Hydrogels for Medical and Related Applications

Joseph D. Andrade, EDITOR

University of Utah

A symposium sponsored
by the Division of
Polymer Chemistry, Inc.
at the 170th Meeting
of the American Chemical
Society, Chicago Ill.,
August 27–28, 1975

ACS SYMPOSIUM SERIES 31

AMERICAN CHEMICAL SOCIETY
WASHINGTON, D. C.   1976

Library of Congress CIP Data

Hydrogels for medical and related applications.
(ACS symposium series; 31 ISSN 0097-6156)

Includes bibliographical references and index.

1. Colloids in medicine—Congresses. 2. Colloids—Congresses.
I. Andrade, Joseph D., 1941-   . II. American Chemical Society. Division of Polymer Chemistry. III. Title. IV. Series: American Chemical Society. ACS symposium series; 31.

R857.C66H9   610'.28   76-28170
ISBN 0-8412-0338-5   ACSMC8 31 1-359

ACS Symposium Series

Robert F. Gould, *Editor*

FOREWORD

The ACS SYMPOSIUM SERIES was founded in 1974 to provide a medium for publishing symposia quickly in book form. The format of the SERIES parallels that of the continuing ADVANCES IN CHEMISTRY SERIES except that in order to save time the papers are not typeset but are reproduced as they are submitted by the authors in camera-ready form. As a further means of saving time, the papers are not edited or reviewed except by the symposium chairman, who becomes editor of the book. Papers published in the ACS SYMPOSIUM SERIES are original contributions not published elsewhere in whole or major part and include reports of research as well as reviews since symposia may embrace both types of presentation.

CONTENTS

PREFACE

There is considerable interest and activity in the application of synthetic and biological polymers in medicine. Synthetic polymers are widely used as surgical and dental implants as well as for blood bags, syringes, tubing, etc. Most of the materials used in medicine have properties greatly different from the tissue with which they are interfaced or replacing. Excepting bones, nails, and the outer layers of skin, mammalian tissues are highly aqueous materials, with water contents ranging up to 90% (blood plasma).

Hydrogels are three dimensional networks of hydrophilic polymers, generally covalently or ionically cross-linked, which interact with aqueous solutions by swelling to some equilibrium value. Aqueous gel networks can be relatively strong (such as in celluose dialysis membranes) or relatively weak, generally becoming weaker as the water content increases, although such variables as the nature of the cross-linker, polymer network, tacticity, and crystallinity can significantly influence the mechanical behavior.

Interest has focused on the utilization of the bulk or the surface properties of hydrogels for biomedical applications. The bulk property of swelling is of particular interest for "swelling implants," i.e., implants which can be implanted in a small dehydrated state via a small incision and which then swell to fill a body cavity and/or to exert a controlled pressure. The swelling of synthetic and natural gels may also help elucidate swelling and osmotic mechanisms in biological tissues.

A related bulk property is the permeability of hydrogels for low molecular weight solutes. Solute diffusivity in gels can be used in sustained drug release applications and in the transport of solutes to gel-entrapped macromolecules, particularly enzymes immobilized in the gel network. Ion interactions or partitioning within the gels are important in bone interfacing applications.

Aqueous gels are relatively subtle systems which equilibrate with and "follow" many environmental changes. The properties of such gels can be highly dependent on cross-linker levels, impurity levels of co-monomers, catalyst residues, and stereoregularity. These variables are discussed for hydrophilic methacrylate ester monomers and their gels. These systems receive the most attention in this volume. Monomer analysis and purification, free radical initiator effects, and tacticity are discussed for the poly(hydroxyethyl methacrylate) system.

The bulk properties of synthetic hydrogels can be widely varied and perhaps tailored to specific end uses. Though the poly(hydroxyethyl methacrylate) system is the most widely studied for medical applications, a large repertoire of gel types are available and have been applied to varying degrees in medicine. Questions of long term biostability or intentional biodegradeability are also very important, but are not discussed here. However, a number of fairly basic studies of gel networks and a number of practical applications based on network properties are presented.

The surface and interfacial properties of hydrogels are somewhat similar to those of natural biological gels and tissues. A number of analogies or comparisons have been made between the hydrogel/water interface and the living cell/physiologic solution interface. Such interfaces are difficult to study and interpret as many of the assumptions of classical surface chemistry cannot be applied. The effect of the gel/water interface on interfacial fluid dynamics and the use of neutral gel coatings to reduce local fluid movement (electroosmosis) in particle electrophoresis experiments are discussed. The interface electrokinetic potential (as measured by the streaming potential) can be varied by use of appropriate comonomer compositions.

The wettability of hydrogel surfaces is rather complicated, perhaps because of the mobility of polymer segments in the interfacial zone. Protein adsorption at certain gel interfaces is also discussed and related to interfacial free energy arguments.

Aqueous gel surfaces can be used in blood-contact applications, as tubing, catheters, and vascular devices. Such applications depend on the nature of the gel/blood interface. As hydrogels generally lack suitable mechanical properties for vascular device applications, there has been considerable activity in coating or grafting the gels to mechanically suitable supports.

In discussing blood compatibility of gels, the application of hydrogels as a substrate for *in vitro* cell or tissue culture studies should be considered. Data on the blood compatibility of hydrogels is available, particularly for polyacrylamide, and there is considerable interest and activity among medical research scientists and a number of commercial firms in hydrogel-coated catheters, drains, and other conduits.

Aqueous gel networks have an important role to play in biomedicine, not only as inert blood conduits and devices, but as biochemically and pharmaceutically active elements. The role of ion and solute partitioning or interactions with gels will prove important for such applications as enzyme entrapment, sustained drug or ion release, and bone induction or calcification, as well as for the more popular applications of blood and tissue interfacing. Suitably equilibrated and solute "stocked" gels should prove very important for cell and tissue culture applications.

Control of hydration/dehydration phenomena, particularly hysteresis effects, is needed in order to ship and store hydrogel products in a dry state, thus reducing bulk and the possibility of microorganism contamination.

This volume is dedicated to the memory of Dr. Willem Prins, a pioneer and leader in the study of gel networks, who died in a boating accident in 1974.

I am particularly grateful to Buddy D. Ratner and Allan S. Hoffman for the review paper which begins this volume and to Donald Gregonis for aid in assembling the papers for publication. The participants in the symposium were also the reviewers of the papers. I thank them all for speedy but critical reviews. The secretarial assistance of Terry Smith is gratefully acknowledged.

University of Utah
Salt Lake City, Utah
March 5, 1976

JOSEPH D. ANDRADE

Synthetic Hydrogels for Biomedical Applications

BUDDY D. RATNER and ALLAN S. HOFFMAN

Departments of Chemical Engineering and Bioengineering,
University of Washington, Seattle, Wash. 98195

It is the intention of this paper to review the literature
concerning the preparation, properties, and biomedical applications of synthetic hydrogel materials. As the literature on this
subject is now voluminous and rapidly expanding (e.g. this
symposium) it is difficult to produce a completely comprehensive
review of the subject. However, a large number of articles will
be examined which should give a useful overview of research
interests and directions in this growing, multidisciplinary
field. This review will be organized as follows: The introduction will discuss general aspects of synthetic hydrogels and
point out similarities between the various distinctly different
chemical structures which fit into this category. A short section is included within the introduction on problems related to
the measurement and description of the biocompatibility of materials. The following six classes of hydrogel materials will then
each be treated individually: poly(hydroxyalkyl methacrylates);
poly(acrylamide), poly(methacrylamide) and derivatives; poly
(N-vinyl-2-pyrrolidone); anionic and cationic hydrogels; polyelectrolyte complexes; and poly(vinyl alcohol). Sections on
surface coated hydrogels, the characterization of imbibed water
within hydrogels and immobilization or entrapment of biologically
active molecules on and within hydrogels for biomedical applications are also included.

1. Introduction

A. General Aspects of Synthetic Hydrogels. A hydrogel can
be defined as a polymeric material which exhibits the ability to
swell in water and retain a significant fraction (e.g., > 20%)
of water within its structure, but which will not dissolve in
water. Included in this definition are a wide variety of
natural materials of both plant and animal origin, materials
prepared by modifying naturally occurring structures, and synthetic polymeric materials. This review article will consider
only synthetic hydrogel systems which are being used as, or have

potential for use, as biomaterials. This constraint is not intended to imply that biomaterials prepared from natural biopolymers are unimportant or uninteresting. Examples of modified
natural biopolymers which are presently receiving attention for
biomedical applications include Cuprophan, cross-linked Dextrans,
and cross-linked, enzymatically treated collagens. For a review
of natural tissues used as biomaterials see the recent article by
Kiraly and Nosé (1).

Hydrogel materials resemble in their physical properties
living tissue more so than any other class of synthetic biomaterial. In particular, their relatively high water contents
and their soft, rubbery consistency give them a strong, superficial resemblance to living soft tissue. Based upon these
properties a number of advantages, some obviously real and
others somewhat speculative can be cited for hydrogel materials.
With respect to the real advantages, two in particular stand out.
First, the expanded nature of the hydrogel structure and its
permeability to small molecules allows polymerization initiator
molecules, initiator decomposition products, polymerization solvent molecules and other extraneous materials to be efficiently
extracted from the gel network before the hydrogel is placed in
contact with a living system. The in vivo leaching of additives
used during the fabrication of polymeric materials has been cited as a cause of inflammation and eventual rejection of implanted
biomaterials (2). Second, the rather soft and rubbery consistency of most hydrogels can contribute to their biocompatibility
by minimizing mechanical (frictional) irritation to surrounding
cells and tissue.

The most intriguing of the potential advantages for hydrogels is the low interfacial tension which may be exhibited between a hydrogel surface and an aqueous solution. This low
interfacial tension should reduce the tendency of the proteins
in body fluids to adsorb and to unfold upon adsorption (3).
Minimal protein interaction may be important for the biological
acceptance of foreign materials as the denaturation of proteins
by surfaces may serve as a trigger mechanism for the initiation
of thrombosis or for other biological rejection mechanisms.

The ability of small molecules to diffuse through hydrogels
may also be advantageous for hydrogel in vivo performance. The
diffusion of important low molecular weight metabolites and ions
through the implant and to the surrounding tissue would occur
with hydrogels, but not with relatively hard, impermeable
plastics.

A number of biomedical applications for hydrogels which have
been mentioned in the literature are listed in Table I. The
wide range of biomedical applications for hydrogels can be attributed both to their satisfactory performance upon in vivo
implantation in either blood contacting or tissue contacting
situations and to their ability to be fabricated into a wide
range of morphologies. The ease with which the physical form

of a hydrogel can be altered allows the physical properties of
the hydrogel to be adjusted specifically for a given application.
Hydrogels can often be prepared in the form of porous sponges,
non-porous gels, optically transparent films, liquids which can
be subsequently crosslinked to form gels and coatings bound by
either covalent bonds or non-covalent forces to a substrate polymer material. It should be emphasized, however, that a particular hydrogel composition suitable for one biomedical application
may have to be significantly modified in composition and form for
a different application. That is, the hydrogel system must be
matched to each biomedical use. Individual classes of hydrogel
biomaterials are discussed in Section II.

Table I
Potential as well as Actual
Biomedical Applications of Synthetic Hydrogels

Coatings	"Homogeneous" Materials	Devices
Sutures	Electrophoresis gels	Enzyme Thera-
Catheters	Contact Lenses	peutic Systems
IUD's	Artificial Corneas	Artificial
Blood Detoxicants	Vitreous Humor Replace-	Organs
Sensors (electrodes)	ments	Drug Delivery
Vascular grafts	Estrous-Inducers	Systems
Electrophoresis cells	Breast or other Soft Tissue	
Cell Culture Sub-	Substitutes	
strates	Burn Dressings	
	Bone Ingrowth Sponges	
	Dentures	
	Ear Drum Plugs	
	Synthetic Cartilages	
	Hemodialysis Membranes	
	Particulate Carriers of	
	Tumor Antibodies	

Although the presence of imbibed water within a polymeric
system is not a guarantee of biocompatibility, it is believed
that the relatively large fraction of water within certain
hydrogel materials is intrinsically related to their high biocompatibility (4), (see Section IV). However, these highly
hydrated and water-plasticized polymer networks are usually
mechanically weak. Furthermore, the higher the water content of
the gel, the poorer the mechanical properties of the gel become.
Fortunately, there are a number of approaches which can be taken
to minimize problems due to the poor mechanical properties of
these gels. Probably the simplest of these approaches consists
of forming the hydrogel over or surrounding a strong polymeric
mesh or woven fabric. Other methods involve coating a device or
material with a layer of hydrogel. In order that the coating

remain intact it must either be used in a non-solvent environment, be crosslinked, or be intimately and/or covalently bound to the support material. Because of the potential importance and complexity of techniques designed to anchor hydrogel coatings, the whole area of hydrogels coated onto other surfaces will be discussed separately in Section III.

Finally, hydrogels usually have a large number of polar reactable sites on which other molecules may be immobilized by a variety of chemical techniques. In addition, biologically active molecules can be entrapped within the network structure of cross-linked hydrogels. Immobilization of biologically active molecules to and within synthetic hydrogels will be discussed in more detail in Section V.

B. <u>Biocompatibility-Operational Definitions</u>. In discussing the biocompatibility of polymers a number of problems are encountered in properly defining the terminology used to describe the response of a living system to an implanted foreign material. Thus, terms such as "non-thrombogenic", "blood compatible", and "biocompatible" are often indiscriminately used to describe a wide range of biological responses. Bruck has itemized a number of factors which might be deleterious to the performance of materials used for long-term internal biomedical applications (<u>5</u>). Based upon his description, the ideal biomaterial (in terms of biological response) could be defined as one which does not cause thrombosis, destruction of cellular elements, alteration of plasma proteins, destruction of enzymes, depletion of electrolytes, adverse immune responses, damage to adjacent tissue, cancer and/ or toxic or allergic reactions. No synthetic material developed fully satisfies these criteria. Also, no single test method for evaluating biomaterials is capable of measuring this wide range of factors relevant to biomaterial response. Based upon the limitations imposed by available testing procedures, discussion in this review paper concerning the biological performance of biomaterials will be oriented towards the test methods which have been used to evaluate the materials. With respect to the most commonly used test methods, the following comments should allow a more critical reading of this review.

<u>Lee White Test (and other related static, in vitro coagulation time assays</u>): The Lee White test compares the coagulation time of recalcified whole blood in a test tube made of or coated with the material to be evaluated with the coagulation time of blood in a standard control tube (usually glass). Variables which can affect the results from this test include changing the donor, changes in the diet or medication of the donor, storage time of the blood, venipuncture technique, and variations in the experimental technique used to measure the clotting times. The test has also been criticized because of the large blood-air interface which is exposed. The validity of results, particularly as they apply to situations involving contact with flowing

blood in an <u>in</u> <u>vivo</u> or <u>ex</u> <u>vivo</u> situation, have frequently been questioned. Methods for the <u>in</u> <u>vitro</u> estimation of the blood compatibility of materials have been critically reviewed (<u>6</u>).

<u>Vena Cava Ring Test</u> (<u>7</u>): This test method involves the implantation of streamlined rings made of or coated with the material to be evaluated into the vena cava of dogs, usually for 2 hour and 2 week test periods. The following terminology has been used to describe the experiment results: "Thrombogenic" is used to designate those materials (formed into test rings) which are completely occluded with thrombus after only 2 hours implantation. "Moderately thromboresistant" is used to describe materials which remain patent after 2 hour implantation but show significant adherent thrombus after the two week test period. The designation "highly thromboresistant" is reserved for materials which show little or no adhering thrombus even after the two week implantation period. A variation of this test using a different procedure for describing the results has recently been published (<u>8</u>). The vena cava ring test does not distinguish between those materials which are truly non-thrombogenic and those which cause thrombus formation but are non-thromboadherent.

<u>Renal Embolus Ring Test</u> (<u>9</u>): This test utilizes rings fabricated from the materials to be evaluated implanted in the canine descending aorta just above the renal arteries. A constriction is made in the aorta below the renal arteries to force a large fraction of the blood flowing through the test ring into the kidneys. After a period of implantation (usually 3-6 days) the rings are examined for adherent thrombi and the kidneys are dissected and examined for infarcts presumably caused by thrombi shed from the ring surface. This test should be able to distinguish between those materials which are truly non-thrombogenic and those which are only non-thromboadherent. Almost all materials examined to date have been shown to cause some infarct damage to the test animal's kidneys.

<u>Soft Tissue Compatibility Tests</u>: For examining the response of the body to materials implanted in soft tissue areas (not in direct contact with the blood stream) there has been little effort extended towards adopting standardized test procedures. Autian has surmised in a literature review that intramuscular implantation may be the most sensitive site for evaluation of this tissue response (<u>10</u>). Coleman, King and Andrade have recently described a comprehensive protocol for the evaluation of tissue response to standardized intramuscular implantations (<u>11</u>). However, the bulk of the papers in the literature do not follow any particular standardized test procedure. The term "tissue compatible", for the purposes of this review paper, will be used to describe those materials which, upon implantation, show a normal acute inflammation reaction and then rapidly "heal in" to a passive state wherein the implant is surrounded by a thin, uniform fibrous capsule in which multinucleated giant cells and other inflammatory cells are generally absent. <u>In</u> <u>vitro</u> tests using cultured cells have

also been utilized with varying degrees of success to evaluate
the toxicity and, to a smaller extent, the biocompatibility of
biomedical polymers. This literature has been covered in a
review by Autian (10).

II. Biomedically Important Hydrogels.
     Hydrogels may be prepared by various polymerization tech-
niques or by conversion of existing polymers. Tables II and III
list examples of monomers and polymers used in preparing these
materials. Although generalizations can be made about hydrogels,
it should be apparent from these tables that this category of
materials covers a wide range of chemical compositions. There
are major differences for each type of material with respect to
synthesis, properties and biocompatibility. Therefore, each of
the more important types of hydrogels will be discussed
individually.

     A. Poly(hydroxyalkyl methacrylates). Included in this class
of compounds are poly(2-hydroxyethyl methacrylate) (P-HEMA), poly
(glyceryl methacrylate) (P-GMA), and poly(hydroxypropyl methacrylate) (P-HPMA). A review article with particular emphasis upon
hydroxyalkyl methacrylate hydrogels has been published (12).
     Poly(2-hydroxyethyl methacrylate) (P-HEMA) hydrogels were
first described and synthesized by Lim and Wichterle in the early
1960's (13). Although this polymer was prepared by DuPont scientists as early as 1936 (14), they did not polymerize the monomer
in the presence of crosslinking agent in aqueous solvent media as
did Lim and Wichterle. To date, P-HEMA hydrogels have probably
been among the most widely studied and used of all the synthetic
hydrogel materials.
     A description of the synthesis and properties of P-HEMA
hydrogels was published in 1965 by Refojo and Yasuda (15). If
HEMA monomer containing crosslinking agent is polymerized in the
presence of a good solvent for both monomer and polymer (e.g.,
ethylene glycol, ethylene glycol-H_2O) an optically transparent
(homogeneous) hydrogel is formed. If the monomer plus crosslinking agent is polymerized in a poor solvent system for the
resulting polymer, an opaque, spongy, white (heterogeneous) hydrogel is formed. As the HEMA monomer is an excellent solvent for
P-HEMA, if the concentration of water (which is by itself a non-
solvent for P-HEMA) in the water-HEMA mixture is 43% or less by
weight, a homogeneous gel will be formed (16)(17). A soluble
P-HEMA polymer which is suitable for dip coating applications can
be prepared by polymerizing to a low degree of conversion in dilute ethanol solution monomer from which most of the contaminating
cross-linking agent has been removed (18,19).
     One of the problems encountered in the preparation of P-HEMA
hydrogels for biomedical applications is the purity of monomer
used in these systems. Typical impurities found in commercial
grades of HEMA monomer are methacrylic acid, ethylene glycol

TABLE II
MONOMERS USED IN HYDROGELS

NEUTRAL	ACIDIC OR ANIONIC

NEUTRAL

HYDROXYALKYL METHACRYLATES $CH_2 = C \diagup{}^{CH_3}_{CO_2-R}$

$R = -CH_2CH_2OH, -CH_2-CH-OH, -CH_2-CH-CH_2-OH$
$\qquad\qquad\qquad\qquad\quad | \qquad\qquad\quad |$
$\qquad\qquad\qquad\qquad CH_3 \qquad\qquad OH$

2,4 PENTADIENE-I-OL $CH_2 = CH-CH=CH-CH_2OH$

ACRYLAMIDE DERIVATIVES $CH_2 = \overset{R}{\underset{}{C}}-CO-N-\overset{R'}{\underset{R''}{|}}$

$(R = -H, -CH_3)$
$(R,R'' = H, -CH_3 - C_2H_5, - CH_2CHOHCH_3$

N-VINYL PYRROLIDONE $CH_2 = CH-N\diagdown$

HYDROPHOBIC ACRYLICS (USED AS COMONOMERS) $CH_2 = C\diagup{}^{R}_{CO_2R'}$

$(R = -H, -CH_3)$
$(R' = -CH_3, -C_4H_9, -OCH_3, -CN, -OCH_2CH_2OCH_3$

ACIDIC OR ANIONIC

ACRYLIC ACID, DERIVATIVES $(R = -H, -CH_3)$ $CH_2 = \overset{R}{C}-CO_2H$

CROTONIC ACID $CH_3-C=CH-CO_2H$ (with VAc)

SODIUM STYRENE SULFONATE $CH_2 = CH-\langle\rangle-SO_3^{\ominus} Na^{\oplus}$

BASIC OR CATIONIC

AMINOETHYL METHACRYLATE, DERIVATIVES $CH_2 = \overset{R}{C}\diagdown{}_{CO_2-C_2H_4 N-R'}^{}$, $\overset{|}{R''}$

$(R, R', R'' = -H, -CH_3, -C_4H_9)$

VINYL PYRIDINE $CH_2 = CH-\langle N\rangle$

CROSSLINKERS

ETHYLENEGLYCOL DIMETHACRYLATE DERIVATIVES $CH_2 = C\diagup{}^{R}$ ... CO / O / $(CH_2CH_2O)_x$ ... $\overset{R}{C}=CH_2$ / CO

METHYLENE-BIS-ACRYLAMIDE $CH_2=CH \qquad CH=CH_2$ / $CO \qquad CO$ / $NH \qquad NH$ / $\diagdown CH_2 \diagup$

Table III

EXAMPLES OF CONVERTED POLYMERS USED AS HYDROGELS

$+CH_2-CH+$ → $+CH_2-CH+$
$\quad\quad | $ $\quad\quad |$
$\quad\quad O$ $\quad\quad OH$
$\quad\quad |$
$\quad\quad C=O$
$\quad\quad |$
$\quad\quad CH_3$

$+CH_2-CH=CH-CH_2+$ → $+CH_2-CH-CH-CH_2+$
$\qquad\qquad\qquad\qquad\qquad\qquad\qquad | \quad |$
$\qquad\qquad\qquad\qquad\qquad\qquad\quad OH \quad OH$

→ POLYELECTROLYTE COMPLEX

$+CH_2-CH+$ → $+CH_2-CH+(CH_2-CH+(CH_2-CH+$
$\qquad | $ $\qquad | \qquad\qquad\quad |$
$\qquad CN$ $\qquad CN \qquad\qquad C=O \qquad CO_2H$
$\qquad\qquad\qquad\qquad\qquad\qquad\qquad | $
$\qquad\qquad\qquad\qquad\qquad\qquad NH_2$

$CH_2 = C\diagup{}^{CH_3}_{CO_2C_2H_4OH}$ + $+CH_2-CH+$ → MIXTURE OF 2 POLYMERS
$\qquad\qquad\qquad\qquad\qquad\qquad\qquad N\diagdown=O$

$OCN-R-NCO + HO+CH_2-CH_2-O)_x H → [\overset{O}{C}-HNR-NH-\overset{O}{C}+CH_2-CH_2-O)_x]$

dimethacrylate, and ethylene glycol. The monomer can be purified
by a complex procedure involving extracting with hexane, distillation under vacuum and treatment with alumina (18,20). Even
after this purification, low levels ($\sim$0.01 - 0.1%) of impurities
remain in the monomer. These impurities could have long-term
effects on the biocompatibility of P-HEMA gels. It has been
suggested that lack of reproducibility in the measurement of the
thrombogenicity of P-HEMA materials in in vivo tests might be due
to as yet unidentified impurities in HEMA monomer (21).

A particularly advantageous property of P-HEMA hydrogels for
biomedical applications is their high degree of chemical stability. P-HEMA gels were found to be resistant to acid hydrolysis
and reaction with amines (22). In another study using a related
crosslinked polymer, poly(diethyleneglycol methacrylate),
alkaline hydrolysis was found to occur to a significant degree
only at high pH and at elevated temperatures (23). P-HEMA gels
in neutral or near neutral aqueous solutions are also relatively
temperature stable and can be steam sterilized with no apparent
damage to the materials (24).

The swelling behavior of P-HEMA gels in various solvents has
been studied in great detail. In water the water content of homogeneous P-HEMA gels is remarkably insensitive to the concentration
of crosslinker in the polymerizing mixture in the 0-4 mole %
crosslinker range (15,25). The water content of heterogeneous
P-HEMA gels strongly depends upon the amount of water (polymer
non-solvent) present during the polymerization reaction (15).
Tables which compare the relative degree of swelling of P-HEMA
gels in a wide variety of organic solvents have been prepared by
two authors (16,22). The swelling behavior of this gel in a
variety of aqueous solutions has also been described in a number
of studies (25-29). Theories relating the swelling of P-HEMA gels
to both hydrophobic and hydrogen bonding interactions within the
gel matrix have been proposed (25,26,30,31). The swelling of
P-HEMA gels has also been studied with respect to the fundamental
equations describing the swelling of crosslinked networks (32).

The mechanical properties of P-HEMA hydrogels have received
some attention (33). These gels are particularly suited for
studies on the fundamental aspects of rubber elasticity, as the
topological aspects of the crosslinked networks can be easily
varied by changing the crosslinking agent, crosslinking agent
concentration, total monomer concentration or polymerization
solvent. Also, co-monomers are easily incorporated into the
system. This literature has recently been reviewed by Janacek
(33).

Biomedical studies utilizing P-HEMA hydrogels are numerous
(34-64). Many of these are listed in Table IV. Fundamental
studies on the healing of disks of P-HEMA subcutaneously implanted in rats and pigs were performed primarily by researchers
at the Institute of Macromolecular Chemistry, Prague (31-36).
Homogeneous P-HEMA and heterogeneous P-HEMA materials with varying

degrees of porosity were examined. In general, most of implants were well tolerated and did not provoke unfavorable reaction. However, P-HEMA gels which were extremely macroporous (those prepared with greater than 70% water in the polymerization mixture) demonstrated poor mechanical properties, and certain undesirable healing characteristics including calcification at the margins or center of the implant. It was therefore concluded that homogeneous or microporous heterogeneous P-HEMA gels might be more suitable for implant purposes.

Table IV
Biomedical Studies Utilizing P-HEMA Hydrogels

Description	Reference
Evaluation of Tissue Response	(34-40)
Antibiotic Delivery:	
From Coated Suture	(41,42)
From Coated Urethral Catheter	(41,43,44)
In Otolaryngology	(45)
Anti-tumor Drug Delivery Device	(46)
Coated Intrauterine Device	(47)
Liver Resection	(48)
Blood Compatibility	(41,49-52)
Coated Sutures	(41,42,50)
Corneal Surgery	(53)
Soft Contact Lens	(54,55)
Ureter Prosthesis	(56)
Breast Augmentation	(57)
Latex Spheres for Cell Surface Studies	(58)
Hemodialysis and Hemoperfusion	(59-62)
Ligament Prosthesis	(63)
Bone Formation in P-HEMA Sponges	(40,64)

Further problems were noted with the healing of porous P-HEMA sponge materials by Winter and Simpson (40). They found that pieces of P-HEMA sponge implanted for 62 days in the skin of young pigs showed evidence of woven bone formation. The effect was found to be readily reproducible. The healing process for these porous materials is unusual in that the sponges are apparently well tolerated for the first month of implantation. However, from 31 days onwards multinucleate giant cells are found in the implant (64). Rubin and Marshall attempted to utilize the observed bone formation in implanted spongy P-HEMA to anchor into the femur and tibia a knitted Dacron-porous P-HEMA anterior cruciate ligament prosthesis (63). Fixation into the bone was not achieved although bone formation was noted in parts of the polymer.
Due to the poor mechanical properties of P-HEMA gels, there have always been experimental difficulties in evaluating the blood compatibility of these materials by accepted in vivo testing

techniques (e.g., the vena cava ring test or the renal embolus ring test). However, the blood compatibility of P-HEMA was observed in a few studies either in vitro, or, using somewhat less conventional techniques, in vivo (41, 49-52). It was found, in all cases, to be a relatively "non-thrombogenic" (or at least non-thromboadherent) material, although the in vivo tests that were used do not allow a systematic grading of thrombogenicity, and do not compare the performance of P-HEMA to a number of other commonly used materials. More systematic studies of the blood compatibility of P-HEMA are discussed in the section on surface coated hydrogels.

The primary clinical use for P-HEMA hydrogels has been for flexible, hydrophilic contact lenses. This is also the only application which has received U.S. Food and Drug Administration approval (specifically for the Bausch and Lomb P-HEMA hydrogel Soflens for the correction of refractive errors and the Warner-Lambert poly(HEMA-co-ethylene dimethacrylate-co-methacrylic acid-co-g-Povidone) Soft-con as a therapeutic 'bandage"). The properties and applications for such contact lenses have been reviewed (54). Studies dealing specifically with oxygen permeability through hydrogel contact lenses (65), and hydration and linear expansion of hydrophilic contact lenses (66) have been published. In another study methyl methacrylate contact lenses and P-HEMA contact lenses were observed with respect to their effect upon the growth of corneal epithelial tissue in culture. The tolerance of the tissue to continuous exposure to the contact lenses was found to be significantly higher for the hydrophilic lenses (55).

A number of problems have been noted with hydrophilic contact lenses. These include low visual acuity as compared to hard lenses, accumulation of material on and within the lenses, susceptibility to mechanical damage, low permeability to oxygen, (reversible) changes in the shape and refractive index of the lenses, and the need for frequent sterilization (54,67). The principle advantages of these hydrogel lenses are that they are easy to fit, well tolerated and can be used for therapeutic purposes other than simple refractive correction. Concerning the degree to which these lenses are tolerated, it has been recently reported that the Bausch and Lomb Soflens can be worn continuously 10 days with few or no problems (67). However, another study reports that it is unadvisable to wear soft contact lenses continuously for long periods of time due to the development of corneal edema (68).

Other hydroxyalkyl methacrylates have not received nearly as much attention as P-HEMA. Poly(hydroxypropyl methacrylate) hydrogels containing a mixture of the two isomers can be readily prepared (25). Although the water content of these hydrogels is fairly low (∿25%) they demonstrate excellent mechanical properties. Poly(glyceryl methacrylate) hydrogels exhibit a considerably higher equilibrium water content than P-HEMA gels (28,60-71). Therefore it has often been thought that this material might show a higher degree of biocompatibility than P-HEMA. This has not been described in the literature to date. It has been suggested that dehydrated P-GMA might be swollen in situ in cases where it is necessary to replace the vitrous humor of the eye (70).

B. Poly(Acrylamide), Poly(Methacrylamide) and Derivatives. Gels of poly(acrylamide) (PAAm) and of some N-substituted derivatives of PAAm can be readily prepared by the free radical polymerization of an aqueous solution of acrylamide containing a small fraction of crosslinking agent (often N, N-methylenebisacrylamide). The gels formed are optically transparent, mechanically weak, and can have extremely high water contents (> 95%). The water content is dependent upon the per cent crosslinker in the system, unlike the homogeneous P-HEMA gel system. Another technique was recently described for preparing gels composed primarily of polyacrylamide with smaller fractions of poly(acrylonitrile) and poly(acrylic acid) (72). This technique involves polymerizing acrylonitrile in concentrated aqueous zinc chloride and then partially hydrolyzing the resulting gel. The high water contents which are attainable with acrylamide gels prepared by solution polymerization or by hydrolysis make them attractive for biomedical applications (see Section IV).

A number of studies have been carried out on the hydrolysis of acrylamide and methacrylamide polymers (73-75). Hydrolysis occurs at significant rates at elevated temperatures if the polymers are in acidic or basic solutions. Thus, polyamide gels should be relatively stable under conditions of physiological pH and temperature. However, problems might occur at steam autoclave temperatures and therefore this procedure might be contraindicated.

The tissue compatibility of poly(N-substituted acrylamides) was investigated by Kopecek, et al. (76). N-substituted acrylamides were used for this study because of their superior hydrolytic stability compared to poly(acrylamide). It was found that discs of poly(N,N-diethyl acrylamide), poly(N-acrylyl morpholine), poly(N-ethyl acrylamide) and also the methacrylamide derivative, poly[N-(2-hydroxypropyl) methacrylamide], implanted subcutaneously in rats were well tolerated by the animals and did not provoke unfavorable reaction. The long term biological interaction of these hydrogels with the test animals was described as being similar to that observed with implanted P-HEMA materials, which were also investigated by this group.

The thrombogenicity of a number of PAAm gels was investigated in vitro using the Lee-White coagulation time test (52). Where the coagulation time of fresh blood samples in glass tubes was approximately 12 minutes, blood in PAAm tubes prepared using singly recrystallized acrylamide monomer showed a clotting time of ~45 minutes. When gels formed from triply recrystallized acrylamide monomer were tested they showed clotting times in excess of 24 hours. The deleterious effects of incomplete removal

of initiator by-products on the thrombogenicity of acrylamide
materials was also noted in this study. The effect of varying
the crosslinking agent concentration in the PAAm gel was found
not to significantly alter the Lee-White clotting times for the
system. Interesting effects of other co-monomers incorporated
into the PAAm gels on the clotting times were noted. Some of
these effects are discussed in Section II F. In general, the
PAAm system prepared with triply recrystallized monomer gave the
best results, although systems containing dimethylaminoethyl
methacrylate were shown to demonstrate excellent Lee-White clotting times. The thrombogenicity of these particular gels was
found to be extremely sensitive to crosslinker concentration and
total gel solids.
    Further information about PAAm gels is contained in a recent-
ly published review covering three types of biomedical hydrogels
(4). In particular the pore size and water content of PAAm is
discussed in relation to various measures of its biocompatibility.

    C. Poly(N-Vinyl-2-Pyrrolidone). Poly(N-vinyl-2-pyrrolidone)
(P-NVP) is a somewhat unique polymer in that, in its uncrosslinked form, it is extremely soluble in water and is also soluble
in many other polar and non-polar solvents. Because of its
strong interaction with water it can be used for preparing gels
which will exhibit high water contents.
    P-NVP gels are also of interest for biomedical applications
as the soluble polymer has had a long history of use in the medical and pharmaceutical fields. One of the most important uses
for P-NVP solutions has been as a plasma expander (77). When
infused intravenously P-NVP is non-toxic and non-thrombogenic and
can be used to maintain circulatory fluid volume in cases of severe injury or trauma. Some retention of P-NVP in the liver,
spleen, lungs and kidneys has been noted (78). In a review of
the world literature on P-NVP in 1962 it was concluded that
P-NVP could be used orally or intravenously with complete safety
(79). However, at the present time P-NVP is no longer used as a
plasma expander in humans because it is not metabolized and is
not retained in circulation as well as other plasma expanders
(e.g., Dextrans). The resistance of P-NVP to digestion by lyso-
somal enzymes has been exploited in a recent study in which the
kinetics of uptake by pinocytosis of I^{125} labelled P-NVP into rat
yolk sac cultured in vitro was measured(80). P-NVP has also been
used in blood volume determinations (81) and in the preservation
of blood and blood components (82). In the pharmaceutical field
it has been used as a tablet binder, a tablet coating and for the
solubilization and stabilization of drugs. An extensive bibliography listing P-NVP medical and pharmaceutical applications is
available (83).
    Hydrogels consisting only of P-NVP have not often been
described in the biomedical literature possible because high
concentrations of crosslinker (5-20%) are needed to produce a

material with useful mechanical properties. P-NVP gels crosslinked with methylene bis(4-phenyl isocyanate) were evaluated for
use as hemodialysis membranes (84). Faster metabolic waste transfer was obtained with these membranes than with conventional cellulose films. P-NVP gels crosslinked with 20%(w/w) methylenebis-
acrylamide have been considered for use as an implantable drug
delivery system (85). In vitro evaluation of the thrombogenicity
of P-NVP gels and P-NVP-PAAm copolymer gels by the Lee-White test
showed some extension of clotting times although problems with
residual NVP monomer in the gels was noted (52). Fibroblast adhesiveness to P-NVP gels has also been measured in vitro (86).
It was determined that increasing the concentration of the gel
from 40% to 96% renders it more adhesive to the cells.

The difficulties involved in preparing homogeneous P-NVP
materials makes NVP an ideal monomer for use in covalent surface
grafting systems. A number of graft P-NVP copolymers are discussed in Section III.

D. Polyelectrolyte Complexes. Polyelectrolyte complexes are
polysalts formed by the coreaction of a cationic polymer such as
poly(vinyl benzyltrimethyl-ammonium chloride) and an anionic polymer such as sodium poly(styrene sulfonate). The complex formed
from these two particular polyelectrolytes was developed by the
Amicon Corporation and is referred to as Ioplex 101. It has
received biomedical evaluation in a number of situations(87,88).

Due to mechanical strength limitations polyelectrolyte complexes have been generally used as coatings on fabrics and other
supports. In order to be used for coatings the gel must be solu-
bilized in a complex, multicomponent solvent system usually containing water, a polar, water soluble organic solvent, and a
strong electrolyte. Ioplex-solvent solutions have generally been
strongly acidic and have been shown to degrade certain plastics
such as nylon-6,6 thus making the requirements for a suitable substrate for the polyelectrolyte complex more stringent (89).

Difficulties have also been found in the sterilization of
Ioplex materials. Autoclave sterilization of these complexes can
result in disintegration of the gel structure (89). Gas sterilization can leave entrapped ethylene oxide in the matrix. The
problems encountered in sterilizing these hydrogels have been
reviewed(4,90).

An advantage of these materials is the ease with which a net
charge (anionic or cationic) can be incorporated in the system.
This is done by adding stoichiometrically greater or lesser amounts
of one of the two polymeric components during formulation. It was
determined using the in vivo vena cava ring test that Ioplex 101
containing 0.5 meq. excess anionic component showed the greatest
thromboresistance (89). However, the performance of the neutral
Ioplex gel in this test indicated that it is perhaps only slightly
less thromboresistant than the anionic gel (4). Cationic Ioplexes
and gels containing an increased number of anionic sites

were significantly more thrombogenic.

E. <u>Poly(vinyl alcohol)</u>. Poly(vinyl alcohol) (PVA) is a
water soluble polymer formed by the hydrolysis of poly(vinyl acetate). Crosslinked gels of PVA have found a number of uses in
the biomedical field.

A crosslinked, highly porous sponge of PVA can be formed by
reacting formaldehyde with soluble PVA and blowing air through
the solution before the polymerization-crosslinking process is
completed. This material was commercially available under the
name Ivalon and had been extensively used in hernia treatment,
duct replacement, cardiac-vascular surgery, plastic surgery, and
reconstructive surgery. Healing problems with Ivalon sponges
were encountered and it was concluded that Ivalon did not meet
up to the early enthusiasm expressed for it, and that other syn-
thetics would be more satisfactory in similar situations (<u>91</u>).

A hydrogel consisting of PVA and the anticoagulant heparin
crosslinked together with glutaraldehyde/formaldehyde mixtures
has been synthesized and demonstrates low thrombogenicity in <u>in</u>
<u>vitro</u> tests (92). Experiments with S^{35} labelled heparin indicate
that heparin does not leach out of the crosslinked gel. A potential problem with this material is that, possibly due to the
presence of heparin, it shows a tendency to adsorb blood platelets. PVA-heparin hydrogels have also been evaluated for use as
hemodialysis membranes (<u>93</u>). They show promise for this application since the permeability to "middle-molecular weight" molecules
such as inulin is much higher for PVA-heparin hydrogels than
for Cuprophan cellophane hemodialysis membranes.

Radiation crosslinked gels of PVA have been proposed for use
as synthetic cartilage in synovial joints (<u>94</u>). The material
prepared for this application is annealed to increase the crystallinity and therefore the physical strength of the hydrogel. This
application for PVA gels is discussed further in Section II F.

A PVA surface with a number of immobilized biomolecules on
it was designed in an effort to simulate the natural blood vessel
intima (<u>95</u>). This material is described in Section V.

F. <u>Anionic and Cationic Hydrogels</u>. Anionic and cationic
hydrogels are usually formed by copolymerizing small amounts of
anionic or cationic monomers with neutral hydrogel monomers (see
Table II). However, they can also be prepared by modifying preformed hydrogels such as by the partial hydrolysis of poly
(hydroxyalkyl methacrylates) or, in the case of polyelectrolyte
complexes, by adding an excess of the polyanion or polycation
component.

The interest in such hydrogel systems stems from observations
on the surface charges on blood cells, blood vessel walls, and
other tissue types. Under normal conditions the blood vessel
walls and blood cells have a negative charge. Sawyer has presented evidence which indicates that when this charge is altered

by pH changes, or by drugs the tendency of the system to throm-
bose is also altered (96). It is generally thought that negatively charged surfaces should be less thrombogenic than positively
charged ones, (97) and experiments have been performed which
support this contention. However, results indicating decreased
thrombogenicity for positively charged surfaces have also been
presented (52,98). The role of charge density (as opposed to
total charge) has been suggested as an important factor with
respect to the thrombogenicity of surfaces (97). At the present
time the importance of surface charge in blood-hydrogel interactions or in in vivo healing is not at all clear.

Cerny, et al., investigated the healing of P-HEMA specimens,
some of which contained charged co-monomers, subcutaneously implanted in several species of laboratory animals (35). He found
that in four groups of rats, for P-HEMA specimens containing 4%
methacrylic acid, calcification (which was found in neutral
P-HEMA specimens) was inhibited. Barvic, et al., found that
P-HEMA gels containing 5 weight % or greater of the co-monomer
diethylaminoethyl methacrylate gave rise to inflammatory reactions after three weeks of implantation in rats (37). Sprincl,
et al., in a recent study, investigated the healing of P-HEMA
specimens containing small amounts of copolymerized anionic
(methacrylic acid) and cationic (N, N, dimethylaminoethyl methacrylate) monomers implanted subcutaneously for long periods
(up to 360 days) in rats (39). They found that the chemical
composition of the gels, within the concentration ranges studied,
showed no apparent effect on long term healing. Thus, the
observation by Barvic, et al., might only be true for short term
implantation. Sprincl, et al., also showed that for P-HEMA gels,
calcification apparently only depended on the physical form of
the gel (i.e., macroporous gels might calcify where microporous
gels wouldn't) rather than on the chemical composition of the gel.
Therefore, the inhibition of calcification noted by Cerny, et al.,
might be due to differences in the physical structure of the gels
he used rather than to the presence of 4% methacrylic acid.

The in vitro thrombogenicity of PAAm gels copolymerized
with dimethylaminoethyl methacrylate, t-butylaminoethyl methacrylate, 2-sulfoethylmethacrylate sodium salt, 2-hydroxy-3-meth-
acryloloxypropyltrimethylammonium chloride, acrylic acid, methacrylic acid, 2-vinyl pyridine, 4-vinyl pyridine and 2-methyl-
5-vinyl pyridine was studied by Halpern, et al., using the Lee-
White technique. Only the dimethylaminoethyl methacrylate-
acrylamide hydrogel showed a significant extension in clotting
times. As was mentioned in Section II D, polyelectrolyte complexes containing 0.5 meq. excess anionic component were found to
have the lowest thrombogenicity in in vivo studies (99). These
two results are somewhat contradictory as the dimethylaminoethyl
methacrylate-acrylamide hydrogel is positively charged at physiological pH while the anionic polyelectrolyte complex has a
negative charge. Other studies on the thrombogenicity of

charged hydrogels will be discussed in Section III.

A unique application for a cationic hydrogel has been proposed by Bray and Merrill (94,99). They constructed a synthetic
articular cartilage material for use in a synovial joint by simultaneously radiation crosslinking poly(vinyl alcohol) and radiation grafting to the PVA chains a cationic monomer. Such a material should strongly adsorb negatively charged hyaluronic acid
and produce an "osmotically-enhanced, viscous gel layer of
boundary lubricant" (94). Cationic monomers which have been used
for this purpose are allyltrimethyl-ammonium bromide and 2-
hydroxy-3-methacryloxypropyltrimethylammonium chloride.

III. Surface Coated Hydrogels

There are a number of techniques which can be used to coat
substrates with hydrophilic polymers or copolymers. (Table V).
Aside from the conventional technique of dipping in a solution
of the polymer, all other methods involve covalent bonding
(grafting) of the hydrophilic polymer to the substrate polymer
chains.

Table V
Techniques for Depositing Hydrogel Coatings

1. Dip-coat in pre-polymer + solvent.
2. Dip in monomer(s) ($\pm$ solvent, polymer)
    then polymerize using catalyst $\pm$ heat.
3. Pre-activate surface ("active vapor,"
    ionizing radiation in air) then contact
    with monomer(s) $\pm$ heat to polymerize.
4. Irradiate with ionizing radiation while
    in contact with vapor or liquid solution
    of monomer(s).

By covalently bonding a hydrogel to the surface of another
polymer a new composite material is formed whose mechanical properties more closely resembles those of the base polymer than the
thin grafted hydrogel layer. The most efficacious technique for
preparing such materials involves generating free radicals on a
plastic surface and then polymerizing a monomer directly on that
surface. A number of techniques have been used for generating
such radicals on surfaces. Some of the more commonly used ones
are listed in Table VI.

There are at least five advantages to using grafting techniques for preparing biomedical hydrogels. The first and most
obvious advantage is the increase in mechanical strength over
the ungrafted hydrogel which can be obtained. A second advantage
is the permanence and durability which should be exhibited by the
covalently bound hydrogel coating as compared to coatings on
devices prepared by dipping techniques. Third, graft polymeriza-

TABLE VI

Techniques and Reactions for Generating Radicals on Surfaces.

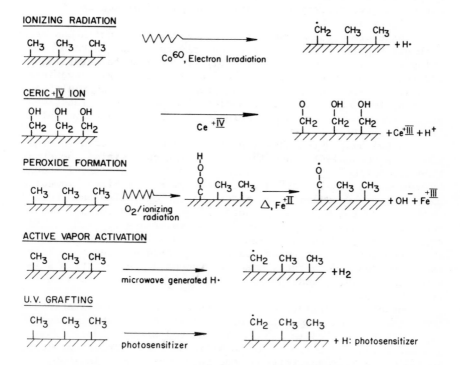

NOTE: The precise nature of the radical intermediates formed has not been elucidated in most cases. Representations in this table show schematically radical species which might be formed.

tion techniques make it possible to prepare complex surfaces formed by successive graftings using different monomers. Fourth, the preparation of hydrogels grafted only on the surface, parti- ally into the substrate, or uniformly throughout a hydrophobic matrix can be effected by varying the polymerization solvent and other grafting parameters. The latter type of grafted hydrogel comprises an interesting class of materials which should have mechanical and surface properties reflecting the characteristics of both the substrate and the hydrogel. Finally, using radiation to prepare a grafted hydrogel, the addition of an initiator is not necessary, thereby eliminating one potential source of contamination in the final product.

There are certain undesirable side reactions which can occur with graft polymerizations, particularly those initiated by ion- izing radiation. These include polymer degradation, crosslinking, and the formation of unwanted chemical species (e.g., peroxides, acids). However, degradation and crosslinking can be minimized by using low doses and the formation of most unwanted functional groups can be eliminated by excluding oxygen and reactive solverts from the grafting system.

As has been noted above, grafting copolymerization techniques provide a convenient means for controlling composition, penetra- tion and morphology of the grafted polymer. Such control should be useful for "tailoring" a grafted polymer to a given biomedical application. Detailed descriptions of methods which have been used to vary the surface character of grafted hydrogels for bio- medical uses have been described in a few papers (100-102). Ex- amples of some of the grafting parameters which can be used to influence the properties of radiation grafted hydrogels include radiation dose, dose rate, monomer concentration, grafting sol- vent, temperature, and the presence of various metal ions in the grafting system.

One of the earliest applications of graft polymerization techniques to the preparation of materials for biomedical appli- cations was reported in 1964 by Yasuda and Refojo (103). They grafted N-vinyl-2-pyrrolidone to silicone rubber in an effort to increase the hydrophilicity of the rubber surface. In 1966, Leininger and co-workers discussed the possibility of grafting vinyl pyridine to a base plastic for use in immobilizing heparin to the surface (104). Laizier and Wajs in 1969 described a transparent polymer prepared by grafting NVP to a silicone resin which might be suitable for contact lens applications (105).

Within the last five years a number of groups have published papers or reports describing grafted hydrogels designed for bio- medical applications. Much of the work in this field is summar- ized in Table VII.

There has been little published on the in vivo evaluation of tissue response to grafted hydrogels. In a preliminary, and as yet unpublished study, radiation grafted P-HEMA and PAAm hydro- gels on silicone rubber were found to be well tolerated when

Table VII

Grafted Hydrogels Prepared For Biomedical Applications

Research Group	Monomer(s) Used	Grafting Technique(s) Used	Applications	References
Yasuda and Refojo	NVP	Electron irradiation	Medical applications	(103)
Battelle, Columbus	Vinyl Pyridine	Radiation grafting	Non-thrombogenic plastic surface	(104)
Laizier and Wais	NVP	Co^{60} (pre-irradiation technique)	Contact lenses	(105)
Miller, et al.	Acrylic Acid, Methacrylic Acid, Acrylamide, Ethylene Sulfonic Acid, Various Esters and Imides	Electron irradiation, chemical treatments	Non-thrombogenic surfaces	(106)
Hoffman, et al.	HEMA, NVP, Acrylamide, Methacrylamide, Methacrylic acid, Ethyl methacrylate	Co^{60} (Mutual irradiation technique)	Artificial heart components, catheters, knitted artery prostheses, blood & tissue compatible interfaces	(100,101, 107-110)
Andrade, et al.	HEMA	Co^{60} (Mutual irradiation technique)	Artificial heart components	(102)
Kearney, et al.	Acrylamide	Co^{60} (Mutual irradiation technique)	Blood compatible interface	(111)
Franklin Institute	HEMA, Acrylamide	Atomized gas plasma grafting technique	Blood & tissue compatible materials, heart assist devices, I.U.D.	(9, 47, 112,113)
Union Carbide Corp.	Vinyl Acetate-co-2% Crotonic Acid, NVP-co-2% Acrylic Acid	Electron irradiation of preformed poly-electrolyte coating, Co^{60} mutual irradiation	Artificial heart and organs	(113-115)
Polysciences, Inc.	Acrylamide, HEMA	Ceric IV ion grafting, radiation grafting (Mutual irradiation tech.)	Blood oxygenator (hollow fiber), intra-aortic assist balloons, assist bladders, aortic implant tubes	(9, 115, 116)
U.S. Army Medical Bioengineering Research & Development Laboratory	HEMA	Co^{60} (Mutual irradiation technique)	Burn dressing	(118)
Univ. of Wisconsin	HEMA	Co^{60} (Mutual irradiation technique)	Dialysis membrane	(117)
Hydromed Sciences	HEMA	Co^{60} (Mutual irradiation technique)	IUD, catheters	(119)

implanted both intraperitoneally and subcutaneously in rats. The films were surrounded by a thin non-adhering fibrous capsule after implantation periods ranging from 5-10 days (120). In an in vitro study on adhesiveness of chick embryo myoblasts to radiation grafted P-HEMA and N-VP hydrogels on silicone rubber it was found that the cells adhere very poorly to grafted hydrogels and that the hydrogel grafted polymers always adhered fewer cells than the ungrafted polymers (110). Whether such results are meaningful in terms of in vivo cell-grafted hydrogel interactions is not yet clear.

    The blood compatibility of grafted hydrogels has been examined in a number of cases by the vena cava ring test and renal embolus ring test. Certain generalizations can be made from these results.

    1. P-HEMA or P-NVP hydrogels grafted to silicone rubber will greatly reduce the thrombogenicity of the silicone rubber as judged by the vena cava ring test (108).

    2. When evaluated by the vena cava ring test several different types of grafted hydrogels have been found to perform well, but some thrombus is usually noted adhering to the ring, particularly after the two week test period. An exception to this is grafted polyacrylamide materials which have, in some cases, shown no thrombus after 14 days (4,121). Recent results, however, using a modification of the vena cava ring test do show thrombus adhering to grafted acrylamide rings in four out of six cases (8). A 60% sodium ionomer of poly(vinyl acetate-co-2% crotonic acid) has, in certain test situations, also showed negligible thrombus accumulation after 14 days (114).

    3. In the renal embolus test, although little thrombus is found within most hydrogel treated test rings, the kidneys almost always show moderate to extensive embolus damage (9).

    Thus, the generally low thrombogenicity rating of synthetic hydrogels based on the vena cava ring test, may, in fact, be due to a low thromboadherance, i.e. a flaking off of thrombus from the hydrogel surface. This aspect of hydrogel-blood interaction is much in need of further investigation.

    Recently platelet adhesiveness to radiation grafted P-HEMA hydrogels on cellulose acetate was measured in an in vitro test cell. The per cent platelets adhering was found to decrease with increasing graft level (figure 1) (117). This trend was very similar to that noted for fibrinogen adsorption to P-HEMA grafted silicone rubber surfaces (figure 2) (109). Decreased adhesiveness of cells to hydrogel grafted surfaces was also noted in the cell adhesion study discussed above (110). Again, these observations tend to support the idea that hydrogels have a low thromboadherence.

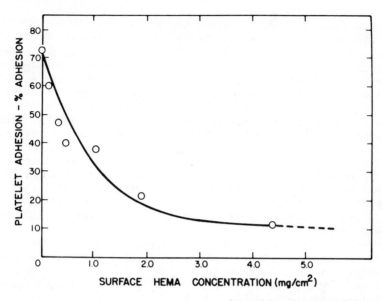

Journal of Biomedical Materials Research

Figure 1.   Platelet adhesion to radiation grafted HEMA hydrogels on cellulose acetate (117)

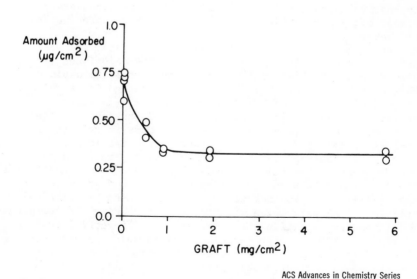

ACS Advances in Chemistry Series

Figure 2.   Fibrinogen adsorption to radiation grafted HEMA hydrogels on silicone rubber (109)

IV. Characterization of the Imbibed Water

Clearly, the presence of large amounts of water within a polymeric network is not the sole factor controlling the biocompatibility and thrombogenicity of such materials. Thus, gelatin or some polysaccharides which often have water contents of 90% or higher are considered to be relatively thrombogenic materials. Also, extremely "open" P-HEMA gels which exhibit high water contents are less tissue compatible than tighter P-HEMA gels with lower water contents (39). Still, the presence of a significant fraction of water is considered important for the biocompatibility and low thrombogenicity of those hydrogels which are biocompatible and "reasonably" non-thrombogenic (or non-thromboadherent) (4). The nature or organization of the water within the hydrated polymeric network may be one of the factors which will influence the interactions that occur between biological systems and the hydrogel.

Problems concerning the organization of water at the molecular level (water structure) are often extremely complex and the literature on the subject is extensive. There are excellent reviews available on water structuring (122,123). Also, a number of review articles have been written on the biological implications of structured water (124-126) and the potential importance of the molecular organization of water to the performance of biomaterials (127,128).

The gross total water contents of swollen hydrogels are most easily measured and most often reported. Information about the molecular nature of water within the network is not as easy to obtain; such water may be (a) polarized around charged ionic groups, (b) oriented around hydrogen bonding groups or other dipoles, (c) structured in "ice-like" configurations around hydrophobic groups, and/or (d) imbibed in large pores as "normal" bulk water.

Attempts have been made to separate the total gel water content into some of these categories using NMR techniques (129). Based upon the NMR data gel water contents would be divided into "bulk water" (category d), "bound water" (categories a, b) and the remaining water, called "interface water" (category c). Results for P-HEMA gels shown in Table VIII suggest that the fractions of bulk, bound and interfacial water vary with the total water content of the gel. Increasing fractions of bulk water and decreasing fractions of bound water are found as the water content of the gel increases. The fraction of interfacial water undergoes only small changes with changing gel water content. Similar conclusions were drawn concerning the state of water in P-HEMA gels which were investigated using the techniques of dilatometry, specific conductivity and differential scanning calorimetry (130). Water sorption studies which provide some insight into the kinetics and thermodynamics of water interaction with hydrogels have also been performed (131 ,132).

Table VIII
The Fraction of Water in P-HEMA Gels of Different
Total Water Content (Wt %)
[Data from Lee, Andrade and Jhon, (114)]

Wt % of Total Water in the Gel	20	25	30	35	40	45	50	55	60
f_w: Fraction of Bulk Water	0	0	0	0.09	0.21	0.30	0.37	0.42	0.47
f_1: Fraction of Interfacial Water	0	0.2	0.33	0.34	0.29	0.26	0.33	0.22	0.20
f_b: Fraction of Bound Water	1.0	0.8	0.67	0.57	0.50	0.44	0.40	0.36	0.33

Studies to date on the organization of water within hydrogels are only preliminary indications; furthermore, the effects
of water organization on biological interactions remain to be
elucidated. It should be noted that the organization and content
of gel water will vary significantly with hydrogel composition,
probably most often in expected directions (i.e., gels with
higher water contents will have lower fractions of bound and
interfacial water).

V. Immobilization and Entrapment of Biologically Active Molecules
on and Within Hydrogels for Biomaterial Applications.
      Hydrogels are, in many respects, eminently suited for use as
a base material for "biologically active" biomaterials. Examples
of classes of biologically active molecules which can be used in
conjunction with hydrogels are listed in Table IX. Examples of
biomedical applications for immobilized enzymes are presented in
Table X. There are a number of distinct advantages for hydrogels
in these types of systems. Small molecules (drugs, enzyme substrates) can diffuse through hydrogels and the rate of permeation
can be controlled by co-polymerizing the hydrogel in varying
ratios with other monomers. Hydrogels may interact less strongly
than more hydrophobic materials with the molecules which are
immobilized to or within them thus leaving a larger fraction of
the molecules active (3). Hydrogels can be left in contact with
blood or tissue for extended periods of time without causing
reaction making them useful for devices to be used in long-term
treatment of various conditions. Finally, hydrogels usually
have a large number of polar reactive sites on which molecules
can be immobilized by relatively simple chemistries.

Table IX
Biologically Active Molecules which may be
Entrapped or Immobilized in Hydrogels

Antibiotics
Anticoagulants
Anti-Cancer Drugs
Antibodies
Drug Antagonists
Enzymes
Contraceptives
Estrous-Inducers
Anti-bacteria Agents

Table X
Biomedical Applications of Immobilized Enzymes

Immobilized Enzyme(s)	Application	Reference
Brinolase	Non-Thrombogenic Surface	(133)
Urokinase	Non-Thrombogenic Surface	(134)
Streptokinase	Non-Thrombogenic Surface	(135)
Asparaginase, Glutaminase	Leukemia Treatment	(136)
Carbonic Anhydrase, Catalase	Membrane Oxygenator	(137)
Urease	Artificial Kidney	(138)
Glucose Oxidase	Glucose Sensor-Artificial Pancreas	(139)
Microsomal Enzymes	Artificial Liver	(140)
Alcohol Oxidase	Blood Alcohol Electrode	(141)
DNase, RNase	Removal of Airborn Infections	(142)

Biologically active molecules can be immobilized within hydrogels permanently, or temporarily. If the hydrogel system is designed to release the entrapped biologically active molecules at a preset rate, these materials are well suited for use as controlled drug delivery devices. In the simplest example of such a system, hydrogels can be saturated with solutions of various antibiotics and other drugs which will leach out to the surrounding tissue upon implantation. A number of papers describing such hydrogel drug delivery systems have already been mentioned (41-46,95). The rate of drug delivery generally decreases rapidly with simple homogeneous hydrogels saturated with a drug solution. By using a biocompatible hydrogel membrane device filled with a drug in the form of a pure liquid or solid, constant drug delivery rates and extended treatment times can be obtained (143,144). A still more sophisticated approach involves the design of a hydrophilic polymer backbone chain onto which alternate "catalyst" groups and labile "drug" groups are

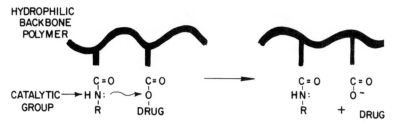

HYDROPHILIC
BACKBONE
POLYMER

CATALYTIC⟶ H N: O
GROUP   R   DRUG

C=O   C=O   C=O   C=O

HN: O⁻
R   + DRUG

Figure 3.   Drug release from a polymeric chain controlled by intramolecular catalysis

bound (e.g. Figure 3). The kinetics of various intramolecularly catalyzed polymeric reactions have been described (145).

Hydrogels which are prepared by the solution polymerization of a monomer in the presence of a crosslinking agent are well suited for entrapping an active biomolecule within the network structure. For the immobilization of an enzyme by the entrapment technique, leakage of the enzyme would be undesirable. Therefore, the "pore" size or average interchain distance of the gel should be smaller than the size of the active enzyme. A "pore" size of 35 Å or smaller should be suitable for retaining most entrapped enzymes (146). For comparison, homogeneous P-HEMA gels have estimated "pore" sizes of approximately 4-5 Å (52,147), while acrylamide gels might have "pore" sizes from 7 Å - 17 Å depending upon the method of preparation (148). However, these pore sizes are probably low with an error estimated at greater than 25% (4). Recent reports on enzyme entrapment include systems involving glucose oxidase entrapped in P-HEMA and P-NVP (149) glucoamylase, invertase, and β-galactosidase entrapped in poly (2-hydroxyethyl acrylate) and poly(dimethylacrylamide) gels (150) and asparaginase and microsomal enzymes entrapped in P-NVP gels (151). The entrapment of heparin in a PVA gel was described in Section II E (92).

Techniques for the covalent immobilization of active molecules to surfaces have been the subject of a number of recent in-depth reviews (140,152,153). A large number of chemical techniques which are particularly applicable for immobilization to hydrogels have been developed. Figure 4 shows chemistries useful for coupling biomolecules to gels containing carboxyl groups. Figure 5 illustrates the probable reactions occurring during the immobilization of a protein to a polymer which contains hydroxyl groups (154). The Ugi reaction is a four component condensation reaction which occurs between an amine, an aldehyde, a carboxylic acid and an isocyanide (155). It is of interest for immobilization to hydrogels because of the many

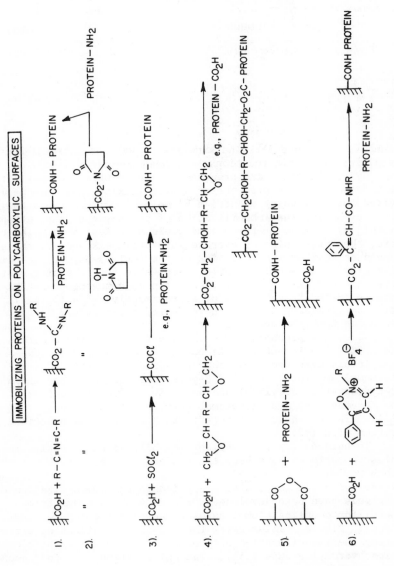

Figure 4. Chemistries useful for coupling biomolecules to gels containing carboxyl groups

potential reactions which might be used depending upon which functional groups are on the hydrogel and which are on the bio-molecule to be immobilized. The Ugi reaction used to couple a protein - NH_2 group to a carboxylic hydrogel is shown in Figure 6. Glutaraldehyde is often used for immobilization of biomolecules to polyacrylamide hydrogels although the precise mechanism of coupling is not yet known (156).

There have been surprisingly few reports on the development of materials intended to be both biocompatible and biologically active. Many active biomolecules have been bound to supports such as Sephadex and Sepharose (modified polysaccharides) which allow large amounts of active biomolecule to be immobilized but which would not be expected to show significant biocompatibility. Devices specifically constructed for the immobilization of en-zymes to be used in contact with blood have, at times, been made of such materials as poly(methyl methacrylate) (136),polyvinyl chloride or polycarbonate (134), all of which are considered to be rather thrombogenic surfaces (although colloidal graphite was used on some of the surfaces, presumably to reduce thrombo-genicity). One of the earliest papers describing an "active" biomaterial prepared by combining radiation graft polymerization plus biochemical and medical concepts is by Hoffman, et al.,(135). In this study streptokinase, albumin and heparin were immobilized on radiation grafted hydrogels based upon P-HEMA and P-NVP. Streptokinase immobilized via an "arm" demonstrated significant fibrinolytic activity. Immobilized heparin, on the other hand, did not seem to retain biological activity when immobilized to these surfaces using either BrCN or carbodiimide chemistries.

Nguyen and Wilkes have described a fibrinolytic surface made by immobilizing brinolase to Enzacryl, a particulate, crosslinked, modified polyacrylamide (133). Significant brinolase activity was maintained. The authors suggest a grafted polymer utilizing acrylic acid and N-acryloyl para-phenylene diamine as a more practical material for both immobilizing brinolase and construct-ing useful devices.

Another approach to preparing a blood compatible surface based upon the immobilization of biomolecules to hydrogels has been taken by Lee, et al. (95). They prepared a three layer support material with PVA at the surface. Onto the PVA they esterified first, half cholesterol esters of dicarboxylic acids and next, the half sialic acid ester of a longer chain dicarbox-ylic acid. The surface was finally treated with tissue culture medium to condition it with salts and proteins found in the blood. The rationale behind the material was to simulate the natural blood vessel intima. Vena cava ring tests indicated generally poor thromboresistance for these complex surfaces (113, 115). Renal embolus rings were relatively free of thrombus, but the kidneys of test animals were often massively infarcted (9).

It has been proposed that an "artificial pancreas" might be constructed by combining a blood glucose sensor with feedback to

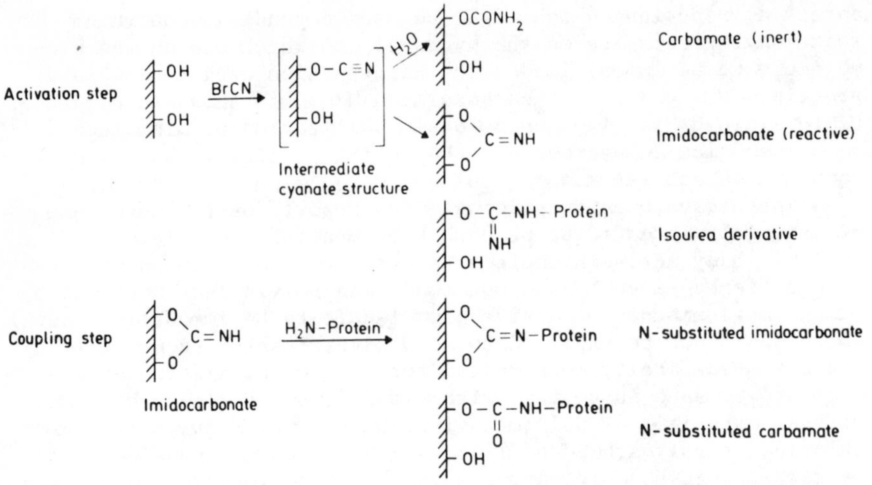

Transactions—American Society for Artificial Internal Organs

Figure 5. Immobilization of a protein to a polymer containing hydroxyl groups (135)

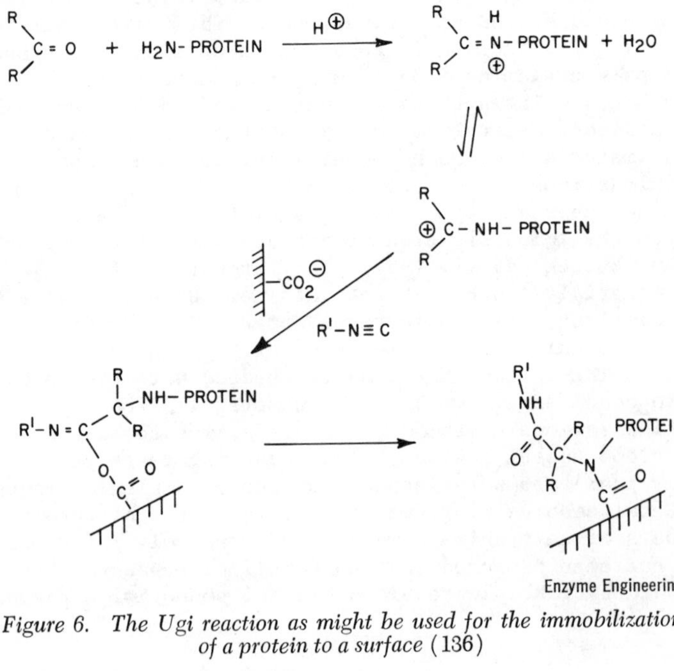

Enzyme Engineering

Figure 6. The Ugi reaction as might be used for the immobilization of a protein to a surface (136)

an insulin delivery system. Potentially, other health problems could also be treated using a combination of a blood sensor and a drug or hormone delivery system. Enzymatic electrodes which can be designed to show high specificity for a given type of molecule might be particularly well suited for use as blood sensors (157). Techniques have been recently described whereby an enzyme is simultaneously entrapped within a crosslinked hydrogel and coated onto a glass electrode. Polyacrylamide hydrogels have been utilized for this application (158). Such electrodes might be expected to show both high specificity and non-thrombogenicity making them particularly well suited for blood sensor applications.

VI. Conclusions

Hydrogels as a class of materials have shown great versatility and excellent performance when used in biomaterial applications. However, in the 15 years since synthetic hydrated polymeric networks were first proposed for biomaterial applications the "surface has barely been scratched" with respect to new types of hydrogels which might by synthesized, fundamental knowledge concerning how and why hydrogels "work", and new biomedical applications for these useful polymers. Specific areas which are clearly in need of further study before the full potential of biomedical hydrogels can be realized are:

(1) Thrombogenicity of hydrogel surfaces, particularly with respect to emboli formation.

(2) Cell adhesion and interaction with hydrated polymer networks, especially with cell types such as platelets, leukocytes and fibroblasts.

(3) The structure of water within hydrogels and its potential relationship to biological interactions.

(4) The behavior of biological molecules immobilized to hydrogels.

(5) The interaction of anionic and cationic hydrogels with blood and other biological systems.

Literature Cited

1. Kiraly, R.J. and Nosé, Y., Biomat., Med. Dev., Art. Org., (1974), 2, 207.
2. Homsy, C.A., J. Biomed. Mat. Res., (1970), 4, 341.
3. Hoffman, A.S., J. Biomed. Mat. Res. Symposium, (1974), No. 5 (Part 1), 77.
4. Bruck, S.D., J. Biomed. Mat. Res., (1973), 7, 387.
5. Bruck, S.D., Biomat. Med. Dev., Artif. Organs, (1973), 1,79.
6. Mason, R.G., Bull. N.Y. Acad. Med., (1972), 48, 407.
7. Gott, V.L. and Furuse, A., Fed. Proc., (1971), 30, 1679.
8. Daniels, A.U. and Mortensen, J.D., Biomat., Med. Dev., Artif. Organs, (1974), 2, 365.
9. Kusserow, B.K., Larrow, R.W. and Nichols, J., "Analysis and Measurement of the Effects of Materials on Blood Leukocytes, Erythrocytes and Platelets," Contract No. PH 43-68-1427, National Heart and Lung Institute, National Institutes of Health, Bethesda , Maryland, Annual Report, (Dec. 1, 1972), PB218-651.
10. Autian, J., Critical Reviews of Toxicology, (1973), 2, 1.
11. Coleman, D.L., King, R.N. and Andrade, J.D., J. Biomed. Mater. Res. Symposium, (1974), No. 5 (Part 1), 65.
12. Wichterle, O., in "Encyl. Polym. Sci. and Technol.", (1971), ed. by H.F. Mark and N.G. Gaylord, Vol. 15, 273.
13. Wichterle, O. and Lim, D., Nature, (1960),185, 117.
14. Ind. and Chem. Eng., (1936), 28, 1160.
15. Refojo, M.F. and Yasuda, H., J. Appl. Polymer Sci., (1965), 9, 2425.
16. Wichterle, O. and Chromecek, R., J. Polymer Sci., (1969), Part C, 16, 4677.
17. Gouda, J.H., Povodator, K., Warren, T.C., and Prins, W., Polymer Letters, (1970), 8, 225.
18. Ratner, B., Ph.D. Thesis, (1972), Polytechnic Institute of Brooklyn.
19. Bohdanecky, M. and Tuzar, Z., Coll. Czech Chem. Commun., (1969), 34, 3318.
20. Halpern, B.D., McGonigal, P.J. and Blessing, H.W., "Polymer Studies Related to Prosthetic Cardiac Materials Which are Non-Clotting at a Blood Interface," Contract No. PH 43-66-1124, National Heart and Lung Institute, N.I.H., Bethesda, Maryland, Annual Report, (Sept. 28, 1972), PB212-724.
21. Bruck, S.D., Personal Communication.
22. Sevcik, S., Stamberg, J. and Schmidt, P., J. Polymer Sci., (1967), Part C, 16, 821.
23. Stamberg, J. and Sevcik, S., Coll. Czech Chem. Commun., (1966), 31, 1009.
24. Wichterle, O. and Lim, D., U.S. Patent 3,220,960, (1965).
25. Ratner, B.D. and Miller, I.F., J. Polymer Sci., (1972), Part A-1, 10, 2425.
26. Refojo, M., J. Polymer Sci., (1967), Part A-1, 5, 3103.

27. Jadwin, T.A., Hoffman, A.S. and Vieth, W.R., J. Appl. Polymer Sci., (1970) 14, 1339.
28. Yasuda, H., Gochin, M. and Stone, Jr., W., J. Polymer Sci., (1966), Part A-1, 4, 2913.
29. Allen, L.F., A.C.S. Polymer Preprints, (1974), 15 (2), 395.
30. Ilavsky, M. and Prins, W., Macromolecules, (1970), 3, 415.
31. Dusek, K., Bohdanecky, M. and Prokopova, E., European Polymer J., (1974), 10, 239.
32. Janacek, J. and Hasa, J., Coll. Czech Chem. Commun., (1966), 31, 2186.
33. Janacek, J., J. Macromol. Sci., (1973), C9, 1.
34. Barvic, M., Kliment, K. and Zavadil, M., J. Biomed. Mater. Res., (1967), 1, 313.
35. Cerny, E., Chromecek, R., Opletal, A., Papousek, F. and Otoupalova, J., Scripta Medica, (1970), 43, 63.
36. Sprincl, L., Kopecek, J. and Lim, D., J. Biomed. Mater. Res., (1971), 5, 447.
37. Barvic, M., Vacik, J., Lim, D. and Zavadil, M., J. Biomed. Mater. Res., (1971), 5, 225.
38. Sprincl, L., Kopecek, J. and Lim, D., Calc. Tissue Res., (1973), 13, 63.
39. Sprincl, L., Vacik, J. and Kopecek, J., J. Biomed. Mater. Res., (1973), 7, 123.
40. Winter, C.D. and Simpson, B.J., Nature, (1969), 223, 88.
41. Levowitz, B.S., LaGuerre, J.N., Calem, W.S., Gould, F.E., Scherrer, J. and Schoenfeld, H., Trans. Amer. Soc. Artif. Int. Organs, (1968), 14, 82.
42. Tollar, M., Stol, M. and Kliment, K., J. Biomed. Mater. Res., (1969), 3, 305.
43. LaGuerre, J.N., Kay, H., Lazarus, S.M., Calem, W.S., Weinberg, S.R. and Levowitz, B.S., Surg. Forum, (1968), 19, 522.
44. Lazarus, S.M., LaGuerre, J.N., Kay, H., Weinberg, S. and Levowitz, B.S., J. Biomed. Mater. Res., (1971), 5, 129.
45. Majkus, V., Horakova, Z., Vymola, F. and Stol, M., J. Biomed. Mater. Res., (1969), 3, 443.
46. Drobnik, J., Spacek, P. and Wichterle, O., J. Biomed. Mater. Res., (1974), 8, 45.
47. Scott, H., Kronick, P.L., May, R.C., Davis, R.H. and Balin, H., Biomat.,Med. Dev., Art. Org., (1973), 1, 681.
48. Michnevic, I. and Kliment, II., K., J. Biomed. Mater. Res., (1971), 5, 17.
49. Warren, A., Gould, F. E., Capulong, R., Glotfelty, E., Boley, S.J., Calem, W.S. and Levowitz, B.S., Surgical Forum, (1967), 18, 183.
50. Singh, M.P. and Melrose, D.G., Biomedical Eng., (1971), 6, 157.
51. Singh, M.P., Biomedical Eng., (1969), 4, 68.

52. Halpern, B.D., Cheng, H., Kuo, S. and Greenberg, H., in
     Artificial Heart Program Conference Proceedings, ed. by
     R.J. Hegyeli, U.S. Government Printing Office, Washington,
     D.C. (1969), 87.
53. Refojo, M.F., J. Biomed. Mater. Res., (1969), 3, 333.
54. Refojo, M.F., Survey of Ophthalmology, (1972), 16, 233.
55. Krejci, L. and Krejcova, H., Brit. J. Ophthal., (1973), 57,
     675.
56. Kocvara, S., Kliment, Ch., Kubat, J., Stol, M., Ott, Z. and
     Dvorak, J., J. Biomed. Mater. Res., (1967), 1, 325.
57. Kliment, K., Stol, M., Fahoun, K. and Stockar, B., J.
     Biomed. Mater. Res., (1968), 2, 237.
58. Molday, R.S., Dreyer, W.J., Rembaum, A. and Yen, S.P.S.,
     J. Cell Biol., (1975), 64, 75.
59. Refojo, M.F., Preprints - Division of Organic Coatings and
     Plastics Chemistry, A.C.S., (1967), 27, (2), 136.
60. Andrade, J.D., Kunitomo, K., Van Wagenen, R., Kastigir, B.,
     Gough, D. and Kolff, W.J., Trans. Amer. Soc. Artif. Int.
     Organs, (1971), 17, 222.
61. Spacek, P. and Kubin, M., J. Biomed. Mater. Res., (1973), 7,
     201.
62. Ratner, B.D. and Miller, I.F., J. Biomed. Mater. Res.,(1973),
     7, 353.
63. Rubin, R.M. and Marshall, J.L., J. Biomed. Mater. Res.,
     (1975), 9, 375.
64. Winter, G.D., Proc. Roy. Soc. Med., (1970), 63, 1111.
65. Holly, F.J. and Refojo, M.F., J. Am. Optom. Assoc., (1972),
     43, 1173.
66. Refojo, M.F., Contact and Intraocular Lens Med. J.,(1975),
     1, 153.
67. Leibowitz, H.M., Laing, R.A. and Sandstrom, M., Arch.Opthal.,
     (1973), 89, 306.
68. Friendly, D.S., Bruner, B.S., Frey, T., Lederman, M.E.,
     Parks, M.M. and Oldt, N., Arch. Ophthal. (1973), 90, 344.
69. Refojo, M.F., J. Appl. Polymer Sci., (1965), 9, 3161.
70. Refojo, M.F., Anales de la Real Sociedad Española de Fisica
     y Quimica, (1972), 68, 697.
71. Nierzwicki, W. and Prins, W., J. Appl. Polym. Sci., (1975),
     19, 1885.
72. Kudela, K., Stoy, A. and Urbanova, R., European Polymer J.,
     (1974), 10, 905.
73. Moens, J. and Smets, G., J. Polymer Sci., (1957), 23, 931.
74. Nagase, K. and Sakaguchi, K., J. Polymer Sci., (1965), A3,
     2475.
75. Pinner, S.H., J. Polymer Sci., (1953), 10, 379.
76. Kopecek, J., Sprincl., L., Bazilova, H. and Vacik, J., J.
     Biomed. Mater. Res., (1973), 7, 111.
77. Jenkins, L.B., Kredel, F.E. and McCord, W.M., A.M.A. Archives
     of Surgery, (1956), 72, 612.
78. Mohn, G., Acta Histochem., (1960), 9, 76.

79. Burnette, L.W., Proc. Sci. Section T.G.A., (Dec. 1962), <u>38</u>,1.
80. Williams, K.E., Kidston, E.M., Beck, F. and Lloyd, J.B.,
     J. Cell Biol. (1975), <u>64</u>, 113.
81. Rivano, R., Lago, T.F. and Biancheri, V., Pathologica(Genoa),
     (May-June, 1957), <u>49</u>, 301.
82. Richards, V. and Persidsky, H., J. Cardiovasc. Surg.,(1964),
     <u>5</u>, 313.
83. PVP - An Annotated Bibliography, 1951-1966, General Aniline
     and Film Corporation, (1967).
84. Luttinger, M. and Cooper, C.W., J. Biomed. Mater. Res.,
     (1967), <u>1</u>, 67.
85. Davis, B.K., Proc. Nat. Acad. Sci. U.S.A., (1974), <u>71</u>,3120.
86. Maroudas, N.G., Nature, (1975), <u>254</u>, 695.
87. Markley, L.L., Bixler, H.J. and Cross, R.A., J. Biomed.
     Mater. Res., (1968), <u>2</u>, 145.
88. Marshall, D.W., Cross, R.A. and Bixler, H.J., J. Biomed.
     Mater. Res., (1970), <u>4</u>, 357.
89. Bixler, H.J., Cross, R.A. and Marshall, D.W., in Artificial
     Heart Program Conference Proceedings, ed. by R.J. Hegyeli,
     U.S. Government Printing Office, Washington, D.C. (1969),
     79.
90. Bruck, S.D., J. Biomed. Mater. Res., (1971), <u>5</u>, 139.
91. Alder, R.H. and Darby, C., U.S. Armed Forces Medical Journal,
     (1960), <u>11</u>, 1349.
92. Merrill, E.W., Salzman, E.W., Wong, P.S.L., Ashford, T.P.,
     Brown, A.D. and Austen, W.G., J. Appl. Physiol., (1970),
     <u>29</u>, 723.
93. Merrill, E.W., Salzman, E.W., Wong, P.S.L. and Silliman, J.,
     ACS Polymer Pre-Prints, (1972), <u>13(1)</u>, 511.
94. Bray, J.C. and Merrill, E.W., J. Biomed. Mater. Res.,(1973),
     <u>7</u>, 431.
95. Lee, H., Stoffey, D.G. and Abroson, F., in Artificial Heart
     Program Conference Proceedings, ed. by R.J. Hegyeli, U.S.
     Government Printing Office, Washington, D. C., (1969), 143.
96. Sawyer, P.N. and Srinivasan, S., in Artificial Heart Program
     Conference Proceedings, ed. by R.J. Hegyeli, U.S. Govern-
     ment Printing Office, Washington, D.C., (1969), 243.
97. Leonard, F., Trans. Amer. Soc. Artific. Int. Organs, (1969),
     <u>15</u>, 15.
98. Walter, C.W., Murphy, W.P., Jessiman, A.G., and Ahara, R.M.,
     Surgical Forum, (1951), <u>2</u>, 289.
99. Bray, J.C. and Merrill, E.W., J. Appl. Polymer Sci., (1973),
     <u>17</u>, 3779.
100. Ratner, B.D. and Hoffman, A.S., in "Biomedical Applications
     of Polymers", (1975), ed. by H.P. Gregor, Plenum Press,
     N.Y., 159.
101. Ratner, B.D. and Hoffman, A.S., J. Appl. Polymer Sci.,(1974),
     <u>18</u>, 3183.
102. Lee, H.B., Shim, H.S. and Andrade, J.D., ACS Polymer Pre-
     prints, (1972), <u>13(2)</u>, 729.

103. Yasuda, H. and Refojo, M.F., J. Polymer Sci., (1964),
      Part A, 2, 5093.
104. Leininger, R.E., Cooper, C.W., Falb, R.D. and Grode, G.A.,
      (1966), Science, 152, 1625.
105. Laizier, J. and Wajs, G., in "Large Radiation Sources for
      Industrial Processes", IAEA, Vienna, (1969), 205.
106. Miller, M.L., Postal, R.H., Sawyer, P.N., Martin, J.G. and
      Kaplit, M.J., J. Appl. Polym. Sci., (1970), 14, 257.
107. a. Hoffman, A.S. and Kraft, W.G., ACS Polymer Preprints,
          (1972), 13(2), 723.
      b. Hoffman, A.S. and Harris, C., ACS Polymer Preprints,
          (1972), 13(2), 740.
108. Ratner, B.D., Hoffman, A.S. and Whiffen, J.D., Biomat.,
      Med. Dev., Artifi. Organs, (1975), 3, 115.
109. Horbett, T.A. and Hoffman, A.S., ACS Advances in Chemistry
      Series, (1975), 145, 230.
110. Ratner, B.D., Horbett, T., Hoffman, A.S. and Hauschka, S.D.,
      J. Biomed. Mater. Res., (1975), 9, 407.
111. Kearney, J.J., Amara, I. and McDevitt, M.B., in "Biomedical
      Applications of Polymers," (1975), ed. by H.P. Gregor,
      Plenum Press, N.Y., 75.
112. Scott, H. and Hillman, E.E., "Active - Vapor Grafting of
      Hydrogels in Medical Prosthesis," Contract No. NIH-HHL1-
      71-2017, Natrional Heart and Lung Institute, National
      Institutes of Health, Bethesda, Maryland, Annual Report,
      (Feb. 1, 1973), PB-221-846.
113. Gott, V. and Baier, R.E., "Evaluation of Materials by Vena
      Cava Rings in Dogs (Vol. 1)", Contract No. PH 43-68-84,
      National Heart and Lung Institute, National Institutes
      of Health, Bethesda, Maryland, Annual Report, (Sept.1972),
      PB-213-109.
114. Kwiatkowski, G.T., Byck, J.S., Camp, R.L. Creasy, W.S., and
      Stewart, D.D., "Blood Compatible Polyelectrolytes for Use
      in Medical Devices", Contract No. NO1-HL3-2950 T, National
      Heart and Lung Institute, National Institutes of Health,
      Bethesda, Maryland, Annual Report, (July 1, 1972-June 30,
      1973), PB-225-636.
115. Gott, V.L. and Baier, R.E., "Twelve Month Progress Report
      on Contract No. PH 43-68-84, National Heart and Lung Insti-
      tute, National Institutes of Health, Bethesda, Maryland,
      Annual Report, (Apr. 30, 1970), PB-197-622.
116. Halpern, B.D. and Blessings, H.W., "Polymer Studies Related
      to Prosthetic Cardiac Materials which are Non-Clotting at
      a Blood Interface", Contract No. PH 43-66-1124, National
      Heart and Lung Institute, National Institutes of Health,
      Bethesda, Maryland, Annual Report, (Feb. 2, 1973),
      PB-215-886.
117. Muzykewicz, K.J., Crowell, Jr., E.B., Hart, A.P., Schultz,M.,
      Hill, Jr., C.G., and Cooper, S.L., J. Biomed. Mater. Res.,
      (1975), 9, 487.

118. Meaburn, G.M., Cole, C.M. Hosszu, J.L., Wade, C.W., and
     Eaton, J., Abstracts - Fifth International Congress of
     Radiation Research, (1974), Seattle, Washington,
     July 14-20, 200.
119. Abrahams, R.A. and Ronel, S.H., Society Plast. Eng. - Tech-
     nical Papers, (1975), 21, 570.
120. Lagunoff, D. and Horbett, T.A., unpublished observations.
121. Ratner, B.D., Hoffman, A.S. and Whiffen, J.D., unpublished
     observations.
122. Eisenberg, D. and Kauzmann, W., "The Structure and Proper-
     ties of Water", Oxford University Press, N.Y.,(1969).
123. "Water - A Comprehensive Treatise", Vol. 1, ed. by F. Franks,
     Plenum Press, New York, (1972).
124. Drost-Hansen, W., in "Chemistry of the Cell Interface",
     Part B, ed. by H.D. Brown, Academic Press, N.Y., (1971),
     p. 1.
125. Ling, G.N., Int. J. Neuroscience, (1970), 1, 129.
126. Klotz, I.M., in "Membranes and Ion Transport", Vol. 1, ed.
     by E.E. Bittar, John Wiley, Inc., N.Y.,(1970), p. 93.
127. Andrade, J.D., Lee, H.B., Jhon, M.S., Kim, S.W., and Hibbs,
     Jr., J.B., Trans. Am. Soc. Artif. Int. Organs, (1973),
     19, 1.
128. Jhon, M.S. and Andrade, J.D., J. Biomed. Mater. Res.,
     (1973), 7, 509.
129. Lee, H.B., Andrade, J.D. and Jhon, M.S., A.C.S. Polymer
     Preprints, (1974), 15(1), 706.
130. Lee, H.B., Jhon, M.S. and Andrade, J.D., J. Colloid Inter-
     face Sci., (1975), 51, 225.
131. Khaw, B., M.S. Thesis, (1973), University of Washington.
132. MacKenzie, A.P. and Rasmussen, D.H., in "Water Structure
     at the Water-Polymer Interface" (1972), ed. by H.H.G.
     Jellinek, Plenum Publishing Corp., N.Y., 146.
133. Nguyen, A. and Wilkes, G.L., J. Biomed. Mater. Res., (1974),
     8, 261.
134. Kusserow, B.K., Larrow, R.W. and Nichols, J.E., Trans. Am.
     Soc. Artif. Int. Organs, (1973), 19, 8.
135. Hoffman, A.S., Schmer, G., Harris, C., Kraft, W.G., Trans.
     Am. Soc. Artif. Int. Organs, (1972), 18, 10.
136. Hersh, L.S., in "Enzyme Engineering, Vol. 2", ed. by E.K.
     Pye and L.B. Wingard, Jr., Plenum Press, N.Y., (1974),
     p. 425.
137. Broun, G., Tran-Minh, C., Thomas, D., Domurado, D. and
     Selegny, E., Trans. Am. Soc. Artif. Int. Organs, (1971),
     17, 341.
138. Chang, T., "Artificial Cells", Charles C. Thomas, Spring-
     field, Ill., (1972), p. 150.
139. Guilbault, G.G. and Lubrano, G., Anal. Chim. Acta., (1973),
     64, 439.
140. Eiseman, B. and Soyer, T., Transplant. Proc., (1971), 3,
     1519.

141. Guilbault, G.G. and Lubrano, G., Anal. Chim. Acta., (1974), 69, 189.

142. Kirwan, D.J., "Removal of Airborn Infections," (1974), presented at Purdue University, Conference on Enzyme and Antibody Engineering - Economics and Directions, Jan. 23-24, 1974, Lafayette, Indiana.

143. Baker, R.W. and Lonsdale, H.K., in "Controlled Release of Biologically Active Agents", Tanquary, A.C. and Lacey, R.E., Eds., Plenum Press, N.Y., (1974), pp. 15-71.

144. Abrahams, R.A. and Ronel, S.H., J. Biomed. Mater. Res., (1975), 9, 355.

145. Morawetz, H., in "Advances in Catalysis and Related Subjects" Vol. 20, (1969), ed. by D.D. Eley, H. Pines and P.B. Weisz, Academic Press, N.Y., 341.

146. Gutcho, S.J., "Immobilized Enzymes - Preparation and Engineering Techniques", Noyes Data Corporation, Park Ridge, N.J., (1974), 141.

147. Refojo, M.F., J. Appl. Polymer Sci., (1965), 9, 3417.

148. White, M.L., J. Phys. Chem., (1960), 64, 1563.

149. Hinberg, I., Kapoulas, A., Korus, R. and O'Driscoll, K., Biotech. Bioeng., (1974), 16, 159.

150. Maeda, H., Suzuki, H., Yamauchi, A. and Sakimae, A., Biotech. Bioeng., (1975), 17, 119.

151. Denti, E., Biomat., Med. Dev., Artif. Organs, (1974), 2,293.

152. Zaborsky, O.R., "Immobilized Enzymes", CRC Press, Cleveland, Ohio, (1973).

153. Brown, H.D. and Hasselberger, F.X., in "Chemistry of the Cell Interface", Part B, ed. by H.D. Brown, Academic Press, N.Y., (1971), p. 185.

154. Axen, R., Porath, J. and Ernback, S., Nature, (1967), 214, 1302.

155. Axen, R., Vretblad, P. and Porath, J., Acta. Chem. Scand., (1971), 25, 1129.

156. Ternynck, T. and Avrameas, S., FEBS Letters, (1972), 23,24.

157. Gough, D.A. and Andrade, J.D., Science, (1973), 180, 380.

158. Guilbault, G.G. and Hrabankova, E., Anal. Chim. Acta., (1971), 56, 285.

Vapor Pressure and Swelling Pressure of Hydrogels

MIGUEL F. REFOJO

Eye Research Institute of Retina Foundation, Boston, Mass. 02114

Hydrogels consist of two components: the polymer network, which is constant in quantity, and the aqueous component, which is variable. At equilibrium swelling, the chemical potentials of the water in the gel and the water surrounding the gel are equal. The addition to the solution surrounding the gel of macromolecules that are too large to penetrate the gel lowers the chemical potential of the water in the solution. Water thus moves out of the gel and the network contracts, decreasing the chemical potential of the water in the network to the value of the water in the solution. Therefore, at equilibrium the osmotic pressure of the macromolecular solution equals the swelling pressure of the hydrogel. The degree of hydration that can be achieved by equilibration with a macromolecular solution can also be obtained by compressing the gel under a mechanical pressure equivalent in magnitude to the osmotic pressure of the macromolecular solution. The mechanical pressure raises the chemical potential of the water in the gel, so water exudes from the gel until equilibrium is reached. Thus the equilibrium water content depends on the mechanical pressure applied to the gel.

The swelling phenomena of gels have been the object of thermodynamic analysis (1,2). The osmotic pressure attributed to the polymer network (π) is the driving force of swelling. The swelling process distends the network, and is counteracted by the elastic contractility of the stretched polymer network (p). Hence the swelling pressure of nonionic hydrogels (P) is the result of the imbibition of solvent driven by an osmotic pressure, counteracted by the contractility of the network which tends to expel the solvent: $P = \pi - p$. At equilibrium, $\pi = p$, and the swelling pressure is zero ($P = 0$).

Swelling pressure can be defined as the pressure exerted by a gel when swelling is constrained but swelling solvent is available. In general, the following empirical relationship (I), developed by Posnjak in 1912 (3), applies to a swelling substance:

$$P = k \times c^n \qquad\qquad (I)$$

where P is the swelling pressure, k and n are constants whose values are usually between 2 and 3, and c is the polymer network concentration. Since n>1, expression (I) indicates the well-known fact that the swelling pressure, P, increases rapidly with the concentration of the polymer in the gel.

In this paper, the swelling pressure of two classes of hydrogels of interest to ophthalmology is investigated. The first class is a group of highly hydrated hydrogels (above about 80% water at swelling equilibrium in water) which is intended for surgical use. The second class consists of hydrogels used to manufacture contact lenses, and these gels are hydrated to about 40 to 75% water at swelling equilibrium in water.

Effect of the External Solution upon the Hydration of Hydrogels. The magnitude of equilibrium swelling of a hydrogel in an aqueous medium is determined by the chemical potential of water in the outside solution. The chemical potential of water in the outside solution is determined by the nature and concentration of the solutes in the solution. The solutes, depending on their molecular size, may or may not penetrate the polymer network. Some solutes which can penetrate the network may interact with the polymer segments, modifying their stretching and contracting properties. Nevertheless, all solutes in the gel water affect the gel by lowering the chemical potential of its water.

Solution tonicity is a biological concept. It is related to osmotic pressure, but it lacks exact physicochemical meaning. Thus, various isotonic solutions may swell or deswell hydrogels depending on the penetration and interaction of the solutes with the network segments (4) (Fig. 1).

Swelling Pressure of High Water-Content Glyceryl Methacrylate Hydrogels and their Ophthalmic Applications. The swelling pressure-volume relationship of poly(glyceryl methacrylate) hydrogels (PGMA) is of practical interest for the development of swelling or expanding surgical implants. A swelling implant is a device that can be placed inside an organ or tissue through a relatively small incision. By imbibing available body fluid it will swell to fill a cavity or to alter the form of the surrounding tissues (5). To be useful, a swelling implant must swell to several times its dry volume under the conditions of implantation. In some applications, the implant must exert sufficient swelling pressure to counteract the constraining pressure of the surrounding tissues. Of interest is a coherent vitreous substitute which could be used to fill the vitreous cavity of the eye (6), and in some cases, the entire eyeball (7).

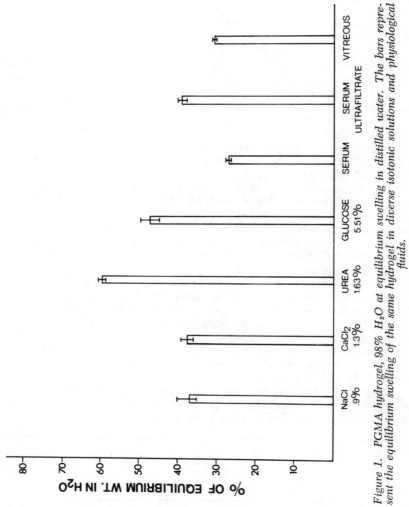

Figure 1. PGMA hydrogel, 98% H_2O at equilibrium swelling in distilled water. The bars represent the equilibrium swelling of the same hydrogel in diverse isotonic solutions and physiological fluids.

Another type of swelling implant is used in ophthalmology in the scleral buckling procedure in retinal detachment surgery (8).

Glyceryl methacrylate monomer (GMA) prepared by hydrolysis of glycidyl methacrylate (9) yields GMA with small amounts of some still unidentified crosslinking agent. However, hydrogels with reproducible physical and chemical properties can be obtained from GMA prepared by this procedure. Polymerization is carried out in glass molds filled with aqueous solutions of GMA with a redox initiator. The monomer dilution in the prepolymer mixture determines the equilibrium degree of swelling of the resulting hydrogel.

The swelling pressure of PGMA hydrogels was obtained by equilibration in solutions of dextran (Pharmacia, mol wt 236,000). The PGMA specimens were placed in dextran solutions and allowed to equilibrate in tightly capped jars at room temperature. Equilibrium swelling was reached after four to ten weeks, depending on the concentration of the solution, and on the size of the specimen. When equilibrium swelling was reached, the dextran concentration was determined from aliquots of the solution. The osmotic pressure of the dextran at different concentrations was determined osmometrically (4).

The dextran molecules in water solution have an ellipsoidal shape. The diameter of the dextran molecules used in these experiments is about 270 Å (10). The dextran molecules are not likely to penetrate into PGMA hydrogels both because a) the average pore size of the hydrogels is smaller than the size of the dextran molecules (for instance, PGMA hydrogels of 94% H_2O have an average pore diameter of about 124 Å)(11), and b) when a highly hydrated gel is placed in a dextran solution, the osmotic dehydration of the gel is faster than the diffusion of the dextran molecules into the gel. As a gel dehydrates, pore size is reduced, further limiting penetration by large dextran molecules.

Figure 2 gives the swelling pressure of several PGMA hydrogels versus the "swelling ratio" (q). The swelling ratio is defined as the volume of the swollen gel over the volume of the same gel in the dry state.

The "swelling ratio" was calculated from the "degree of swelling" (γ), the density of the swollen gel (d), and the density of the dry gel (d_0), according to (II):

$$q = \gamma \frac{d_0}{d} \qquad (II)$$

γ is the ratio of the weights of the swollen gel to the dry gel (4). The densities were obtained from Figure 3, which gives the density of a PGMA hydrogel versus the weight fraction of water in the swollen gel (Cw = % H_2O/100). Densities were determined by the hydrostatic weighing method. Gels were weighed both in air,

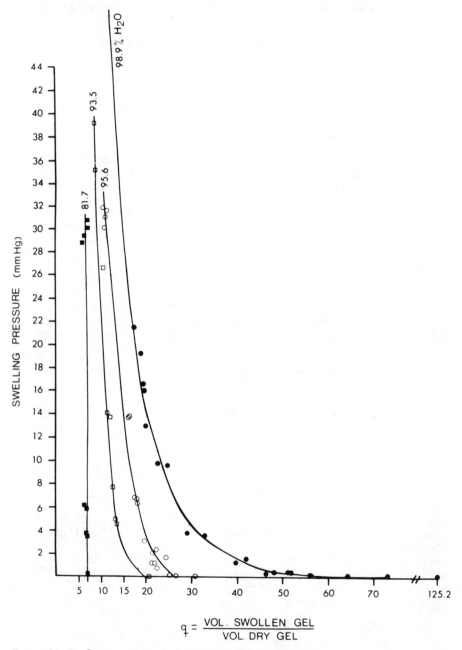

$$q = \frac{\text{VOL. SWOLLEN GEL}}{\text{VOL. DRY GEL}}$$

Figure 2. Swelling pressure vs. "swelling ratio" of several PGMA hydrogels. The equilibrium swelling of the PGMA hydrogels in distilled water is given for each swelling pressure curve in the graph.

and immersed in n-heptane at room temperature. γ is related to Cw by:

$$\gamma = \frac{1}{1-Cw} \qquad\qquad (III)$$

The density of the different PGMA hydrogels was obtained from Figure 3 assuming that the densities of the dry gels (xerogels) and hydrogels were not affected appreciably by the amount of crosslinkage.

The results given in Figure 2 show that a substantial amount of water is removed from highly hydrated hydrogels (jellies) when they are subjected to a slight compression. Thus, the volume of PGMA hydrogel containing 98.9% H_2O by weight (q = 125.6) was more than halved in volume (q = 45) under about 1 mm Hg osmotic pressure. These jellies exude liquid water when they are allowed to stand under their own weight in the air. As the water content in the hydrogel decreases, the pressure required to compress water out of the gel increases. Hence, PGMA with 95% H_2O by weight (q = 30) at equilibrium in water lost about one third of its volume of water (q = 20) under only about 4 mm Hg, but PGMA hydrogel with 82% H_2O by weight (q = 6.9) lost practically no water under ten times as much osmotic pressure. The relationship of hydration by weight, Cw, and swelling ratio by volume (q) is given by:

$$q = (\frac{1}{1-Cw})\frac{d_o}{d} \qquad\qquad (IV)$$

Hence, when the value of Cw is near one, such as in jellies, small differences in hydration represent substantial volume changes.

The swelling pressure properties of jellies and hydrogels are important from the point of view of the two kinds of swelling implants mentioned above, a vitreous substitute and a scleral buckling device. A vitreous substitute must mimic the natural vitreous body, which is a highly hydrated jelly. The gel is implanted in its dry state into the eyeball through the smallest possible incision. Then it absorbs available intraocular fluid, swelling freely as long as it does not adjoin the walls of the eye, until it fills the vitreous cavity. Fully swollen, the implant must occupy the vitreous cavity while exerting a minimum of pressure against the extremely sensitive retina.

The second type of swelling implant is placed on the outside of the eyeball, in the sclera. This implant, upon swelling, exerts pressure to buckle the wall of the eye inward, thereby approximating the choroid that carries the blood supply to a detached retina. Such an implant must be soft to avoid pressure necrosis in the sclera, but not so fragile that it will crumble under pressure. About a five-fold swelling against the constraining tissue is sufficient. For this application, a hydrogel with 80 to 85% water content at equilibrium swelling in water appears to be most useful.

Vapor Pressure and Swelling Pressure of Hydrogel Contact Lens Materials. Since Wichterle and Lim (12) first proposed the use of hydrogels for contact lenses and other medical devices, many new hydrogels have been developed (13). The first commercial hydrogel contact lenses were made of slightly crosslinked poly(2-hydroxyethyl methacrylate) (PHEMA), which is still the material most often used in the soft lens industry. A second compound used to make hydrogel contact lenses is vinylpyrrolidone (VP), in the form of a copolymer of HEMA and VP, P(HEMA/VP) (14). Lenses which contain VP, but no HEMA are also made, such as a copolymer of methyl methacrylate and VP, P(MMA/VP) (15).

As different hydrogel contact lenses become available, it is of interest to investigate and to compare their relative water retention. Differences in hydration at swelling equilibrium are important in the evaluation of the optical and physiological performance of the lenses. This study determined the equilibrium swelling of several hydrogel contact lens materials under osmotic and mechanical pressure, as well as the water activity of the hydrogels under various swelling conditions.

1. Determination of Swelling Pressure of a PHEMA Hydrogel by Equilibrium Swelling in Dextran Solution. PHEMA I was obtained by solution polymerization (16). At equilibrium swelling in distilled water, it contains 40% H_2O by weight on a wet basis. PHEMA I specimens were placed in dextran-40 solutions and allowed to equilibrate in tightly capped jars at room temperature [Different values have been reported for average pore diameter of PHEMA I hydrogels; the maximum is 35 Å (13). Dextran-40 in water has a molecular diameter of about 105 Å (10)]. Equilibrium swelling was reached after two months. After equilibration, the dextran concentration in the jar was determined gravimetrically. The osmotic pressure (π, in atm.) of dextran (mol wt 26,000) was calculated according to equation (V):

$$\pi = A_1c + A_2c^2 + A_3c^3 \qquad (V)$$

where c is the concentration (in $g \cdot cm^{-3}$) and $A_1 = 0.852$ atm·cm³. g^{-1}, $A_2 = 13.52$ atm·cm⁶g⁻², and $A_3 = 66.8$ atm·cm⁹·g⁻³ are the virial coefficients according to Vink (17). The results of the swelling pressure of PHEMA I obtained by this procedure are given in Figs. 4,5.

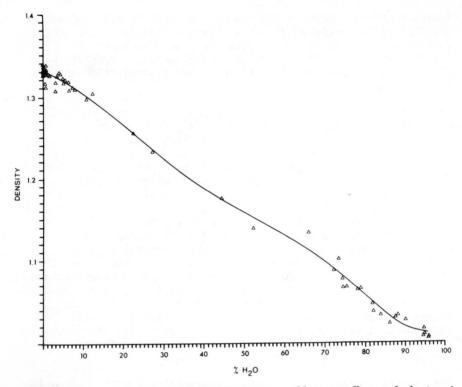

Figure 3.   Density of PGMA hydrogel, 96% H₂O at equilibrium swelling, vs. hydration of the gel.   The curve was printed by a computer using the least squares method applied to the data points.

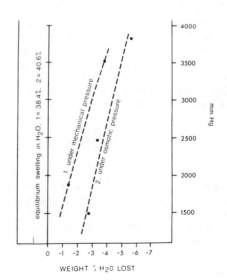

Figure 4.   Dehydration of two PHEMA hydrogels under osmotic and mechanical pressure

   2. <u>Determination of Swelling Pressure of a PHEMA Hydrogel by Equilibrium Swelling Under Mechanical Compression.</u> PHEMA
II was obtained by bulk polymerization (16). It contains 38.5%
H_2O at equilibrium swelling in distilled water. Circular pieces
of PHEMA II hydrogel 20 mm in diameter (1 to 2 mm thick) were
placed between two sintered glass disks in a water bath under
iron weights (five to ten kilograms), which were separated from
the upper sintered glass disc by a plastic cylinder protruding
above the surface of the water bath. The water bath containing
the hydrogel disc and the weights on top of it were kept in a
closed chamber to prevent evaporation. Equilibrium swelling of
the hydrogel was obtained after several weeks of compression.
   The relationship between hydration and swelling pressure of
PHEMA II obtained by this procedure is shown in Figs. 4,5.
   While a slight pressure easily removes water from the high
water-content hydrogels (Fig. 2), it is more difficult to expel
water by osmotic or mechanical means from hydrogels having low
water-content, such as the commonly used PHEMA contact lens
hydrogels, which contain about 40% water at equilibrium swelling
(Figs. 4,5). Similar results are expected from hydrogels containing less than 80% water at equilibrium (Fig. 2). Most hydrogel contact lens materials contain about 35 to 75% water at
equilibrium in physiological saline solution. It requires substantial pressure to remove water from hydrogel lenses, which is
advantageous because if lid pressure were to squeeze water from
the lenses in the eye their performance would suffer. Of course,
the optical properties, the shape, and the size of a hydrogel
lens are all dependent on its water-content.

   3. <u>Determination of Water Activity in Hydrogels.</u> One way
to facilitate water loss from hydrogels is to decrease the relative humidity; this decreases the chemical potential of the water
vapor in the surrounding atmosphere to a low value. It is well
known that hydrogels can lose water rapidly by evaporation. When
this happens, the network, which is under elastic tension, will
contract. As the hydrogel dehydrates, the chemical potential of
the water remaining in the gel decreases and is manifested as an
imbibition pressure, which is equal in magnitude to the osmotic
or mechanical pressure needed to compress the gel to the same
partially dehydrated state.
   The water retention, or water escaping tendency (fugacity)
of hydrogels was determined (Figs. 6,7) by measuring the equilibrium relative humidity (% ERH) of the hydrogels at 32°C, which
is approximately the surface temperature of the eye.
   Four different hydrogels were used in these experiments:
(a) a PHEMA hydrogel with 42.5% H_2O at equilibrium swelling; (b)
a copolymer of methyl methacrylate and vinyl pyrrolidone, P(MMA/
VP), used in the manufacture of hydrogel contact lenses under the
trade name Sauflon (containing about 70% H_2O at equilibrium
swelling in distilled water); (c) a copolymer of HEMA and VP

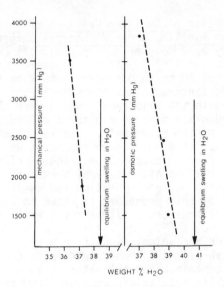

Figure 5. Swelling pressure of two PHEMA hydrogels equilibrated under mechanical pressure (PHEMA II, 38.5% H₂O) and osmotic pressure (PHEMA I, 40% H₂O), respectively

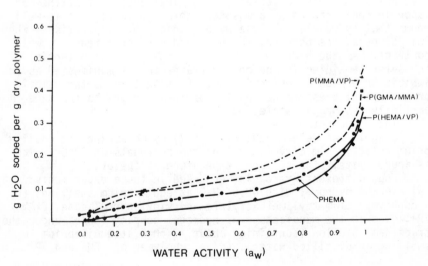

Figure 6. Water sorption isotherms of hydrogel contact lens materials

[P(HEMA/VP)], known as PHP and also used in contact lenses. It contains about 45% water at equilibrium swelling in water; and (d) a copolymer of MMA and GMA [P(GMA/MMA)] which contains 41% water at equilibrium swelling.

The equilibrium relative humidity (water activity) was determined with the hydrogel in a closed container having a calibrated humidity and temperature sensor connected by cable to a receiver (Hygrodynamics Universal Hygrometer Indicator, American Instrument Co., Silver Springs, Maryland). The chamber containing the hydrogel and the sensor was maintained in a constant temperature dry incubator at 32°C. When relative humidity reached equilibrium, the weight of the gel was recorded. The operation was repeated for different states of hydration to obtain the isotherms in which the weight of water sorbed per unit of dry polymer weight was plotted with reference to water activity (Fig. 6). The water activity was obtained under desorption conditions, except for P(HEMA/VP), which was tested under resorption conditions. The isotherms obtained have the standard sigmoid shape of water sorption in polymers. Figure 7 shows the same results of water activity versus water content in the hydrogels, expressed in percent hydration on a wet basis, which is conventionally used in hydrogel literature.

ERH (relative humidity of the space around the hydrogel, when moisture will neither leave nor enter the hydrogel) is a ratio of existing partial vapor pressure of water in the hydrogel (P_e) to the saturation water vapor pressure in the air (P_s). It is given by:

$$\% \text{ ERH} = \frac{P_e}{P_s} \times 100 \qquad (VI)$$

P_e/P_s is, of course, the "water activity" (a_w) in the hydrogel at the given temperature. Thus, the water activity in hydrogels at different levels of hydration can be determined directly and could be used to calculate the swelling pressure (P) of the hydrogels at different hydrations according to:

$$P = \frac{RT}{V_w} \ln \frac{1}{a_w} \qquad (VII)$$

where R is the universal gas constant, T the absolute temperature, and V_w the partial molar volume of water. Under the experimental conditions (Fig. 7), small differences in water activity in the hydrogels were not detectable at hydrations above about 30% water. Not only the hydrogels at equilibrium swelling in water, but liquid water as well, gave an ERH just below the true value of 100%. This is a limitation of the hygrosensor used.

Because of the large value of RT/V_W (about 1390 atm. at 32°C) a very small vapor-pressure decrease can result in a substantial increase in swelling pressure. Thus, the hygrometer technique cannot be used to determine the swelling pressure of hydrogels near equilibrium hydration. The sorption isotherms of four hydrogels are given in Figure 6. The sigmoid shape of the curves may be an indication of the three classes of water which, according to Lee, Jhon, and Andrade (18), may be present in these hydrogels. Because of the difficulty of obtaining the equilibrium relative humidity of a xerogel, the section of the curves at lower hydration are not as sharply defined as the other two sections.

The ERH is a measure of "free water-vapor pressure" which is, in essence, a measurement of the "freedom of water" or its "escaping tendency" (19). Therefore, the hygrodynamics experiments do give a good indication of the proportion of "free" water of hydration in hydrogels, that is, water in the aqueous phase of the hydrogel with the same vapor pressure as liquid water at the same temperature. Water activity versus hydrogel hydration (Fig. 7) shows that the water of hydration in hydrogels above about 25 to 30% has approximately the same vapor pressure as liquid water. Figure 8, thus, represents the amounts of "free" water ($a_W \approx 1$) and somewhat "bound" water ($a_W < 1$) in hydrogels as replotted from Figure 6. The amount of "bound" water, about 30% of water in the hydrogels, is similar to the amounts that Lee, Jhon, and Andrade (18) assigned as "bound" plus "interfacial" water. The rest of the water of hydration in the hydrogels is "free" or "bulk" water.

Thus, the contact lens materials examined all seem subject to losing substantial amounts of water by evaporation. The fact that about 30% of the water is retained more tenaciously in the lens materials does not seem to have any practical importance from the point of view of water retention of hydrogel lenses and optical performance. Of course, frequent blinking and good tear supply are essential for good results with all hydrogel contact lenses.

Osmotic Effects Due to the Aqueous Phase of a Hydrogel Contact Lens. In addition to the swelling pressure (or imbibition pressure) of hydrogels, which is a property of the polymer phase of gels, there are other osmotic effects attributable to the aqueous phase that are of particular interest in the contact lens field.

Most of the aqueous phase of a hydrogel can be freely exchanged with the surrounding aqueous media by diffusion. The movement of water, ions and other dissolved substances is restricted to some degree by friction with the polymer network. However, due to the large surface area of contact lenses relative to their thickness, most of the aqueous phase will interchange with the tears in a few minutes.

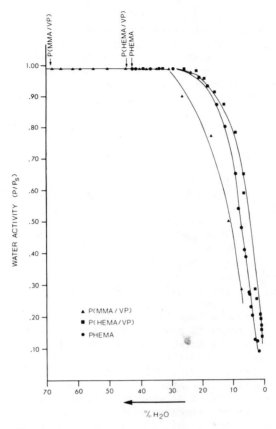

Figure 7. Water activity in hydrogel contact lens materials vs. hydration

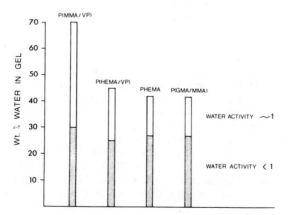

Figure 8. Amounts of "free" water ($a_w \simeq 1$) and "bound" water ($a_w < 1$) in hydrogel contact lens materials

The aqueous phase in a contact lens can be isotonic, hypotonic or hypertonic with respect to tears. When a hydrogel lens is equilibrated in isotonic saline solution (0.9% NaCl), the aqueous phase of the hydrogel is isotonic to the normal tears. When such a lens is placed in the eye, there will be an interchange of its aqueous phase and the tear film, but the tear tonicity will not change in the process of equilibration of the lens. The possible change in size and optics of a lens from equilibrium swelling in isotonic solution and in tears is negligible.

If the hydrogel lens is wetted with tap or distilled water prior to placing it in the eye, the aqueous phase of the lens will be hypotonic to tears. Water will move from the lens to the tears. The lens will then contract, often adhering tenaciously to the cornea and causing much discomfort. Exchange of tears and the aqueous phase of the gel will take place until a state of isotonicity is reached and then the lens will release from its adhesion to the cornea.

If the lens is placed in a sodium chloride solution of more than 0.9%, the aqueous phase of the hydrogel will be hypertonic to tear film. As the lens is placed in the eye, water will be drawn osmotically from the tear film into the lens, but salt will also diffuse from the lens into the tear. This will change the tonicity of the tears to a hypertonic state. The lens effect can be quite large as the total volume of the tear film and the tear meniscus is roughly comparable to the water content of a hydrogel lens. The hypertonic tears will dehydrate the corneal epithelium resulting in ocular discomfort (itching) to the patient. As the isotonicity of the tears is re-established through dilution, the sensation of comfort is again restored.

The existence of a thin aqueous film between a hydrogel lens and the corneal epithelium is a matter of controversy. If the lens is in direct contact with the cornea, the osmotic effect due to the aqueous phase of the hydrogel lens will be acting directly upon the corneal epithelium with the same results discussed above.

Abstract

The swelling pressure of two types of hydrogels, which were classified according to application and hydration, were investigated. 1. Poly(glyceryl methacrylate) (PGMA) hydrogels which are intended for surgical uses in the eye, are divided into two subgroups, a) hydrogels of about 80 to 85% H_2O at equilibrium swelling, for scleral buckling procedures in retinal detachment surgery, and b) hydrogels of above 98% H_2O at equilibrium swelling, for vitreous implantation. The swelling pressure-volume relationship of these hydrogels was determined with dextran solutions. 2. Hydrogel contact lens materials (40-75% H_2O at equilibrium). The swelling pressure of PHEMA hydrogels was determined by equilibration under osmotic and mechanical pressure.

Although slight pressure will remove water from highly hydrated hydrogels, it is very difficult to expel water from medium hydration (40-80% H_2O) hydrogels. The equilibrium relative humidity at different states of hydration of PHEMA, P(HEMA/VP), P(MMA/VP), and P(GMA/MMA) was investigated. The results show that up to 30% H_2O, the water activity in the hydrogels is below the activity for pure water. However, all water above 30% hydration up to equilibrium swelling has the same vapor pressure as pure water.

Acknowledgements

The author is indebted to Dr. F. Holly for his helpful discussion. Ms. F.L. Leong provided technical assistance. A. Seidman provided editorial assistance.

This study was supported by PHS Grant EY-00327 of the National Eye Institute, National Institutes of Health.

Literature Cited

1. Katchalsky, A.: Prog. Bioph. (1954) 4:1
2. Rijke, A.M., and Prins, W.: J. Polym. Sci. (1962) 59:171
3. Posnjak, E.: Kolloidchem. Beih. (1912) 3:417, in M.R. Kruyt, Ed. Colloid Science II. Elsevier Publishing Co., New York, 1949, p 557
4. Refojo, M.F.: Anales Qui. (Madrid) (1972) 68:697
5. Refojo, M.F.: J. Biomed. Mater. Res. Symposium (1971) 1:179
6. Daniele, S., Refojo, M.F., Schepens, C.L., and Freeman, H.M.: Arch. Ophthal. (1968) 80:120
7. Dohlman, C.H., Refojo, M.F., Webster, R.G., Pfister, R.R., and Doane, M.G.: Arch. Ophthal. (Paris) (1969) 29:849
8. Calabria, G.A., Pruett, R.C., and Refojo, M.F.: Arch. Ophthal. (1971) 86:77
9. Refojo, M.F.: J. Appl. Polym. Sci. (1965) 9:3161
10. Senti, F.R., Hellman, N.N., Ludwig, N.M., Babcock, G.E., Tobin, R., Glass, C.A., and Lamberts, B.L.: J. Polym. Sci. (1955) 17:527
11. Refojo, M.F.: J. Appl. Polym. Sci. (1965) 9:3417
12. Wichterle, O. and Lim, D.: Nature (1960) 185:117
13. Refojo, M.F.: Encycl. Polymer. Sci. Technol. Supplement (in press)
14. Seiderman, M.: U.S. Patent 3,721,657 (1973)
15. Frankland, J.D., and Highgate, D.J.: Ger. Offen. (1973) No. 2,312,470 [in Chem. Abstr. 80:27872z]
16. Refojo, M.F., and Yasuda, H.: J. Appl. Polymer Sci. (1965) 9:2425
17. Vink, H.: European Polym. J. (1971) 7:1411
18. Lee, M.B., Jhon, M.S., and Andrade, J.D.: J. Colloid and Interface Sci. (1975) 51:225
19. Quinn, F.C.: Reprint No. 472, Am. Instrument Co., Div. Travenol Labs, Silver Springs, Md.

3

The Role of Proteoglycans and Collagen in the Swelling of Connective Tissue

F. A. MEYER, R. A. GELMAN, and A. SILBERBERG

Weizmann Institute of Science, Rehovot, Israel

Connective tissue provides the chemical environment for most body cells and the mechanical content in which they function. It varies in composition according to this function and characteristically involves a fibrillar skeletal network composed mainly of collagen fibrils and the so-called ground substance, a highly disperse system of proteoglycans, macromolecules involving both protein and carbohydrate chains.

Bone, cartilage, vessel walls, skin, tendon, umbilical cord and synovial fluid are typical examples of connective tissues. All of them are characterized by a low cell content. The major biological roles fulfilled by connective tissues in the body are physical in nature. They involve mechanical function such as motion and transport and (chemical) communication between cells. In its mechanical role connective tissue transmits energy from the muscles permitting the performance of work and the absorbance of stresses from the environment thereby protecting delicate structures. Clearly, the physico-chemical properties and the structure of connective tissue are the important factors fixing these features.

The structure and chemistry of the individual connective tissue components viz. collagen, elastin and proteoglycans can be summarized as follows:

Collagen (1). Collagen is a fibrous protein which is formed by the assembly of monomer units of tropocollagen. Tropocollagen is a rod shaped molecule ~300 nm long and 1.4 nm in diameter with a molecular weight of 300,000. It consists of three chains of equal length. Five tropocollagen molecules can pack to give a microfibril, the units are longitudinally staggared around the microfibril axis. Microfibrils pack in a tetragonal lattice to give fibrils. There is a gap between consecutive tropocollagens in the fibril which together with the staggering gives collagen a characteristic 65 nm repeat distance. Stability of the structure is due in part to the quarter staggering, which optimizes secondary bond interactions and covalent crosslinks as well

occur between and within tropocollagen units. The highly
ordered compact arrangement within the fiber gives it a rigid
character.

Large variations are seen in the degree of crosslinking,
fibril diameter, fibril organization and fibril concentration
in various tissues. Moreover, four genetically distinct types
of collagen exhibiting some tissue specificity have been characterized which in part may explain the variations of collagen
morphology seen.

Elastin (2). Elastin is generally involved with the
fibrillar network. It contains an unusually high proportion of
non-polar amino acids and the amino acids, desmosine and iodo-
desmosine, which may be unique to this protein. A corpuscular
structure has been proposed for elastin where a large part of
the peptide chain is folded forming globular centers. Crosslinking occurs at the surface between such units. The basis of
the   elastic nature of this protein arises from the work required to unfold the hydrophobic regions.

Ground Substance Components (3). The major ground substance components are proteoglycans which consist of glycosaminoglycans attached to protein. The most abundant glycosaminoglycans are hyaluronic acid, chondroitin sulfate, keratan
sulfate, and dermatan sulfate. The glycosaminoglycans are
linear chains consisting of repeating negatively charged di-
saccharide units. The proteoglycan, hyaluronic acid, is of
high molecular weight and in solution adopts a very voluminous
random coil configuration. It is associated with a small amount
of protein (<0.5%) which is not important for the integrity of
the molecule. This is the dominant proteoglycan found in loose
connective tissue, e.g. skin, vessel walls, heart valves and
umbilical cord and is present at concentrations of less than 1%
based on tissue volume.

In loose connective tissue the system is bathed in a
medium which contains about 3% serum protein. The tissue is
maintained in a steady state in contact with the microcircula-
tion of the blood and the lymph system. Exchange with blood
involves the non-selective filtration of water and low molecular weight constituents (salt, metabolites, etc.) through the
endothelial wall of the capillary system, their reabsorption
on the venous side and a one way transfer of serum albumin and
other macromolecular plasma constituents to the tissue. These
macromolecules are recovered predominantly through the lymph
system, which is formed in open channels in direct and open
contact with tissue and is returned to the blood through an
active pumping system. The blood, on the other hand, communi-
cates with the tissue through a partially selective membrane
represented by the vessel wall. In response to pathological

changes, tissue will swell and deswell, passing from one steady state to another.

We have investigated the structural organization of loose connective tissue and have established the roles of various connective tissue components in effecting tissue volume changes. These results could be of interest in the development of hydrogels for medical application and will here be reviewed.

Experimental

Umbilical cord (Wharton's Jelly) was used in these studies. The methods have been described (4,5,6) and will here only be summarized. Tissue slices (some 300-600 mg in weight) were incubated in small volumes of swelling medium (saline) containing added macromolecules or other constituents. Weights both of the tissue and the external solution were monitored as was the concentration of added macromolecules outside. In this way both weight (volume) changes could be calculated and the distribution of the macromolecular species determined. Using various enzymes, specific modifications into tissue structure could be introduced and evaluated.

Results

The most important results can be reviewed as follows:

1. **Tissue Slices Incubated in Serum Albumin Solutions (up to about 12%) in Ringer's Physiological Saline**. The final equilibrium volume of tissue is larger than the initial volume at all concentrations (Figure 1). The dependence of the equilibrium swelling ratio on serum albumin concentration is given in Figure 2. If extrapolation of the data is permitted, no swelling would occur at about 15% serum albumin. Only a small fraction of the hyaluronic acid (as measured by uronic acid release) is lost during incubation.

2. **Treatment of the Samples with Testicular Hyaluronidase**. This produces a tissue of abolished swelling tendency and practically no volume changes occur against any of the serum albumin concentration used (Figure 2). If previously swollen samples are enzymatically treated, they deswell to the original volume (Figure 3).

3. **Treatment of Tissue with Proteolytic Enzymes, For Example, Trypsin**. This produces a final equilibrium volume larger than the control where no trypsin treatment has been used (Figure 4).

A trypsin treatment followed by a hyaluronidase treatment, or vice versa, results in the same final equilibrium volume which is, however, larger than the original volume of the tissue sample. The same final volume results irrespective of when the enzymes were added during the swelling process (Figure 4).

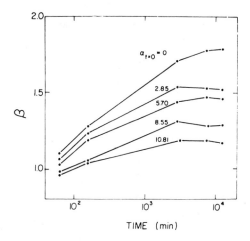

TIME (min)

Figure 1. The degree of swelling (β) with respect to the initial tissue volume as a function of time. Incubated at 4° at neutral pH in solutions of bovine serum albumin whose initial concentration ($\alpha_{t=0}$) is given in percent (w/v). In the case of the more concentrated solutions there is an initial volume decrease. Eventually all samples are swelling, and the final equilibrium volume is larger than the initial.

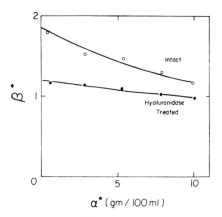

α^* (gm / 100 ml)

Figure 2. The equilibrium degree of swelling (β*) plotted against external serum albumin concentration (α*) at equilibrium for the intact (○) and hyaluronidase-treated (●) tissue. Note that in contrast to the intact tissue, treatment with hyaluronidase produces a tissue of practically no swelling tendency. The dependence of 1/β* on α* is linear. So plotted, the data show no swelling tendency of intact tissue at about α* = 15%.

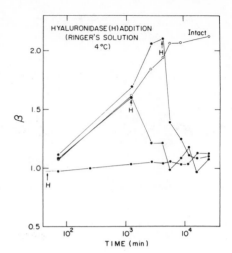

Figure 3. The swelling (β) as a function of time for intact tissue (○) and for tissue treated with hyaluronidase (●). Treatment with the enzyme occurs at the times indicated by the arrows. Such treatment returns the tissue to its original volume irrespective of the degree of swelling of the tissue.

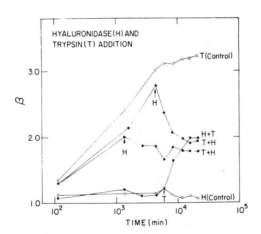

Figure 4. The swelling of tissue (β) as a function of time for tissue treated with either trypsin or hyaluronidase (○) or a combination of the two (●) at the times indicated by the arrows. Note that the equilibrium value of β after trypsin treatment is 3.2 as compared with a value of about 2.1 for intact tissue (see Figure 3). Trypsin treatment followed by hyaluronidase treatment or vice versa gives the same equilibrium value of β independent of the time and order of treatment.

Trypsin, chymotrypsin and pronase each results in a specific final equilibrium volume.

4. Treatment of the Tissue with Glutaraldehyde. This abolishes the tendency of tissue to swell. The tissue volume is fixed at the volume at which the treatment was applied.

5. Interaction of the Tissue with High Molecular Weight Poly(L-Lysine). At physiological salt concentrations this causes one-to-one binding of the polybase by the acid groups of the proteoglycans. In this condition the proteoglycans are extensively protected from subsequent hyaluronidase action (5). The interaction also serves to freeze in the configuration in which the tissue finds itself at the time of interaction. Thus when incubating the tissue in a poly(L-lysine) solution, normal swelling (due to saline penetration) and tissue fixation (due to polybase penetration) take place. The higher the poly(L-lysine) concentration, the faster is the penetration and the closer is the ultimate volume, to which the tissue swells, to the original volume of the tissue sample (Figure 5). As the ionic strength of the suspending medium is increased to about 0.5 M, poly(L-lysine) is excluded totally from the tissue and no effect whatsoever on tissue volume is experienced (Figure 6).

An investigation of the tissue by electron microscopy (4) shows that the collagen fibers are randomly distributed in space and are composed of bundles of parallel fibrils some 54 nm in diameter and some 54 nm apart.

Discussion

These results are consistent with the following functional model (4): In the usual kind of gel, the network acts both as the swelling agent and the source of the elastic restraining force; in connective tissue these two functions are separated. The collagen based fibrillar network is the mechanical spring and is thermodynamically not active, whereas the proteoglycan component acts as the active swelling agent but is not directly forming the network. If the potential of the solvent medium is too low, fluid will enter the tissue and dilute the proteoglycans. Since the proteoglycans, however, are mechanically enmeshed with the fibrillar network, the driving force is transferred to the fiber network and thus produces its expansion. When the proteoglycans are digested away the collagen fiber system returns the tissue to its original volume and water content. Hence the unstressed state of the fiber network, i.e. the mechanical reference state, is the native state of the tissue.

Treatment with proteolytic enzymes alters the fibrillar system and shifts its forceless state to one of higher volume. Reaction with gluteraldehyde "freezes" in the particular deformation state of the system and no volume change at all

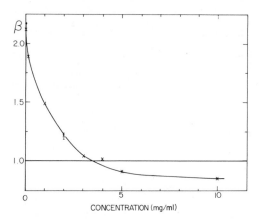

Figure 5.   Dependence of the ultimate value of the degree of swelling (β) on poly(L-lysine) concentration in 0.15M NaCl at neutral pH. Note that poly(L-lysine) binds strongly to proteoglycans at this sodium chloride concentration and freezes in the volume of the tissue at which it interacts with the polysaccharide. Hence, only at high poly(L-lysine) concentrations (fast polybase penetration into the tissue) is the poly(L-lysine) able to abolish swelling.

Figure 6.   The ultimate degree of swelling (β) as a function of salt concentration (A) and salt concentration in the presence of 1 mg/ml of poly(L-lysine) (B) at neutral pH. Note that there is interaction with poly(L-lysine) when the sodium chloride concentration is less than 0.5M. At a concentration of 1 mg/ml poly(L-lysine) the freezing in occurs only after some swelling has already taken place (see Figure 5).

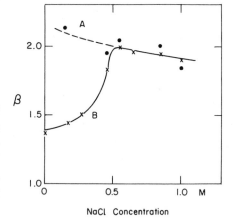

occurs. Interacting the tissue with poly(L-lysine) produces a
one-to-one complex with the proteoglycans and swelling is restricted. This is probably because the proteoglycans being
complexed are no longer osmotically active. Simultaneously
the interaction stiffens the proteoglycans and crosslinks them
into a network so that no further volume change can occur.

It is of interest, using non-interacting macromolecular
probes (6,7) that most of the tissue space, outside that excluded by the collagen fibrils, is accessible.

Tissue is thus a collagen fibril "gel" which is osmotically
active because of the presence of the proteoglycans. These may
form an independent network, but are more likely linked through
their physical involvement with the collagen system.

The finding that tissue is not in equilibrium with a protein
solution of composition in any way comparable with that of
interstitial fluid, confirms that mechanisms operate to maintain tissue, in vivo, in a "dehydrated" state. It may be
assumed that this "dehydrated" state of tissue is maintained
by the blood flow through exchange processes across the endothelial barrier which acts as a semipermeable membrane with
respect to macromolecules. The swelling tendency of tissue
due to the proteoglycans is thus offset permitting the collagen
fiber network, as it grows and assembles, to be in a force-free
state.

The results obtained in the present study describe the
behavior of umbilical cord. There are indications that other
types of connective tissue tend to exhibit similar behavior
(8), but not all forms of connective tissue are necessarily
expected to fall into this pattern. The division into an
osmotically active and inactive network system is believed,
however, to be generally correct.

Literature Cited

1. Traub, W. and Piez, K.A. Adv.Prot.Chem. (1971) 25, 243.
2. Partridge, S.M., in: "The Comparative Molecular Biology
    of Extracellular Matrices", H.C. Slavkin ed. (1972),
    Academic Press, New York and London.
3. Laurent, T.C. Pflügers Arch. (1972), 336 (Suppl.) S21.
4. Meyer, F.A. and Silberberg, A. Microvascular Res. (1974)
    8, 263.
5. Gelman, R.A. and Silberberg, A. Connective Tissue Res.,
    accepted for publication.
6. Meyer, F.A., Koblenz, M. and Silberberg, A., to be
    published.
7. Meyer, F.A. and Silberberg, A. Pflügers Arch. (1972)
    336 (Suppl.), S37.
8. Hedbys, B.O. Exp.Eye Res. (1961) 1, 81.

4

The Role of Water in the Osmotic and Viscoelastic Behavior of Gel Networks

MU SHIK JHON,* SHAO MU MA, SACHIKO HATTORI, DONALD E. GREGONIS, and JOSEPH D. ANDRADE

Department of Materials Science and Engineering, University of Utah, Salt Lake City, Utah 84112

The existence of an ordered structure at water/solid inter-faces has been generally accepted. A structured molecule possess certain preferred orientations and cannot move independent of its neighboring molecules. In the case of water the word "structured" should not be misinterpreted; we do not mean that water possess long range orderness. By "structure" we mean the orderness rela-tive to that in bulk water. In a strict sense, even the molecules in bulk water are structured because of hydrogen bonding and other near neighbor interactions.

Drost-Hansen (1) has discussed a three-layer model for the structure of water near certain water/solid interfaces. According to this model, water molecules near the solid surface are struc-tured; those sufficiently far away from the surface have bulk water structure and those in between have decreasing orderness as a function of distance from the interface. Others (2,3) have indicated the existence of three states of water in natural macro-molecular gels (2) or in membranes of cellulose acetate (3). Jhon and Andrade (4) proposed a three-state model of water in hydrogel systems. They suggested that three classes of water exist in hy-drogels, namely X water (bulk water), Z water (bound water), and Y water (intermediate forms or interfacial water). Following this, Lee, Jhon and Andrade (5,6) tested the model by thermal expansion, specific conductivity, differential scanning calorimetry, and proton spin-lattice nuclear magnetic relaxation studies for poly-(hydroxyethyl methacrylate) (PHEMA gels). Very recently, Choi, Jhon and Andrade (7) again extended the model by means of thermal expansion, specific conductivity, and dielectric relaxation studies for (2,3-dihydroxypropyl methacrylate) gels (DHPMA gels).

In this paper theories for the osmotic and viscoelastic behavior of hydrogels are developed in terms of water structure. Some experimental results on the viscoelastic behavior of hydrogels are presented.

*On leave from Korea Advanced Institute of Science, Seoul, Korea; to whom correspondence should be addressed.

Swelling and Osmotic Pressure

According to Flory (8), the network structure may have several roles. In a solvent the network dissolves and takes the role of a solute. In a solution it permits the passage of solvent molecules and keeps out other dissolved materials and, hence, acts as a membrane. As the network swells the polymer chains are elongated and exert a force in opposition to the swelling. In this case, the network acts as a pressure generating device. His theory of swelling is based on the balancing of the osmotic pressure by the mechanical contraction. To obtain his formal expression, Flory utilizes the Flory-Huggins polymer solution theory assuming random mixing between solute and solvent and a rigid lattice. The expression includes two terms, an entropy term (combinational) and a heat term due to intermolecular forces (non-combinational). Later, Prigogine et al. (9) developed a corresponding state theory for polymer solutions with a more rigorous expression for the non-combinational contribution than the Flory-Huggins approach. However, the combinational term is still used in its original form.

In the swelling of hydrogels, the random mixing assumption between high polymer and water is not generally valid, because of the high degree of structuring of water in some gel networks. The Prigogine theory, which is limited to non-polar and moderately polar systems with no hydrogen bonding, is also not generally applicable.

In this paper, a new semi-empirical interaction parameter, Δ_2, is introduced and used in an equation of swelling based on a "Solubility Model" (10) to avoid the above-mentioned difficulties. The equilibrium condition for isotropic swelling (9) classically requires that

$$\left(\frac{\partial \Delta F_m}{\partial N_1}\right)_{T,P} + \left(\frac{\partial \Delta F_{el}}{\partial N_1}\right)_{T,P} = 0, \qquad [1]$$

where ΔF_m and ΔF_{el} are the free energy of mixing and elastic behavior, respectively, and N_1 is the mole fraction of solvent. The "Solubility Model" (10) still carries the same elastic term, but the first term is modified. The process of transferring one mole of water from bulk to gel network involves three steps (Figure 1):

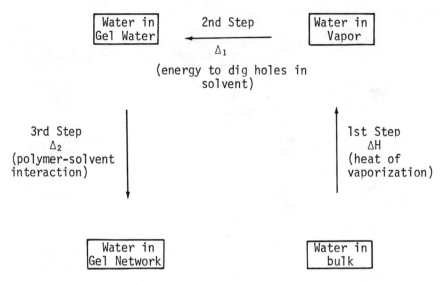

Figure 1. A schematic of the process of transferring one mole of water from bulk to gel network

First Step. To bring a mole of water from the bulk phase to the vapor phase: This requires an energy which equals the heat of vaporization of water, ΔH_v:

$$\Delta H_v = \Delta H_v^{\,o} + \int_0^T Cp \; dT, \qquad\qquad [2]$$

where $\Delta H_v^{\,o}$ is the heat of vaporization at some reference temperature, T, and Cp is the specific heat. Second step. To
bring water molecules from the vapor phase to the existing gel
water: At or near equilbrium the gel network is expanded with
water, which consists of X, Y and Z types, filling the empty
spaces. The energy, Δ_1, required in this step is for creating a
mole of cavity of the size of the solute molecule (water in
vapor) ag ainst the solvent (gel water) surface tension:

$$\Delta_1 = 4\pi r^2 (\gamma_z Z + \gamma_y Y + \gamma_x X)/(1 + K), \qquad [3]$$

where $4\pi r^2$ is the surface area of the water molecule; γ_x, γ_y,
γ_z, are the surface tensions and X, Y, Z are the weight
fractions of the three types of water, X, Y, Z respectively.
     The term $1/(1 + K)$ needs a little more explanation. Jhon,
Grosh, Ree and Eyring (11) proposed a model in which (bulk)

water is visualized as containing at least two solid-like structures, the ice-I-like open structure and the ice-III-like closed structure, in equilibrium with each other and with the gas-like molecules (A gas-like molecule refers to a molecule which is surrounded by holes). A term $1/(1 + K)$ is introduced in equation [3] on the assumption that the energy of cavity formation is neglible in the ice-I-like part of the liquid structure, and K is the equilibrium constant between ice-I-like and ice-III-like domains. Third step. To bring water molecules in gel water to gel network: This step requires an energy of interaction between the polymer molecules and the surrounding water molecules, Δ_2 (12):

$$\Delta_2 = fn \int_{AB}^{\infty} U(R) <S_{A+R+B}>_{Av} dR, \qquad [4]$$

where A and B denote water and high polymer, respectively; n is the number density of solvent molecules; f is the quantity which takes account of the fact that the distribution of solvent molecules of the potential minimum is denser than the average density; $U(R)$ is the intermolecular potential of A and B molecules at a separation distance, R; and $<S_{A+R+B}>_{Av}$ is the average value of the surface area. To evaluate $U(R)$ and $<S_{A+R+B}>_{Av}$, the reader should consult the cited references (10, 12). Although the new interaction parameter Δ_2 is difficult to evaluate directly, it can be obtained for available systems in the input data in Equation [2], Equation [3] and second terms in Equation [1] are provided.

Experiments to support this semi-empirical solubility theory would be first, to determine the interaction parameter, Δ_2, for various gel-water systems. These can be obtained by measuring swelling degrees as the only input data, the rest being available in the literature. Swelling experiments on PHEMA networks made of controlled purity are underway. The obtained Δ_2 will be compared to the Flory interaction parameter χ (8). Other solvent systems such as alcohols should also be studied. Second, a phase diagram may be constructed for the water-polymer system. Perfect symmetry at the consolute temperature will support the hypothesis of random mixing of water and polymer molecules. Deviations from it would indicate support of the solubility theory.

The three-state model of water structure can also be applied to osmotic pressure of polymer solutions. The usual expression for osmotic pressure can be used except that the volume should be replaced by $V(X fx)$, where fx is the ratio between the solubility of solute in X or bulk water to that in gel water. We believe that only the portion of water available for osmotic pressure should be included in the osmotic pressure equation. Tying both swelling and osmotic pressure experiments together with the three-state theory, one may

perform a swelling and/or osmotic pressure experiment of pure
water as a function of temperature. As water is heated from
- 30°C to 100°C, any change in swelling and/or osmotic pressure
would be partly due to melting of ice-I-like clusters (11).

Viscoelastic Properties

The effect of solvent on the viscoelastic behavior of
hydrogels has been widely reported in the literature (13,14,15).
In creep and stress relaxation measurements, the retardation
time and relaxation time are functions of solvent. In dynamic
studies the solvent reduces the γ relaxation process and
shifts the β process to lower temperatures. Furthermore, the
concentration and the nature of low molecular weight compounds
affect the size and shape of the secondary loss maximum as
well as the apparent activation energy. In equilibrium
studies, C_2 in the Mooney-Rivlin equation is affected by the
solvent.

Hydrogels, in the rubbery region, behave much like rubber.
Therefore, in this study, a theory for the stress-strain
relation in hydrogels was developed by modifying the theory
of rubber elasticity. Consider a freely orienting chain
which contains n segments. The force F needed to maintain
the chain at an average elongation $\bar{L}$ is given by the
expression:

$$F = \frac{kT}{\ell_0} L^* \left(\frac{\bar{L}}{n\ell_0}\right) \quad \text{or} \quad \bar{L} = n\ell_0 L\left(\frac{\ell_0 F}{kT}\right). \qquad [5]$$

The stress σ needed to maintain a rubber network at high
elongation is given by (16)

$$\sigma/\nu kT = \frac{1}{3} n^{\frac{1}{2}} L^*(\alpha/n^{\frac{1}{2}}) - \alpha^{-2}, \qquad [6]$$

where ν, α, ℓ_0, k and T are, respectively, the number of
chains per unit volume, the extension ratio, the segment
length, Boltzmann's constant, and the absolute temperature.
L and L* are the Langevin function and inverse Langevin
function, respectively which are defined by $x = L(y) = \coth$
$y - 1/y$ and $y = L^*(x)$, respectively. For a hydrogel with
mainly X water, it is reasonable to assume that the polymer
chains can rotate freely and Equations [5] and [6] apply.
However, for hydrogels with mostly Z water, due to the constrained state, only two limited conformations can occur,
i.e., internal isomerization between two conformers keeping
the position of each chain end fixed.

In magnetic theory, it is known that the magnetic field
shifts the relative amount of two orientations. For M magnetic dipoles each of which can exist either in the direction
of the magnetic field, H, or against the field, the relation
between the magnetization, I, and the magnetic field takes

the form [17] $I/m \ M = \tanh \left(\frac{mH}{kT}\right)$, where m is the magnetic moment. In polymer elasticity the force plays a corresponding role. The equivalent expression for polymer elasticity can be written as:

$$Z\left(\frac{\ell_0 F}{kT}\right) = \frac{e^{\ell_0 F/kT} - e^{-\ell_0 F/kT}}{e^{\ell_0 F/kT} + e^{\ell_0 F/kT}} = \tanh \left(\ell_0 F/kT\right). \qquad [7]$$

We shall refer to $Z\left(\frac{\ell_0 F}{kT}\right)$ as the Z function.

For hydrogels with mostly Y water structure, we expect that both the L function and the Z function should fail to apply and a Y function should govern:

$$Y(\ell_0 F/kT) = \int_0^{\pi} \sin\theta \ \cos\theta \ \tanh \left(\ell_0 F \cos\theta/kT\right) d\theta . \qquad [8]$$

In this case, not only two orientations (conformer) are permitted, but the position of each chain end is not fixed. With the same argument the numerical value of a Y function lies between the corresponding X function and Z function. Hence, equation [6] can be modified as follows:

$$\alpha/\nu kT = \frac{1}{3} n^{\frac{1}{2}} \left[XL^*(\alpha/n^{\frac{1}{2}}) + YY^*(\alpha/n^{\frac{1}{2}}) + ZZ^*(\alpha/n^{\frac{1}{2}}) - \alpha^{-2}\right], \qquad [9]$$

where Y^*, Z^* are the inverse Y function and inverse Z function, respectively.

   To test our hypothesis, the stress-elongation curves for three poly(hydroxyethyl methacrylate) gels with different water contents were obtained from stress-strain measurements at room temperature (23°C) (Figure 2). The observed data are given by the solid lines in the figure. The values of X, Y and Z are taken from Lee (5,6). Choosing n = 100 (we found that results obtained for n = 100 and n = 1000 do not differ significantly), the values of σ can be calculated as a function of α. The constants, νkT, needed for fitting the experimental curves are 1.09×10^7 dynes/cm^2 (Gel I, 45% water); 6.03×10^7 dynes/cm^2 (Gel II, 31% water); and 2.59×10^7 dynes/cm^2 (Gel III, 29.9% water). The calculated points are indicated in the figure. Gels I and II were prepared with the indicated amount of water in the polymerization mix while Gel III was prepared with 100% hydroxyethyl methacrylate monomer and then swelled to 29.9% water content. Although the water content in Gel II and Gel III are about the same, the former is considerably tougher than the latter.

   It is possible that, at 23°C, the behavior of Gel II is not within its rubbery region, since the agreement between the calculated and measured σ values for this gel is less

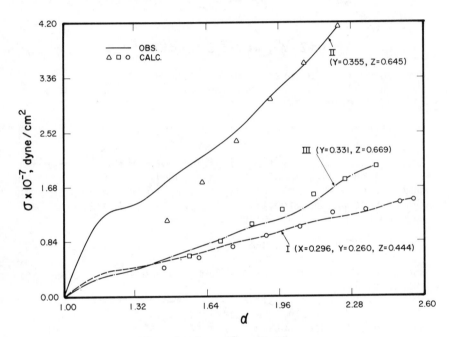

Figure 2.   Stress–elongation curve

satisfactory than those for the other two gels.

Summary

A theory is developed to interpret the osmotic, swelling and viscoelastic behavior of hydrogel networks in terms of a three-state model of water structure. For isotropic swelling under equilibrium conditions, Flory assumed random mixing between the solvent molecules and the polymer molecules. Since water molecules in hydrogels possess higher degrees of orderness than those in the bulk, it is believed that the solubility theory (10) should be used instead of the classical Flory theory. This is because solubility theory considers the free energy of mixing of water with the gel network, which includes the heat of vaporization of water, the energy required to create holes in the gel water and the energy of interaction between water and polymer. In our theory only the portion of water available for osmotic pressure was included in the osmotic pressure equation. For the viscoelastic behavior of hydrogels, the theory of rubber elasticity was modified to accommodate the effect of three types of water on their stress-strain relationship. Polymer chains with mainly X water can rotate freely, those with mainly Z water can only have two restricted conformations and those with mainly Y water can have intermediate behavior. In actual cases, the contributions from all three types of water should be considered since they are coexisting in any polymer-water system. A few experiments are proposed. According to some experimental results, the theory provides good agreement.

Acknowledgement

We gratefully acknowledge support of this work by NIH Grant HL16921-01 and NSF Grant GH38996X.

Abstract

A three-state model of water structure in hydrogels has been extended to describe the osmotic, swelling and viscoelastic behavior of gel networks. The solubility theory modification of the classical Flory theory is proposed to explain the osmotic and swelling behavior of gel networks. In describing the viscoelastic behavior of hydrogels, three functions, governed by the three types of water, are used to explain the stress-strain relations in the rubbery region.

Literature Cited

1. Drost-Hansen, W., Indust. Eng. Chem. (1969) 61, 10.
2. Aizawa, M., and Suzuki, S., Bull. Chem. Soc. Japan (1971) 44, 2907.
3. Krishnamurthy, S., McIntyre, D., and Santee, E. R., Jr., J. Polym. Sci. (1973) 11, 427.
4. Jhon, M. S., and Andrade, J. D., J. Biomed. Mater. Res. (1973) 7, 509.
5. Lee, H. B., Andrade, J. D., and Jhon, M. S., Polymer Preprints (1974) 15, 706.
6. Lee, H. B., Jhon, M. S., and Andrade, J. D., J. Colloid and Interface Sci. (1974) 51, 225.
7. Choi, S. H., Jhon, M. S., and Andrade, J. D., Submitted for publication.
8. Flory, P. J. "Principles of Polymer Chemistry," Cornell University Press, Ithaca, New York (1953).
9. Prigogine, I., Belleman, A., and Englert-Chowles, A., J. Chem. Phys. (1956) 24, 518.
10. Jhon, M. S., Eyring, H., and Sung, Y. K., Chem. Phys. Letters (1972) 13, 36.
11. Jhon, M. S., Grosh, J., Ree, T., and Eyring, H., J. Chem. Phys (1966) 44, 1465.
12. Kihara, T., and Jhon, M. S., Chem. Phys. Letters (1970) 7, 559.
13. Janacek, S., and Kolarik, J., Coll. Czech. Chem. Comm. (1965) 30, 1597.
14. Janacek, J., and Ferry., J. D., J. Poly. Sci. (1969) Part A-2, 7, 1681.
15. Ilavsky, M., and Prins, W., Macromolecules (1970) 3, 415.
16. Bueche, F., J. Appl. Poly. Sci. (1960) 4, 107.
17. Hill, T. L., "An Introduction to Statistical Thermodynamics," Addison-Wesley Publishing Company, Inc., Reading and London (1960).

Permeability as a Means to Study the Structure of Gels

N. WEISS and A. SILBERBERG

Weizmann Institute of Science, Rehovot, Israel

Permeation flow of the suspending medium past a non-moving gel matrix can also be interpreted as the translation of the gel substance through the stationary suspending medium. Hence the permeation coefficient bears some relationship to the sedimentation coefficient of an isolated macromolecule of the same chemical build-up as the gel substance. There should, in fact, be a one-to-one relationship between the two if the macromolecule and the gel network are both freely draining, i.e. if the hydrodynamic effects are linearly additive and the resistance to flow per gram of gel substance is the same as the translational friction coefficient per gram of the free macromolecule. Non-linear (in segment concentration) hydrodynamic interactions between the parts of the macromolecular chains will, however, affect the flow resistance and the permeation rates will depend upon the structural arrangements of the network strands. Hence measurements of permeability can be used for an analysis of gel structure.

The main difficulty in the way of using this approach has been the lack of suitable experimental equipment which would avoid large pressures and large pressure gradients in the gel.

A device will be described and results obtained with it discussed.

Experimental

The principle of the apparatus is illustrated in Figure 1. Two compartments are involved. Flow is from 1 to 2. Volume displacement is achieved by bending of a thin metal diaphragm by means of a piston driven by a linear actuator. Volume displacement is measured by means of a LVDT (linear variable differential transformer) connected to the piston and calibrated in terms of volume displacements in a separate set up involving a fine glass capillary. Pressure is measured in each compartment by means of sensitive, very small volume change pressure transducers. The cells are filled to completion by rotatory valves on each compartment.

Volume displacements of the order of 10^{-5} ml are aimed at
and such small volume changes correspond approximately to the
volume changes that might be expected from the response of the
pressure transducer, the finite compliance of the walls and the
compressibility of the fluid. To overcome this difficulty the
pressure in compartment 1 is maintained equal to ambient by means
of a feedback circuit which is driven by pressure transducer 1
and actuates the volume displacement device 1. Hence there is no
compression of the fluid in compartment 1 and the walls of cell 1
are not under stress. A similar feedback circuit keeps the
pressure in compartment 2 at a selected level $\Delta P = P_2 - P_1$ over
ambient by suitably adjusting the volume displacement device 2.
While volume changes deriving from the causes listed above do
arise in compartment 2; they are not important for the volume
measurement since only the pressure in this compartment has to
be known.

An outline drawing of the instrument is given in Figure 2
and the feedback control circuits are described in Figure 3.

The gels studied in the present investigation (1) are based
on acrylamide co-polymerized in water with N-N' methylene bis
acrylamide as crosslinking agent, ammonium persulphate as
initiator and N,N,N',N'-tetramethyl ethylenediamine as regu-
lator. Since this reaction is strongly influenced by oxygen the
reaction was carried out under nitrogen and in solutions through
which nitrogen had been bubbled to saturation. Nevertheless,
even if these precautions were taken, the upper layers of any gel
produced tended to be of different structure. To avoid such
effects interfering with the gel samples actually tested and in
order to produce well defined and well anchored gel blocks the
following procedure was adopted.

The special mould illustrated in Figure 4 was used. The
reaction mixture was filled into the mould under nitrogen, the
thickness of the final gel layer being controlled by spacer as
shown. The upper and lower parts was then removed as explained
in the text to the Figure and the cup with the gel layer in place
was clamped between the two cells of the instrument.

The gel mixtures prepared are characterized by the concentration of monomer and the percentage of crosslinking reagent in
the monomer mixture.

Measurements at one pressure were completed within one
minute in general and could be reproduced with great accuracy.

Sedimentation results were obtained on fractions of linear
polyacrylamide in a Model E Ultracentrifuge at the concentrations
shown and were corrected for pressure by extrapolation using the
method of Elias (2). The polymers were synthesized using the
same reaction procedure as for the gels but leaving out the multi-
functional monomer.

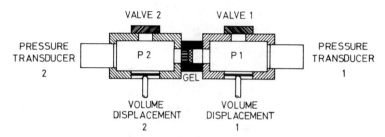

Figure 1.   Schematic outline of permeability apparatus

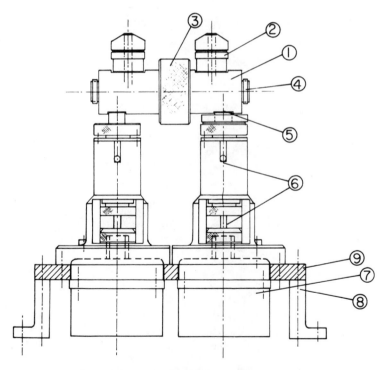

Figure 2.   Outline drawing of permeability apparatus. 1, half-cell; 2, rotatory valve; 3, connecting ring; 4, connection for pressure transducer; 5, volume displacement device; 6, LVDT; 7, linear actuator; 8,9, base.

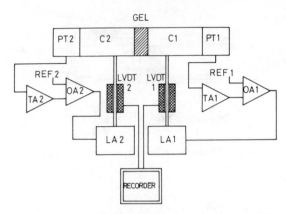

Figure 3. Feedback control circuit. PT, Statham P23D[d] pressure transducer; TA, Hewlett Packard transducer amplifier/indicator 311-A; OA, Kepco bipolar operational power supply BOP 36-1.5 (M); LA, Derritron VP.2 MM vibrator; LVDT, Sanborn Linearsyn differential transformer 585 DT-050; C, permeability half-cell.

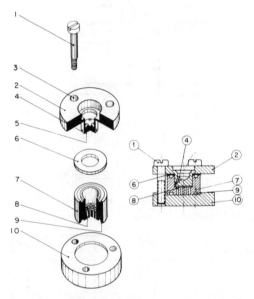

Figure 4. Special mould for preparation of gel blocks. 1, Clamping screws; 2, upper clamp; 3, holes for screws; 4, escape hole for excess gel mixture; 5, upper defining surface for gel block; 6, spacer ring defining block thickness; 7, cell with perforated bottom for gel block; 8, perforation; 9, seal; 10, lower clamp.

Theoretical Relationship between Permeation Coefficient K_s and the Sedimentation Constant ($\underline{3}$)

The flux J per unit cross-section of the gel is given by

$$J = (K_s/\eta)(-\nabla p) \quad , \tag{1}$$

where ∇p is the applied pressure gradient, K_s is the permeation coefficient and η is the viscosity of the suspending medium of the gel. It should be noted that

$$K_s/\eta = L_p \quad , \tag{2}$$

where L_p is the hydraulic volume flow permeation coefficient used in membrane analysis. In terms of compliance to flow L_p can be defined as the reciprocal of the friction coefficient, $\hat{f}_w$, per unit volume of suspending medium, i.e.

$$K_s/\eta = L_p = 1/\hat{f}_w. \tag{3}$$

The friction coefficient $\hat{f}_w$ is defined as the force exerted on the immobile gel matrix, per unit volume of the suspending medium, at unit relative velocity.

If the suspending medium is at rest (in the mean) and the gel substance (in the same configuration as before) is moving relative to it, we can define a friction factor $\hat{f}_g$ which is the force which has to act on unit volume of gel substance for it to move with unit velocity relative to the suspending medium. Since the force which would be required to move the gel substance through the suspending medium at a certain velocity must equal the force to move the suspending medium through the gel substance at the same (mean) velocity, we can write ($\underline{4}$):

$$(1/\hat{f}_w) = (1/\hat{f}_g)((1-\phi)/\phi) \quad . \tag{5}$$

If, in the case of movement of the gel substance through the suspending medium, the force acting upon the gel substance derives from a centrifugal field (ultracentrifuge) it may be questionable whether the gel membrane will not change configuration (the centrifugal but not the hydrodynamic forces are uniformly distributed). If, nevertheless, we make this identification

$$v/\omega^2 r = s = (1-\rho\bar{v})/\hat{f}_g \bar{v} \quad , \tag{6}$$

where v is the velocity of movement, ω is the angular velocity, r is the location, from the center, of the phase boundary in the ultracentrifuge, ρ is the density of the solution and $\bar{v}$ is the solute partial specific volume. Whereas s calculated from eqn. (6) may not therefore match s measured in an actual experiment, it nevertheless provides a rendition of the permeability data in a form which can be compared with sedimentation values. Hence combining equations (3), (5) and (6),

$$K_s/\eta = L_p = 1/\hat{f}_w = (1/\hat{f}_g)(1-\phi)/\phi$$

$$= (s\bar{v}/(1-\rho\bar{v}))((1-\phi)/\phi) . \qquad (7)$$

Equation (7) can be used to calculate an effective K_s, L_p or $\hat{f}_w$ value from a measurement of s, or an effective s (or $\hat{f}_g$) value from a measurement of K_s or L_p.

Since the volume fraction ϕ can be written

$$\phi = c\bar{v} , \qquad (8)$$

where c is the concentration of the gel substance in weight per unit volume, we can rewrite equation (7) as follows

$$cK_s = \eta (s) (1-c\bar{v})/(1-\rho\bar{v}) . \qquad (9)$$

An equation identical to this has been derived by Mijnlieff and Jaspers (5). It differs from equation (9) only in that the factor $(1-c\bar{v})/(1-\rho\bar{v})$ is written in the equivalent form $(1-\bar{v}/\bar{v}_w)$ where $\bar{v}_w$ is the partial specific volume of the solvent.

Mijnlieff and Jaspers attached physical significance to their result believing that permeation and sedimentation results could be compared in this way. It is important to note, however, that equation (9) can hold only in the unlikely event that exactly the same system is used both in determining K_s and s. In particular, data obtained from sedimentation studies, of uncrosslinked material, should not, when used in (9) be expected to match K_s, measured on a crosslinked system even of the same overall concentration and even if as in our case the experimental s-values were corrected for the high pressure in the ultracentrifuge.

On the other hand, as already pointed out, equation (9) represents a useful means of translating permeation into effective sedimentation results and vice versa. In particular, one may look upon sedimentation data of independent macromolecules as corresponding to zero crosslink density.

Results

The results calculated as cK_s either from equation (1), in the case of permeation data, or from equation (9) in the case of sedimentation are represented in Figure 5.

Sedimentation results depended on molecular weight. The data shown is for a sample of 1.57×10^5 M.W. At concentrations

above the minimum in the curve, however, the molecular weight
dependence of the sedimentation data was negligible.

Discussion

Molecular, Structural Interpretation of K_s

Since permeation flow occurs relative to the gel matrix,
the system is hydrodynamically two phase: the flowing suspending
medium and the immobile gel substance. If all the suspending
medium is flowing equally freely around all parts of the gel
substance we can replace the latter by a system of equally sized
spheres, evenly strung out along the strands of the network. If
there are ν such spheres per unit volume of gel substance and
each has a friction coefficient f_o,

$$\hat{f}_g = \nu f_o \tag{10}$$

and combining equations (7), (8) and (10)

$$cK_s = \eta \; (1/\nu f_o)(1-c\bar{v})/\bar{v} \; . \tag{11}$$

Since in the extreme free draining case envisaged, f_o by definition does not depend upon the density of crosslinks cK_s,
should be independent of crosslink density as well.

On the other hand, if we assume some clustering of the segments around points, for example, in regions where the crosslink
density locally is above average, these regions will act to
hinder flow through them (become non-draining) and a more appropriate picture to use is that of a number of oversized spheres,
equal to the number of clustered regions, acting as friction
centers (Figure 6). If there are again ν such spheres per unit
volume of gel substance each containing a length ℓ of gel network
chain and each of effective Stokes radius a,

$$f_o = 6\pi\eta a \tag{12}$$

and

$$\nu = 1/\ell A \; , \tag{13}$$

where A is the molecular cross-section of the gel network chains.
Combining equations (11), (12) and (13)

$$cK_s = [(A/6\pi)(1-c\bar{v})/\bar{v}](\ell/a) . \tag{14}$$

In equation (14) the term in square brackets, but not (ℓ/a), is
independent of the degree of crosslinking. Hence K_s as a function of the degree of crosslinking is determined solely by the
ratio (ℓ/a).

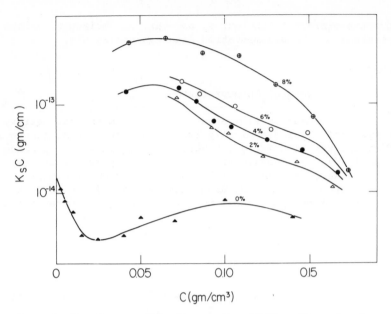

Figure 5. Specific permeability $K_sc = s_\eta(1 - (\bar{v})/(1 - \rho\bar{v})$ as a function of concentration c for aqueous polyacrylamide gels. Parameter: percent crosslinking monomer per ordinary monomer added. 0% corresponds to non-crosslinked material, and the data represent sedimentation results corrected for pressure.

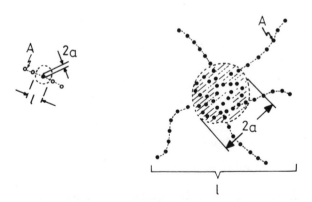

Figure 6. Freely and partially draining gel. $V = (Al/\bar{cv})$ $(1 - \bar{cv})$; V, effective volume of gel containing one hydrodynamic resistance center of radius a and a length of polymer chain l. (i) Freely draining case; (ii) partially draining case.

This ratio is a measure of the segment distribution around each point of resistance. If one can pack a large amount of chain (large ℓ) into a small volume (small a) permeation rates will be high. If, on the other hand, the length ℓ approaches the dimensions of the chain diameter, a, as well, will tend towards this value. For a given concentration, c, therefore, the lowest permeation rates will occur when $\ell/a \simeq 1$ and the chains are freely draining. Where there are structural inhomogeneities, however, for example a local clustering of crosslinks, $(\ell/a) \gg 1$ (many segments grouped in non-draining spaces) permeation rates will be high. This can be seen even better if equation (14) is rearranged to give

$$K_s = V/6\pi a \; . \tag{15}$$

Here

$$V = (1-c\bar{v}) /(c\bar{v}/A\ell) \tag{16}$$

is the volume of suspending medium associated with each effective sphere of hydrodynamic radius a (Figure 6). Clearly K_s will be the larger, the larger V and the smaller a, or in other words, K_s will be large if there is local crowding of the gel network around an effective crosslink leaving large volumes of suspending medium free of segments, or, trivially, if we go from a micro-porous network to a macro-porous network keeping the amount of gel substance fixed.

The results of Figure 5 interpreted as ℓ/a from equation (14) are used in Figure 7. The following relationship was used to evaluate $A/\bar{v}$:

$$A/\bar{v} = M_o/N_A b \tag{17}$$

where M_o is the molecular weight, b the length of the monomer unit and N_A is Avogadro's number. With $M_o = 71$ gm and $\underline{b} = 2.5 \times 10^{-8}$ cm, $A/\bar{v}$ is found to be 4.71×10^{-15} gm/cm. Since $\bar{v} = 0.701$ ml/gm the cross-section $A = 3.3 \times 10^{-15}$ cm^2.

If we put $V = \frac{4\pi}{3} R^3$ and assume that the volume fraction of polymer in the non-draining region of radius a is ϕ_{cl} we find that

$$\frac{\ell_o}{\ell} = \frac{a^3}{R^3} \frac{\phi_{cl}}{\phi} \tag{18}$$

where $\ell_o \simeq \ell - 4(R-a)$ is the actual length of chain involved with the non-draining region. It follows that

$$(\frac{R}{a})^3 [1-4 (\frac{R}{a} - 1)/ \frac{\ell}{a}] = \frac{\phi_{cl}}{\phi} \tag{19}$$

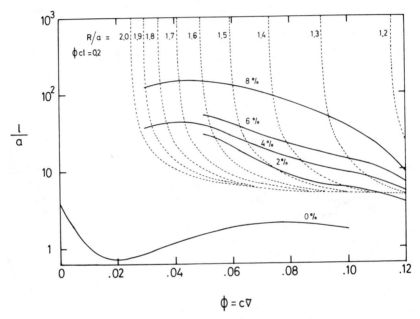

*Figure 7. Permeability data of Figure 5 recalculated according to Equation (14)
in relation to the model of Equation (19)*

We will assume that in the region where the results in Figure 5 come together at higher concentrations, the overall concentration ϕ just matches the concentration ϕ_{cl} in the cross-link regions. This suggests that $\phi_{cl} \simeq 0.2$ and we make the further assumption that this will now also be the concentration in the crosslink regions at all lower overall concentrations as well. With this we can calculate the dependence of ℓ/a on ϕ for given R/a using equation (19). The results are indicated by the dotted lines in Figure 7. The full lines in this figure are the results from Figure 5 redrawn using equation (14) for ℓ/a and equation (8) for ϕ.

It will be seen, interestingly, that $\ell/a \simeq 1$ for the non-crosslinked polymers but that it lies between 10 and 100 for the crosslinked systems. From the roster of lines at fixed R/a values we find that porosity indeed goes up as permeability is increased. At the same time, however, as larger regions become freely draining (increasing R/a) more and more polymer segments will be found outside the effective crosslink region. This will eventually tend to decrease permeability and accounts for the maximum in the curves.

Since $\ell A = \phi (4\pi/3) R^3$, we find using equation (19),

$$a^2 = A[\ell/a - 4 \, (\frac{R}{a} - 1)] / \frac{4\pi}{3} \, \phi_{cl}. \qquad (20)$$

We have calculated that $A = 3.3 \times 10^{-15}$ cm^2 and assumed that $\phi_{cl} = 0.2$. Hence equation (20) can be used together with the data in Figure 7 to calculate a for each experimental point. It turns out that **a**, the radius of the effective crosslink region, ranges from about 1.5 nm for the gels of low degree of cross-linking to 7 nm for the gels at the high degree of crosslinking which we have measured.

Hence permeability measurements can be used to characterize the internal arrangement of gels.

Acknowledgment

The aid in part from a grant by the Israel National Academy of Sciences and Humanities in support of this research is gratefully acknowledged.

Literature Cited

1. Weiss N., Ph.D. Thesis, Feinberg Graduate School, Weizmann Institute of Science, Rehovot 1975.
2. Elias H.G., Makromol.Chem. (1959) 29, 30.
3. Weiss N. and Silberberg A., Second International Congress of Biorheology, Rehovot, 1975. Biorheology, (1975) 12 107.
4. Spiegler K.S., Trans.Farad.Soc. (1958) 54 1408.
5. Mijnlieff P.F. and Jaspers W.J.R., Trans.Farad.Soc. (1971) 67 1837.

6

Diffusion through Hydrogel Membranes

I. Permeation of Water Through Poly(2-hydroxyethyl methacrylate) and Related Polymers

S. J. WISNIEWSKI, D. E. GREGONIS, S. W. KIM, and J. D. ANDRADE

Departments of Applied Pharmaceutical Sciences and Materials Science and Engineering, University of Utah, Salt Lake City, Utah 84112

Hydrogels are known for their unique physical properties and potential biomedical applications. The transport phenomena of some poly(2-hydroxyethyl methacrylate) (poly HEMA) membranes have been previously investigated (1-4). Ratner and Miller (5) have shown that poly HEMA membranes have a high permeability to urea due to interaction between the solute and membrane, and, in addition, consider the existence of pores in the poly HEMA gel structure, with the water regions of the gel acting as "pores" for solute transport.

Recently Chen found (6) that the absorption of water by dehydrated poly HEMA was a function of crosslinker content. In his study diffusion coefficients were calculated from absorption kinetic data using Fick's law. Several factors influence the structure of hydrogels and subsequently their diffusion characteristics. These factors include initiator, crosslinking agents, crosslinker content, equilibrium water content and impurities.

Yasuda et al. (1,11) have utilized a theory, based on a free-volume concept of diffusive transport, for hydrated homogeneous membranes. They concluded that this concept excellently explains the diffusive transport parameters as a function of water content over a wide hydration range.

In this study, we have been concerned with varying crosslinker concentrations and water content. It is hoped that this data will provide additional information on the basic transport mechanisms for water in different gel systems and on the role of water in the transport of other solutes.

Experimental

Hydroxyethyl methacrylate was obtained as a gift from Hydron Laboratories (New Brunswick, New Jersey) and was used without additional purification. Methoxyethyl methacrylate and methoxyethoxyethyl methacrylate were prepared in our laboratories by base catalyzed transesterification of methyl methacrylate with the corresponding alcohol. Ethylene glycol dimethacrylate (Monomer Polymer Laboratories, Philadelphia, Pennsylvania) and tetraethylene glycol dimethacrylate (Polysciences, Warrington, Pennsylvania) were purified by base extraction to remove inhibitor followed by distillation. Azobismethylisobutyrate, prepared by the method of Mortimer (7), was used to initiate polymerization at a concentration of 7.84 m moles/liter monomer. Crosslinker concentration was calculated on a mole-to-mole basis with HEMA. The desired monomer solution was mixed with 45% v/v distilled water and polymerized between glass plates at 60°C for 24 hours.

A glass diffusion cell, which contains two compartments of equal volume (175 ml), was designed. Each chamber was stirred at a constant rate to reduce boundary layer effects. At the beginning of each experiment, one compartment was filled with tritiated water, the other was filled with distilled water. The increase in tritiated water was monitored by removing 50μℓ aliquots at various times. These aliquots were placed in 10 ml of liquid scintillation "cocktail" (Aquasol, New England Nuclear Company) and counted in a Packard Scintillation counter. The thickness of the wet membrane was measured using a light micrometer (Van Kauren Company). Membrane densities were measured by weighing sections of wet membrane of known volume. Water content was obtained from the difference in weight of wet and dry membrane (dried to constant weight under vacuum at about 60°C). All experiments were run at 23°C ± 1°C.

Results and Discussion

The equation used to treat our data was derived elsewhere (8) and is given as follows:

$$\ln (1 - 2 C_t/C_0) = - (\frac{1}{V_1} + \frac{1}{V_2}) AUt,$$

where C_t = ^3H_2O count at time t; C_0 = ^3H_2O count at time 0; V_1 = V_2 = compartment volume = 175 ml; A = membrane contact area = 14.9 cm^2; U = permeability (cm/sec); and t = time (seconds).

A plot of $\ln (1 - 2 C_t/C_0)$ vs. time will yield a straight line with slope = $-(1/V_1 + 1/V_2)AU$. Substituting our values of A, V_1 and V_2 gives U = -5.87 x slope (cm/sec). The diffusion coefficients are given by D = Ud/K_D, where d = wet membrane thickness and K_D is the partition coefficient. By definition K_D = water concentration in membrane/water concentration in bulk, which for our case reduces to $K_D = (P_M/P_W)W_f$. P_M and P_W are the wet membrane density and the density of water at 23°C, respectively. W_f is the weight fraction of water in the wet membrane and is equal to W_W/W_M, where W_W is the weight of water and W_M is the weight of the wet membrane.

Table 1 lists the obtained parameters for the various gels
used. The data for the ~ 0.2 mole-% diester crosslinker poly
HEMA gels and unreported data on gels approximately 3-4 times
thicker indicate that the permeability U is roughly proportional
to 1/d, while D is essentially independent of thickness. Variations in D for the same gels appear to be mainly due to errors
in the thickness measurement and a small degree of non-uniformity
in thickness.

W_f and K_D decrease as crosslinking content increases, as
expected. However, K_D seems to increase again at approximately
6 mole-% crosslinker. The increasing K_D of the higher crosslinked gels may be the result of greater interaction between
water and a highly crosslinked polymer network. This water, a
relatively small fraction of the total gel water, is not available
for transport or is involved in a much slower diffusion process.
(13,14).

A plot of diffusion coefficients vs. percent EGDMA and
TEGDMA crosslinkers is shown in Figure 1. The diffusion coefficients decrease as the crosslinker content increases, approach-
ing a limiting value at about 6 mole-% crosslinker. This behavior
may indicate a "partition" type membrane. Diffusion coefficients
rapidly increase with decreasing crosslinker content below about
2.5 mole-% crosslinker, suggesting the development of "loose
pores."

Two basic mechanisms have been considered in explaining
solute transport through a polymer membrane: 1) a microporous
type, which can act as a sieve with the solute molecules being
transported thourgh the minute pores of the membrane, and 2) a
"partition" type membrane, which further acts to slow the diffusion process due to the interaction between diffusing solute and
membrane matrix or membrane water. Craig's work (9) showed
that, in general, a linear relationship exists between molecular
weight and half time rates of transfer for porous membranes.
However, the effect of molecular weight on the half time rates
through certain "partition" membranes indicated no such trend
(10). Recently, Chen (6) has described three different diffusion
mechanisms from his water absorption studies: 1) a dissolution
mechanism for higher crosslinking content, 2) a pore flow
mechanism for low crosslinking content, and 3) an intermediate
mechanism at intermediate crosslinker concentrations. We have
found similar results from our study. Water diffuses through the
crosslinked poly HEMA membranes via a predominantly pore mechanism
from 0% crosslinker to approximately 2.5 mole-% crosslinker.
Above 4 mole-% crosslinker, water transport is mainly controlled
by the interaction of water with the gel matrix. The intermediate region lies between 2.5 to 4 mole-%.

A description of the homogeneous membrane model and the
theoretical assumptions involved in the derivation of the relation
between diffusion coefficients and water content is found

TABLE I.

Diffusion Coefficients, Permeabilities and Hydrogel Composition

45% H$_2$O-HEMA EGDMA-mole-%	d(cm)	P(g/cc)	W$_f$	K$_D$	D x 10^6 (cm^2/sec)	U x 10^5 (cm/sec)
7.5	0.0777	1.26	0.381	0.481	2.3	1.4
5.3	0.0713	1.22	0.348	0.426	2.3	1.4
3.8	0.0717	1.20	0.355	0.427	2.4	1.4
2.3	0.0696	1.20	0.376	0.452	2.6	1.7
0.75	0.0739	1.25	0.403	0.505	3.0	2.1
0.38	0.0679	1.27	0.417	0.531	3.3	2.6
≈0.02	0.0717	1.20	0.415	0.499	3.6	2.5
	0.0583	1.25	0.419	0.525	3.4	3.1
	0.0638	1.30	0.420	0.547	3.4	2.9
	0.0682	1.22	0.418	0.511	3.3	2.5

45% H$_2$O-HEMA TEGDMA-mole-%	d(cm)	P(g/cc)	W$_f$	K$_D$	D x 10^6 (cm^2/sec)	U x 10^5 (cm/sec)
7.5	0.0776	1.20	0.405	0.488	2.5	1.6
4.6	0.0736	1.22	0.368	0.450	2.6	1.6
2.5	0.0702	1.23	0.377	0.465	2.7	1.8
1.4	0.0697	1.23	0.402	0.496	3.1	2.2
0.46	0.0793	1.33	0.418	0.557	3.4	2.4

Copolymers: Volume-%	d(cm)	P(g/cc)	W$_f$	K$_D$	D x 10^6 (cm^2/sec)	U x 10^5 (cm/sec)
MEMA	0.0660	1.22	0.0348	0.0426	0.23	0.015
33% HEMA-67% MEMA	0.0694	1.25	0.173	0.217	0.58	0.18
67% HEMA-33% MEMA	0.0686	1.29	0.308	0.397	1.6	0.94
MEEMA	0.0901	1.11	0.631	0.702	7.6	5.9
33% HEMA-67% MEEMA	0.0733	1.20	0.504	0.605	5.1	4.2
67% HEMA-33% MEEMA	0.0747	1.20	0.444	0.534	3.9	2.8

MEMA — methoxyethyl methacrylate

MEEMA — methoxyethoxyethyl methacrylate

HEMA — hydroxyethyl methacrylate

elsewhere (11), and that relation is given as follows:

$$\ln D/D_0 = \beta x(1 - \alpha)/(1 + x\alpha),$$

where $x = (1 - K_D)/K_D$, $\alpha = V_p/V_w$, $\beta = V^*/V_w$, D = experimental
diffusion coefficient with partition coefficient K_D, D_0 = diffusion coefficient of water in pure water, V_w = free volume in
unit volume of pure water, V_p = free volume in unit volume of
polymer phase, and V* = a characteristic volume parameter describing the diffusion of a permeant molecule in the medium.

It is possible to rearrange this equation in order to
obtain a linear plot. That is,

$$(\ln D/D_0)^{-1} = - \frac{1}{\beta(1 - \alpha)} x^{-1} - \frac{\alpha}{\beta(1 - \alpha)}.$$

A plot of $(\ln D/D_0)^{-1}$ vs. x^{-1} should yield a straight line with
slope = $-1/\beta(1-\alpha)$ and intercept = $-\alpha/\beta(1-\alpha)$. Table II gives values
of $(\ln D/D_0)^{-1}$ and x^{-1} for the gels studied, using $D_0 = 2.4 \times 10^{-5}$ cm²/sec from (12). The values for poly HEMA, uncrosslinked
($\sim$ 0.02 mole-%), are an average of four different gels. Figure
2 is a plot of $(\ln D/D_0)^{-1}$ vs. x^{-1} for all but the crosslinked
gels. Least squares fit yields values of $\alpha = 0.69$ and $\beta = 11$
with a correlation coefficient of -0.998. The intercept yields
a value of 1.6×10^{-7} cm²/sec for the hypothetical zero water
content polymer. Inclusion of crosslinked poly HEMA gels yields
values of $\alpha = 0.72$ and $\beta = 13$ with a correlation coefficient of
-0.988. Even though a fair correlation is obtained with the
inclusion of the crosslinked gels, these gels seem to exhibit a
trend of their own. These results and previous results (11)
indicate that structural differences of the monomers used may
play some role in the diffusion process other than just determining water content. Changing of the water content of swollen
gels for pure systems by varying polymerization water concentration may shed some light on structural effects in the diffusion
process. This study is being continued.

Acknowledgements

This work was supported by NHLI Grant No. HL 16921-01. The
donations of generous quantities of hydroxyethyl methacrylate
from Hydro Med Sciences, Inc., is gratefully acknowledged.

Abstract

Water transport through fully swollen poly (2-hydroxyethyl
methacrylate) (poly HEMA) hydrogels, containing varying concentrations of ethylene glycol dimethacrylate (EGDMA) and tetraethylene

TABLE II.

Values of $(\ln D/D_0)^{-1}$ and x^{-1} for Hydrogels Studied

(Using $D_0 = 2.4 \times 10^{-5}$ cm^2/sec)[12]

Hydrogel (Copolymers in volume-%)	$\ln(D/D_0)^{-1}$	$x^{-1} = K_D/1-K_D$
MEMA	-0.22	0.0445
33% HEMA-67% MEMA	-0.27	0.277
67% HEMA-33% MEMA	-0.37	0.658
MEEMA	-0.87	2.36
33% HEMA-67% MEEMA	-0.55	1.15
67% HEMA-33% MEEMA	-0.65	1.53
HEMA	-0.51	1.09
Mole-% EDGMA		
7.5	-0.43	0.927
5.3	-0.43	0.742
3.8	-0.43	0.745
2.3	-0.45	0.825
0.75	-0.48	1.02
0.38	-0.50	1.13
Mole-% TEGDMA		
7.5	-0.44	0.953
4.6	-0.45	0.818
2.5	-0.46	0.869
1.4	-0.49	0.984
0.46	-0.51	1.26

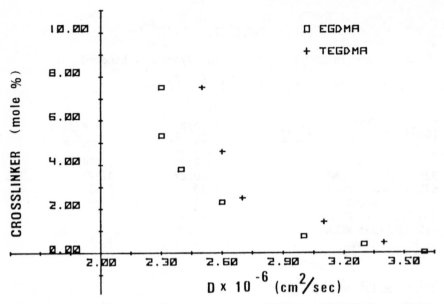

Figure 1.   Plot of diffusion coefficients vs. mole % EGDMA and TEGDMA in PHEMA hydrogel

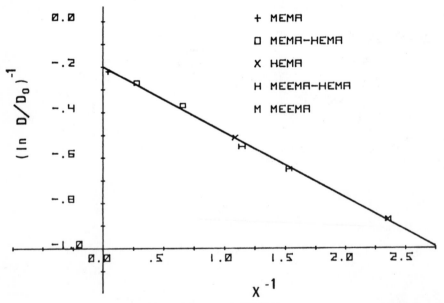

Figure 2.   Plot of $(\ln D/D_o)^{-1}$ vs. X^{-1} for hydrogels swollen to different equilibrium water concentrations

glycol dimethacrylate (TEGDMA) crosslinkers, was investigated using tritiated water and a designed diffusion cell. Diffusion coefficients obtained in these experiments decrease as the concentration of crosslinker increases. This decrease is very sharp at crosslinker concentrations of 0-2.5 mole-%. At higher crosslinker concentrations (2.5-7.5 mole-%), the diffusion coefficients appear to reach a limiting value. This study indicates that a crosslinked poly HEMA membrane may provide both "partition" and "pore" mechanisms of solute transport based on crosslinker concentration.

Hydrogels of varying water content were prepared to investigate the relationship between water content and diffusion coefficient. Poly(methoxyethyl methacrylate) (poly MEMA), poly(methoxyethoxyethyl methacrylate) (poly MEEMA) and copolymers of MEMA and MEEMA with HEMA were formulated to give hydrogels with water content ranging from approximately 3% to 63%. The results of the diffusion experiments were examined in light of a free-volume model of diffusive transport. Excellent theoretical-experimental correlation was obtained for the hydrogels used in this study.

Literature Cited

1. Yasuda, H., Lamaze, C., and Ikenberry, L., Makromol. Chem. (1968) 118, 19.
2. Spacek, P., and Kubin, M., J. Poly. Sci., Part C, (1967) 16, 705.
3. Ikenberry, L., Yasuda, H., and Clark, H., Chem. Eng. Prog. Sym. Ser. (1968) 64 (84) 69.
4. Refojo, M. F., J. Appl. Poly. Sci. (1964) 9, 3417.
5. Ratner, B. D., and Miller, I. F., J. Biomed. Matl. Res. (1973) 7, 353.
6. Chen, R. Y. S., Polymer Preprints (1974) 15, No. 2, 387.
7. Mortimer, G. A., J. Org. Chem. (1964) 30, 1632.
8. Mah, M. Y., Master's Thesis, University of Utah, 1972.
9. Craig, L. C., and Konigsberg, W., J. Phys. Chem. (1961) 65 116.
10. Lyman, D. J., Trans. Am. Soc. Art. Int. Org. (1964) 10, 17.
11. Yasuda, H., Lamaze, C. E., and Peterlin, A., J. Poly. Sci., A-2 (1971) 9, 1117.
12. Wang, J. H., Robinson, C. V., and Edelman, I. S., J. Amer. Chem. Soc. (1953) 75, 466.
13. Chang, Y. J., Chen, C. T., and Tobolsky, J. Poly. Sci., Poly. Phys. Ed., (1974), 12, 1.
14. Hoffman, A. S., Modell, M., and Pan, P., J. Appl. Poly. Sci., 14, 285 (1970).

7

The Chemistry of Some Selected Methacrylate Hydrogels

DONALD E. GREGONIS, CHWEN M. CHEN, and JOSEPH D. ANDRADE

Department of Materials Science and Engineering, University of Utah,
Salt Lake City, Utah 84112

Hydrogels have been described as biocompatible materials(1). Polymers of hydroxyethyl methacrylate (HEMA) are the most studied synthetic hydrogels (2). At present, polymers of this material are widely used in corrective contact lenses (3); it has been used as a coating for catheters and other medical devices (4-6). In our investigations, we wanted to study a variety of hydrophilic methacrylate polymers to evaluate their biological behavior in relation to the concentration and type of groups incorporated into the polymer. In order to pursue this goal, a thorough understanding of the bulk properties of the hydrogels is required. In this work the equilibrium water content of the gels is regulated by varying copolymer ratios. Charged groups are incorporated into the polymer by copolymerization with acidic or basic methacrylates. The monomers that are being investigated are shown in Table I. The amount of water in the equilibrated polymers covered the range from 3.5% to greater than 90% (Table II). For this study, hydroxyethyl methacrylate was selected as the fundamental monomer because of past work and availability.

TABLE I.

Hydrophilic Methacrylate Monomers

$$\begin{array}{c} O \\ \parallel \\ -C \\ | \\ O-CH_2CH-R_1 \end{array} \quad R_2$$

1. R_1 = OH, R_2 = H; hydroxyethyl methacrylate (HEMA).
2. R_1 = OCH_2CH_2OH, R_2 = H; hydroxyethoxyethyl methacrylate(HEEMA).
3. R_1 = $OCH_2CH_2OCH_2CH_2OH$, R_2 = H; hydroxydiethoxyethyl methacrylate (HDEEMA).
4. R_1 = $-OCH_3$, R_2 = H; methoxyethyl methacrylate (MEMA).
5. R_1 = $OCH_2CH_2OCH_3$, R_2 = H; methoxyethoxyethyl methacrylate (MEEMA).
6. R_1 = $OCH_2CH_2OCH_2CH_2OCH_3$, R_2 = H; methoxydiethoxyethyl methacrylate (MDEEMA).
7. R_1 = CH_2OH, R_2 = OH; 2,3-dihydroxypropyl methacrylate(DHPMA).

88

TABLE II.
Equilibrium Water Swelling of Pure Homopolymers

1. pHEMA   -   0.40*   5. pMEEMA   -   0.62

2. pHEEMA -   0.80   6. pMDEEMA -   >0.90

3. pHDEEMA -   >0.90   7. pDHPMA   -   0.70 to >0.90 (7)

4. pMEMA   -   0.035

---

*Water fraction (W_f) = weight of water in polymer/weight of hydrated polymer

Monomers (See Table I)

The laboratory synthesis of these monomers is accomplished by the following classical chemical reactions:
1. Reaction of methacryl chloride with the corresponding alcohol:

(Monomers 1-7)

2. Transesterification of methyl methacrylate with the corresponding alcohol:

(Monomers 1-7)

3. Acid catalyzed hydrolysis of glycidyl methacrylate (7):

(Monomer 7)

4. Acid catalyzed hydrolysis of isopropylidineglyceryl methacrylate (8-9).

(Monomer 7)

A thorough survey of hydrophilic methacrylate monomers
covering both laboratory and industrial synthesis is available
(10). Moderate amounts of monomer (0.1 to 1 ℓ.) satisfies most
of our requirements. A description of our purification scheme
is presented below. Normal preparative chromatographic techniques
are impractical for the purification of this quantity of monomer.

The preparation of monomers 1, 2, and 3 is complicated by
the formation of dimethacrylate esters. These diesters are
undesired in the product because they act as crosslinking agents.
They may be removed from the product by extraction techniques.
Since the diesters are less soluble in water than the corresponding monesters, the impure monomer is dissolved in water (∿1:1
ratio, v/v) and extracted with a non-polar solvent, such as
carbon tetrachloride, benzene or petroleum ether (11). The less
water soluble diesters are extracted into the organic phase. The
desired monomer is isolated by saturation of the aqueous solution
with sodium chloride and extraction with a more polar solvent,
usually methylene chloride. Ether solvents are not used for any
extraction unless they have been rigorously purified. Anhydrous
ethers readily air oxidize to form peroxides, which are good
polymerization initiators. The monomer is carefully extracted
to remove acidic and basic impurities and then eluted through
Grade II neutral alumina to remove other polar impurities.

The monomers should finally be purified by vacuum distillation. The vacuum system must be able to carry out distillations
as low as 0.1 mm Hg. pressure. Inhibitor is required to deter
polymerization when the product is distilled. Cuprous chloride
or copper powder is routinely added to the distillation flask and
offers the advantage of not co-distilling with the monomer.
Hydroquinone or p-methoxyphenol are more effective inhibitors
and are required in distillations above 100°C, but they co-distill
with the product. A nitrogen bleed into the distillation flask
largely eliminates the deleterious effects of oxygen. Table III
gives the boiling points of the monomers.

TABLE III.
Boiling Points of Hydrophilic Methacrylate Monomers

	Monomer	Boiling Point, °C
1.	Hydroxyethyl methacrylate(HEMA)	69° @ 0.1 mm Hg
2.	Hydroxyethoxyethyl methacrylate (HEEMA)	97-99° @ 1 mm Hg (2)
3.	Hydroxydiethoxyethyl methacryl- ate (HDEEMA)	120-122° @ 0.4 mm Hg (2)
4.	Methoxyethyl methacrylate(MEMA)	57° @ 8 mm Hg
5.	Methoxyethoxyethyl methacrylate (MEEMA)	65-70° @ 0.1 mm Hg
6.	Methoxydiethoxyethyl methacryl- ate (MDEEMA)	95-101° @ 0.1 mm Hg
7.	2,3-dihydroxypropyl methacrylate (DHPMA)	140° @ 0.6 mm Hg

Gas chromatography (GC), high pressure liquid chromatography (HPLC) and thin-layer chromatography (TLC) (12) are used for analysis of purity. Certain impurities that cannot be detected by one technique are easily detected by another technique. Both UV absorbance and refractive index are monitored with the high pressure liquid chromatograph.

The methacrylate system offers a wide variety of monomers, which we call gel modifiers, to copolymerize with the hydrophilic monomers to change the characteristics of the gel. For example, copolymerization with methacrylic acid introduces negatively charged groups into the gel. Dimethylaminoethyl methacrylate may be used to introduce positive charges into the gel. Some of the commercially available monomers used for gel modification are listed in Figure 1. Modifiers 11-14 in Figure 1 are not discussed further in this paper.

Methacrylate diesters are also commercially available. These diesters function as crosslinking agents. The crosslinkers that we have used are the dimethacrylates of ethylene glycol(EGDMA) and tetraethylene glycol(TEGDMA)(2,13). The latter is more soluble in water.

Initiators

Methacrylates may be polymerized by either radical or anionic initiators (14). For radical initiation we selected a class of azo initiators. These azo initiators are readily available and decompose at a uniform rate unaffected by solvent or induced decomposition (15). The azo initiators may be modified so that the end groups introduced into the polymer chain are very similar to the mer unit. They thermally decompose at a reasonable temperature (16) and can be decomposed at low temperatures by UV light. In this study, esters from azobisisobutyronitrile (AIBN) were prepared by the following method (17). Mild acid alcoholysis of AIBN converts the nitrile group to iminoester salts, which precipitate from the reaction solution. The product is isolated by filtration and the imino-ester group is hydrolyzed to form the ester (Figure 2).

Azobis(methyl isobutyrate) (Compound 16) has been previously prepared (17-19). More water soluble azo initiators were prepared from the monomethoxyglycols (Compounds 17 and 18). Initiator azobis (methoxydiethoxyethyl isobutyrate) (Compound 18) is completely water soluble. In other work we have shown that all three AIBN derivatives initiate polymerization of HEMA at an equal rate, which is slightly faster than AIBN (20). For bulk polymerization, 7.84 micromoles of initiator are used per milliliter of monomer. This ratio is kept constant and does not vary with water content. For polymerization of high water content gels, the more water soluble derivatives of AIBN are used.

Providing no active hydrogens are present in the monomer or solvent, anionic initiators may be used. These conditions prohibit polymerization of the hydroxylic monomers 1, 2, 3, 7 or

the use of water or alcohols as solvents. The anionic initiators
are organometallic compounds, such as n-butyl lithium, or strong
bases, such as lithium t-butoxide. Although there are disadvan-
tages to these initiators, they are used to polymerize methacrylates in a wide range of tacticities from chains having high
syndiotactic triads to chains having high isotactic triads. Our
group is investigating such polymers to determine the effect of
tacticity on the properties of the hydrogels (21).

For certain studies soluble polymers were required (22).
Radical polymerization of the methacrylate system without solvent
gives a polymer that swells to a high degree in good solvents,
but does not dissolve. It is felt that some crosslinking of the
polymer occurs by radical chain transfer mechanisms (23). Soluble
methacrylate polymers may be obtained with a radical initiator
at high dilutions of the monomer. We routinely polymerize the
monomers in ethanol (1:10,v/v). These solutions have been used
for solvent casting (24) and gel coatings (22).

Preparation of Gel Membranes

Gel membranes were required for our diffusion studies (13)
and were used for our work on water swelling of gels. Polymerization of the monomer was initiated by azobis (methyl isobutyrate)
using the same concentration used for the polymerization of bulk
gels, 7.84 micromoles per milliliter of monomer. This ratio is
independent of water concentration. All gel modifiers and crosslinking agents are expressed as molar quantities in terms of the
monomer volume. The membranes were prepared by polymerization
of the monomer solution between flat plates using a silicone
rubber spacer to regulate thickness. The polymerization conditions were standarized at 60°C for 24 hr. At low water concentrations (W_f <0.20), polyethylene or polypropylene mold plates are
used; at higher water concentrations, glass plates are used.
This is because gels polymerized at low water concentration adhere
strongly to the glass surface. Polymerization of membranes on a
glass surface is prefered since it has a more uniform surface.
The thickness of the gel membranes is 0.75 mm unless otherwise
stated.

Synthesis and Polymerization of C^{14}-labeled HEMA

To determine the relative stability of the methacrylate gels
in various media, C^{14}-labeled HEMA was synthesized by reacting
methacryl chloride and 1,2-C^{14}-ethylene glycol, using excess
carrier ethylene glycol to prevent the formation of excessive
amounts of C^{14}-labeled diester (ethylene glycol dimethacrylate).
The C^{14}-HEMA was purified utilizing the "salting out" technique
described earlier. The product was diluted with carrier HEMA and
was then distilled by micro-vacuum techniques. The distillate
was determined to have a specific activity of 9.1 x 10^5 dpm/ml.

To check the radioisotopic purity, a 1 ml aliquot was chromato-
graphed on 100g Grade II alumina. The column was eluted with a
linear gradient from 100% petroleum ether (250ml) to 5% methanol
in diethyl ether (250ml). The radioactivity eluted in one peak.
the fractions were combined and evaporated. The product exhibited
an identical infra-red spectrum to control HEMA.
    To initiate the polymerization of C^{14}-HEMA, azobis(methyl
isobutyrate) was added (7.84 micromole/ml). The monomer was
polymerized in a polypropylene mold. The mold dimensions were
50mm x 50mm x 1mm. The standard polymerization conditions were
60°C for 24 hours unless otherwise noted. After this time the
mold was cooled to 0°C, and the C^{14}-pHEMA was removed and weighed.
The C^{14}-pHEMA was then placed in various solvents at 37°C. Aliquots
were removed at various time intervals and counted using liquid
scintillation techniques. The amount of radioactivity found in
the solvent was calculated as a percentage of the overall radioactivity in the polymer. Figure 3 shows the percent radioactivity
extracted with time for some typical gels. Gels 1 and 2 are dup-
licates. Gel 6 was polymerized under identical conditions but
was extracted with 95% ethanol. Gel 9 was polymerized with 45%
water and extracted with water. The radioactivity of all gels
extracted with water leveled out after one day, but the radioactivity extracted with ethanol (Gel 6) appeared to increase
after this time at about 0.5%/week. Table IV provides a complete
list of the C^{14}-HEMA polymerization conditions, extraction solvents, and the percent radioactivity extracted. The percent
radioactivity extracted was calculated as the average of the
isotope found in solution at points in time from 4 days to 5
weeks.

TABLE IV.
Elution of Radioactivity from C^{14}-pHEMA Gels
Equilibrated in Various Solvents

Gel#	Polymerization Conditions	Solvent	% Radioactivity in Solvent after 5 wks.
1	24 hours @ 60°C	Distilled H_2O	4.7% ± 0.13
2	24 hours @ 60°C	Distilled H_2O	4.8% ± 0.08
3	24 hours @ 60°C	Human Reference Serum	3.6% ± 0.27
4	3 1/2 hours @ 60°C	Distilled H_2O	4.8% ± 0.14
5	66 hours @ 60°C	Distilled H_2O	2.6% ± 0.41
6	24 hours @ 60°C	95% Ethanol	9.1% ± 0.90
7	24 hours @ 60°C (Gel contained 1% unlabeled EGDMA)	Distilled H_2O	2.9% ± 0.30
8	24 hours @ 60°C (Gel contained 5% unlabeled EGDMA)	Distilled H_2O	2.3% ± 0.11
9	24 hours @ 60°C (Gel contained 45% H_2O v/v)	Distilled H_2O	1.7% ± 0.09

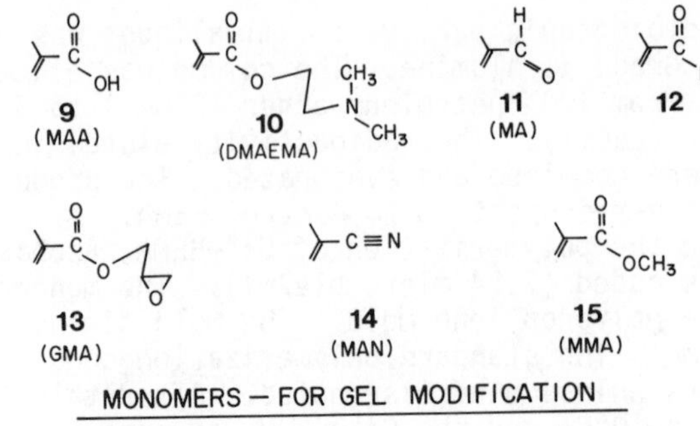

9
(MAA)
10
(DMAEMA)
11
(MA)
12
13
(GMA)
14
(MAN)
15
(MMA)
MONOMERS — FOR GEL MODIFICATION

Figure 1. Commercially available gel modifiers

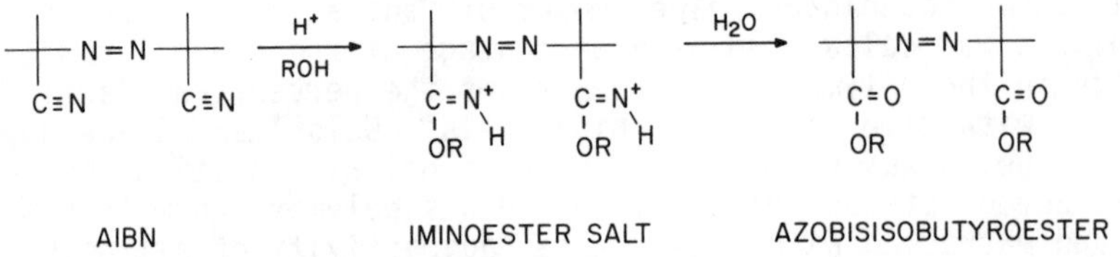

AIBN
IMINOESTER SALT
AZOBISISOBUTYROESTER

16. R = CH_3 AZOBISMETHYLISOBUTYRATE

17. R = $CH_2CH_2OCH_3$ AZOBISMETHOXYETHYLISOBUTYRATE

18. R = $CH_2CH_2OCH_2CH_2OCH_2CH_2OCH_3$ AZOBISMETHOXYDIETHOXYETHYLISOBUTYRATE

Figure 2. Preparation of azobisisobutyrate esters

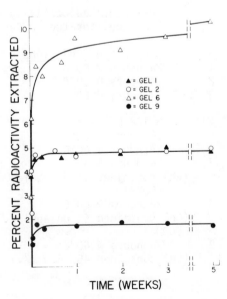

PERCENT RADIOACTIVITY EXTRACTED
TIME (WEEKS)
▲ = GEL 1
○ = GEL 2
△ = GEL 6
● = GEL 9

Figure 3. Radioactivity extracted at given time for ^{14}C-HEMA gel

The amount of unpolymerized C^{14}-HEMA or C^{14}-pHEMA that dissolved in the solutions is given in Table IV. The better solvent, ethanol (#6) contained more radioactivity than the poorer solvent, water (#1,2). At shorter polymerization times, more radioactivity was found in the solvent than after long polymerization times (#4 vs. #5). There did not appear to be any significant difference, however, between Gels 1, 2 and 4. With higher percent crosslinker, the amount of radioactivity that is found in the solvent decreased (#1, 7 and 8). It was somewhat surprising to find that the least radioactivity was found in the solvent when C^{14}-HEMA was polymerized with 45% water (#9). These preliminary studies are continuing. Isolation and identification of the extracted radioactivity will be necessary to determine if it is due to unreacted monomer or to polymers.

Water Swelling of pHEMA and pMEEMA Membranes

Isotropic swelling is assumed in the gels. Linear swelling (ℓs) and water fraction (W_f) are defined as follows:

$$\ell s = \frac{\text{length of equilibrium swollen gel}}{\text{length of dry gel}}$$

$$W_f = \frac{\text{weight of water in gel}}{\text{weight of hydrated gel}}$$

$$\%\ell s = \frac{\text{wet length - dry length}}{\text{wet length}} \times 100$$

$$\%W_f = W_f \times 100$$

Experiments to determine the kinetics of linear swelling and water fraction of the gel were performed. The pHEMA gels polymerized with different water concentrations reached equilibrium water content after 24 hours. At equilibrium the water fraction (W_f) in the pHEMA gel was 0.40 ± 0.05 (Figure 4), whether or not the HEMA was polymerized with water. The same experiments were not done with MEEMA since this monomer is not very soluble in water, however, anhydrous pMEEMA reached an equilibrium water fraction of 0.62 after 48 hours.

The kinetics of linear swelling (ℓs) do not follow the same pattern as the kinetics of water uptake as measured by the water fraction (W_f). This is found in both pHEMA (Figure 5) and pMEEMA gels (Figure 6). For a pHEMA gel polymerized without water and then allowed to equilibrate in water, the water fraction (W_f) increased to 0.20, half the final equilibrium water fraction, but the linear swelling (ℓs) was measured to be 1.015, only 5% of the final equilibrium linear swelling (Figure 5). An analogous result is obtained for pMEEMA gels (Figure 6). In effect, there is a much greater initial amount of water taken into the gel than would be anticipated from just the linear swelling measurement.

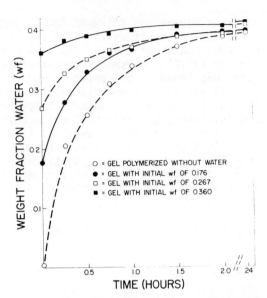

*Figure 4.   Water uptake of pHEMA polymerized
at different water contents*

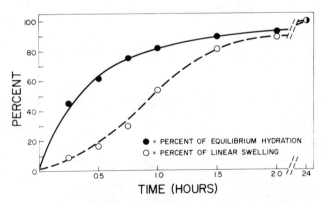

*Figure 5.   Percent equilibrium hydration (0.40 $w_f = 100\%$)
and percent equilibrium linear swelling (1.165 $l_s = 100\%$)
plotted vs. time for pHEMA gel*

Similar results are observed by Refojo for pHEMA and other gels (25). Absence of linear swelling at low water fractions is observed by us in an alternative experiment with pHEMA gels. HEMA was polymerized at various water concentrations and then allowed to equilibrate in distilled water (Figure 7). It is found that when HEMA is polymerized with a water fraction (W_f) of 0.1 or less (W_f of 0.1 is 25% of the equilibrium W_f), the gel still swells to the same degree as an anhydrous pHEMA gel.

These results indicate that there is free volume or "voids" in the hydrogels, and this volume is filled with water molecules before the gel is able to exhibit any linear expansion. This concept has been used to explain the swelling and mechanical behavior of β-keratin (26) and other biopolymers (27). Similar studies on other hydrophilic gels are in progress.

Copolymers

In order to obtain gels that would swell in water a speci-fied amount, copolymers of the hydrophilic monomers were investi-gated. At the same time the water solubility of the comonomers was also studied. It was hoped that monomer solubility behavior would be useful to explain some aspects of swelling of the polymer. Figure 8 shows HEMA-MEMA comonomers and copolymers and their relationship to solubility and swelling in water. Water has a maximum solubility of 3.5% (v/v) in MEMA monomers. HEMA monomer is infinitely soluble in water. The comonomer solutions exhibit a water solubility that increases slightly as the amount of HEMA is increased up to 40% MEMA-60% HEMA. At that point the water solubility increases greatly until 20% MEMA-80% HEMA where the comonomers become infinitely soluble. The copolymers of MEMA-HEMA appear to have a linear water fraction (W_f) relation-ship from 0.035 water for pure pMEMA to 0.40 for pure pHEMA.

Figure 9 shows the relationship between MEEMA-HEMA and water solubility and between their copolymers and the degree of swelling in water. In this case, MEEMA monomer dissolves only about 8% water and as HEMA monomer is added, the amount of water that it dissolves increases sharply. Near a 60:40 HEMA:MEEMA ratio, water becomes infinitely soluble in the comonomers. The water swelling of the copolymers show the opposite relationship. pMEEMA swells to 63% H_2O, and as increasing amounts of HEMA are incorporated in the copolymer, the degree of swelling decreases until a 40% H_2O uptake is reached for pHEMA.

These above results indicate that the hydrophilicity of the monomer and the degree of crosslinking of the polymer (28) are not the only factors that determine the degree of water swelling of the polymer. In the methacrylate system, we have rationalized that the length of the ester chain is a third factor that should be taken into account when discussing swelling. This is probably due to the fact that the long ester functionality decreases the packing energy of the methacrylate polymer.

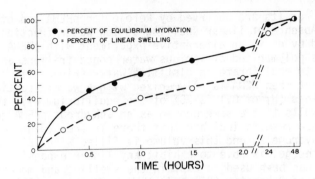

*Figure 6.   Percent equilibrium hydration (0.62 w_f = 100%)
and percent equilibrium linear swelling (1.37 l_s = 100%)
plotted vs. time for pMEEMA gel*

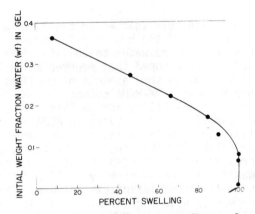

*Figure 7.   Plot of percent linear swelling vs. the
initial amount of water in HEMA gel. Gel
polymerized without water swells to 100%.*

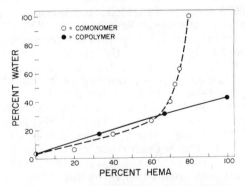

*Figure 8.   Water solubility of MEMA–HEMA
comonomers (v/v) and the equilibrium water
weight fraction, w_f, of MEMA–HEMA copolymers*

HEMA-Ionic Methacrylate Copolymers

It has previously been shown that marked swelling in water of crosslinked poly(methacrylic acid) gels is a function of the degree of neutralization (29, 30, 31). We needed to know the effect of swelling in hydrogels as a function of charge incorporated into the gel. For this investigation, methacrylic acid (MAA) is used to introduce anionic groups and dimethylaminoethyl methacrylate (DMAEMA) is used to introduce cationic groups in pHEMA. The gels were prepared by adding molar amounts of MAA and/or DMAEMA to the HEMA monomer along with 40% v/v water and were polymerized as membranes using standard conditions. The gels were equilibrated for two days in distilled water before the equilibrium water fraction (W_f) was determined.

As shown in Figures 10 and 11, pHEMA gels polymerized with up to 5 molar percent MAA or DMAEMA show little change in equilibrium water content from that of pure pHEMA gel polymerized at identical conditions (W_f = 0.40). This is not surprising since carboxylic acids and tertary amines are weak acids and bases and do not ionize to an appreciable extent at neutral pH. To form charged groups, the MAA-HEMA copolymers were equilibrated in pH 10 NaOH solution for twelve hours and the DMAEMA-HEMA copolymers were equilibrated in pH 3 HCl solution for twelve hours. After this time they are equilibrated in distilled water for two days, changing the water several times in this period to remove excess ions from the gels.

The conversion of MAA to its carboxylate salt in the gels dramatically increases the equilibrium water content of the gel (Figure 10). For example, the HEMA gel containing only 2 molar percent MAA equilibrated with a water fraction greater than 0.90. Converting the HEMA-DMAEMA gel to its hydrochloride salt did not show the same dramatic swelling effect, even though the water fraction increased with increasing concentration of DMAEMA. The 5 molar percent DMAEMA gel water fraction measured 0.46 (Figure 11). It is known that tertary amines act as chain transfer agents with the methacrylate system (32,33). In these experiments, the tertary amine would also become part of the polymer chain. The effect of this would give the DMAEMA-HEMA gels a higher crosslink density, possibly accounting for its lower degree of swelling in water.

Figure 12 shows the equilibrium water content of HEMA gels containing equal molar quantities of MAA and DMAEMA. When equilibrated in distilled water the water fraction of the gel increased slightly with increasing molar percent MAA-DMAEMA to about 0.5 water fraction for the 5 molar percent gel. The carboxylate groups were then converted to their sodium salt and the amino groups were converted to free amines by equilibrating the gel in pH 10 NaOH solution for one day. The gels were re-equilibrated in distilled water for two days and a much greater swelling degree was observed. The 5 percent MAA-DMAEMA gel had

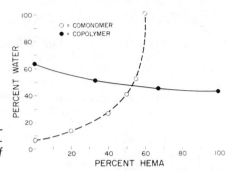

Figure 9. Water solubility of MEEMA–HEMA comonomers (v/v) and the equilibrium water weight fraction, w_f, of MEEMA–HEMA copolymers

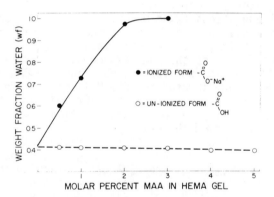

Figure 10. Equilibrium water weight fraction, w_f, of HEMA–MAA copolymers

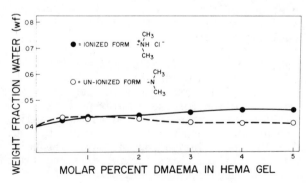

Figure 11. Equilibrium water weight fraction, w_f, of HEMA–DMAEMA copolymers

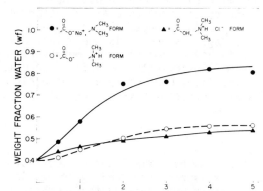

Figure 12. Equilibrium water weight fraction, w_f, of HEMA–MAA–DMAEMA terpolymers. MAA and DMAEMA are polymerized at equal molar concentrations.

a water fraction measuring 0.81 (Figure 12). This same effect was not observed when the amine was converted to its hydrochloride salt by equilibrating the gel in pH 3 hydrochloric acid followed by two days equilibration in distilled water. The 5 percent MAA-DMAEMA had a water fraction of only 0.53 (Figure 12). It is not known at present why the carboxylate salt gel swells to a greater degree than the amine hydrochloride gel. Polymerization of the amine salt of DMAEMA with HEMA would possibly reduce the crosslink density due to chain transfer mechanisms found in free tertary amine groups. Careful experiments have to be designed in this regard since it is known that changing the pH changes the reactivity ratios of ionic methacrylates (34). Future experiments will be directed towards a correlation of equilibrium water fraction, monomer type, concentration of charge and groups and degree of covalent crosslinking in the gels.

Acknowledgements

We gratefully acknowledge NIH Grant HL 16921-01 for financial support of this research, and Hydro Med Sciences, Inc., for generous donations of pure hydroxyethyl methacrylate.

Abstract

A summary of hydrophilic methacrylate monomer synthesis is discussed. Radical polymerization of the methacrylate system was accomplished by azobisisobutyronitrile (AIBN) derivatives. Some new water soluble AIBN derivatives were prepared and used. Copolymers of various hydrophilic methacrylate monomers afford a method to control water uptake. Some water swelling parameters of poly(hydroxyethyl methacrylate) and poly(methoxyethoxyethyl methacrylate) are described. The water uptake as measured by the water fraction is shown not to be a linear function of the degree of swelling. The water swelling behavior of poly(hydroxyethyl methacrylate) containing ionic groups is described using methacrylic acid for anionic groups and dimethylaminoethyl methacrylate for cationic groups. The synthesis of C^{14}-hydroxyethyl methacrylate is reported. After the radiolabeled monomer was polymerized, the amount of radioactivity in various solvents was measured.

Literature Cited

1.   Wichterle, O., and Lim, D., Nature (1960) 185, 117.
2.   Wichterle, O., Encyclopedia of Polymer Sci., (1971) 15, 273.
3.   Wichterle, O., U. S. Pat. 3,361,858 (1968).
4.   Majkus, V., Horakova, Z., Vymula, F., and Stol, M., J. Biomed. Mat. Res. (1969) 3, 443.
5.   Refojo, M. F., J. Biomed, Mat. Res. (1969) 3, 333.

6. Taylor, G. R., Warren, T. C., Murray, D. G., and Prins, W., J. Surgical Res. (1971) 11, 401.
7. Refojo, M. F., J. Appl. Polymer Sci. (1965) 9, 3161.
8. Fegley, V. W., and Roland, S. P., U. S. Patent 2,680,735.
9. Rohm & Haas Co., British Patent 852,384 (1965).
10. Yocum, R. H. and Nyquist, E. B., "Functional Monomers," Marcel Dekker, New York, Vol. 1, 1973, pp 299-487 and Vol. 2 (1974).
11. Horsley, L. H., and Sheetz, D. P., U. S. Patent 3,162,677.
12. Brinkman, U. A. Th., Van Schaik, T. A. M., deVries, G., and deVisser, A. C., this symposium.
13. Wisniewski, S. J., Gregonis, D. E., Kim, S. W., and Andrade, J. D., this symposium.
14. Lenz, R. W., "Organic Chemistry of Synthetic High Polymers," John Wiley & Sons, New York, 1967, p. 244.
15. Pryor, W. A., "Free Radicals," McGraw-Hill, New York, 1966, pp. 128-133.
16. Walling, C., "Free Radicals in Solution," John Wiley and Sons, New York, 1957, p. 511.
17. Mortimer, G. A., J. Org. Chem. (1965) 30, 1632.
18. Thiele, J., and Hauser, K., Ann. (1896) 290, 1.
19. Lim, D., Coll. Czech. Chem. Comm. (1968) 33, 1122.
20. Ma, S. M., Gregonis, D. E., Chen, C. M., and Andrade, J. D., this symposium.
21. Russell, G. A., Dalling, D. K., Gregonis, D. E., deVisser, A. C., and Andrade, J. D., this symposium.
22. Ma, S. M., Gregonis, D. E., Van Wagenen, R., and Andrade, J. D., this symposium.
23. Flory, P. J.,"Principles of Polymer Chemistry," Cornell University Press, Ithaca, New York, 1953, p. 384.
24. Sorenson, W. R., and Campbell, T. W., "Preparative Methods of Polymer Chemistry," Interscience Publications, New York, 1961.
25. Refojo, M. F., Contact and Intraocular Lens Med. J., (1975) 1, 153.
26. Bradbury, J. H., and Leeder, J. D., J. Appl. Polymer Sci., (1963) 7, 533.
27. Hiltner, A., Case Western Reserve University, Cleveland, Ohio, personal communication.
28. Flory, P. J., "Principles of Polymer Chemistry," Cornell University Press, Ithaca, New York, 1953, p. 581.
29. Schaefgen, J. R. and Trivisonno, C. F., J. Am. Chem. Soc., (1951) 73, 4580.
30. Schaefgen, J. R. and Trivisonno, C. F., J. Am. Chem. Soc., (1952) 74, 2715.
31. Katchalsky, A. and Eisenberg, H., J. Polymer Sci., (1951) 6, 145.
32. Lenz, R. W., "Organic Chemistry of Synthetic High Polymers," John Wiley and Sons, New York, 1967, p. 355.

33.   Shalati, M. D. and Scott, R. M., Macromolecules, (1975) $\underline{8}$, 127.

34.   Alfrey, M. D., Overberger, C. G., and Penner, S. H., J. Am. Chem. Soc., (1953) 75, 4221.

8

Analysis and Purification of 2-Hydroxyethyl Methacrylate by Means of Thin-Layer Chromatography

U. A. TH. BRINKMAN, T. A. M. VAN SCHAIK, and G. DeVRIES
Department of Analytical Chemistry, Free Reformed University,
Amsterdam, The Netherlands

A. C. DeVISSER
Department of Materials Science, Free Reformed University,
Amsterdam, The Netherlands

Three-dimensional networks of hydrophilic methacrylate polymers, have been recognized (1,2) as useful materials for biomedical and surgical applications. Implants and other protheses are being prepared from such hydrogels, often reinforced with a fabric structure. The research program of our departments aims at the development of a hydrogel that 1) combined with a filler can serve to substitute hard tissue (bone), and 2) promotes calcification and/or bone growth in its matrix, so that the initially soft material is filled up by the body itself with hard tissue. As hydrogel, poly(hydroxyethyl methacrylate) (PHEMA), has been selected which can be prepared from the commercially available monomer, 2-hydroxyethyl methacrylate (HEMA) by various methods (3-5).

PHEMA, which has been widely studied and discussed for medical applications, has largely been accepted (1,6) as a biocompatible, safe and stable material for medical use. In order to prepare a gel that meets such requirements, the initial monomer should be of high purity. The commercially available HEMA contains relatively large proportions of methacrylic acid (MAA) and the diester ethylene dimethacrylate (EDMA); the latter can be formed from HEMA by transesterification.

Other contaminants may also be present, as is discussed below. The removal of EDMA from HEMA, as well as MAA and its other esters, is particularly desirable, because the proportion of diester in the polymerizing mixture affects the nature of the network formed (7). Since the rate of both the polymerization and the transesterification reaction strongly increases with increasing temperature, techniques such as distillation and gas chromatography do not appear to be entirely favorable for the ultimate purification and analysis of HEMA samples.

Therefore, in the present study, thin-layer chromatography (tlc) has been selected for both qualitative analysis and small-scale preparative work.

Materials and Methods

2-Hydroxyethyl methacrylate was purchased from BDH (Poole, England) and Merck (Darmstadt, W. Germany); these products have a yellow color. Colorless highly pure (> 99%) HEMA was obtained as a gift from Hydro Med Sciences (New Brunswick, N.J., U.S.A.).

Ethylene dimethacrylate was obtained from BDH (Poole, England) and Koch and Light (Colnbrook, England). Methacrylic acid, ethylene glycol, 2-hydroxypropyl methacrylate, methyl methacrylate, hydroquinone and p-methoxyphenol, were purchased from Merck. Diethyleneglycol methacrylate was synthesised (8) by mixing appropriate amounts of diethylene glycol (855 g) and methyl methacrylate (500 g) at a temperature of $60^{\circ}C$, adding a 4 N solution of sodium methanolate in methanol (10 g) and heating the reaction mixture for a period of time of 30 min. Subsequently, the mixture was poured into water (400 g), washed with n-hexane and extracted twice with diethyl ether. After repeated washings with water, the ether extract was dried and diethyleneglycol methacrylate was distilled at reduced pressure.

All further chemicals used in this study were reagent-grade products, and were used without further purification.

Thin-layer chromatography was carried out in Hellendahl staining jars, using precoated silica gel plates (Kieselgel 60 F_{254}, Merck) cut into the appropriate size (4 x 8 cm^2).

Spots were applied with a pointed paper wick and chromatography is carried out in a non-saturated atmosphere. After development over a length of run of approximately 7 cm, detection is done using the methods reported below. Reproducibility of the R_F values in the solvent systems used in the present study was satisfactory.

Preparative-scale thin-layer chromatography was carried out on 20 x 20 cm^2, 2-mm thick precoated silica gel plates (Kieselgel 60 F_{254}, Merck). 100-150 mg HEMA, as a 20% (v/v) solution in ethanol, were applied with a Camag chromatocharger equipped with a disposable plastic syringe. Development over a distance of 15-17 cm, without prior acclimatisation, is done in a normal rectangular tank.

Refractive indices were measured on a thermostat-
ted (20 $\pm$ 0.1°C) Abbe refractometer. IR-spectra were
run on a Shimadzu IR 400 spectrometer, using cells
with NaCl windows.

Results and Discussion

Reference Substances. As has been stated in this
introduction, HEMA generally contains several impurities, e.g. EDMA, methacrylic acid and ethylene glycol.
Since methacrylic acid and its derivatives tend to
polymerize even at room temperature, an inhibitor
such as hydroquinone, p-methoxyphenol, phenothiazine
or octylpyrocatechol is usually added to these products (9). According to the specifications, the HEMA
samples obtained from BDH and Merck contain 200 ppm
p-methoxyphenol, and 100 ppm hydroquinone plus 100
ppm p-methoxyphenol, respectively; HEMA from Hydro
Med Sciences, Inc., contains 36 ppm p-methoxyphenol.
Hydroquinone is also present in EDMA and methacrylic
acid samples.

Methyl methacrylate may also be present as an
impurity in commercially available HEMA, since it is
one of the starting materials in its synthesis.
However, during tlc, it evaporates off from the chromatoplate (b.p., 100°C) and escapes detection in
qualitative analysis. In preparative-scale work, such
evaporation during development ensures its removal.
By developing a sample of pure methyl methacrylate and
spraying with water (10) immediately after the run,
we have shown its R_F value to be considerably higher
than that of HEMA in both solvent systems recommended
below. As regards the presence of water, one is
referred to Fig. 3 and the accompanying text.

As a consequence of the above, in addition to
HEMA, 5 compounds were included in our study, viz.
EDMA, methacrylic acid, ethylene glycol, hydroquinone
and p-methoxyphenol (cf. Table I). The purity of the
samples obtained as reference substances was tested,
using two or more of the solvent systems discussed
below. The samples were found to be chromatographically pure, apart from the presence of an inhibitor,
with the exception of EDMA. Here, tlc with n-hexane -
diethyl ether (1:1) as mobile phase revealed the
presence of some eight additional spots (Figs. 1a and
1b). These contaminants were removed by distillation
at reduced pressure (b.p. of EDMA at 4 mm Hg, 90°C)
or, preferably, by preparative-scale tlc (Figs. 1c
and d).

Table I. Detection limits of HEMA and some of its contaminants.

Name	Detection limit (µg/µl) * with				
	UV	I_2	$KMnO_4$	Echtblausalz B	Diazotised sulpha-nilic acid
2-Hydroxyethyl methacrylate (HEMA)	30	100	0.6		
Ethylene dimethacrylate (EDMA)	30	15	0.6		
Methacrylic acid	30	60	0.6		
Ethylene glycol	-	15	0.2		
Hydroquinone	15	0.6	0.6	0.2	0.2
p-Methoxyphenol	15	0.6	0.6	0.2	0.2

*For HEMA, d_4^{20} has a value of 1.070-1.074 (14). Therefore, 1µg contaminant/µl of HEMA roughly corresponds with 1000 ppm or 0.1 wt.%.

With the former technique, addition of hydroquinone or of a mixture of KCl, BiCl$_3$ and BiI$_3$ (11) is necessary in order to prevent polymerization.

Detection. From among a large number of non - specific methods of identification, three detection methods were selected for further research. Under U.V.-light (254 nm) all compounds but ethylene glycol are visible as dark spots on a fluorescent background. Treatment with iodine vapor (red-brown spots on a yellow-white background) is a suitable means to detect the glycol too. Unfortunately, the sensitivity of the combined U.V./I$_2$ procedure is rather low (cf. Table I). Therefore as an alternative, detection has been done by spraying with a 0.2% solution of KMnO$_4$ in acetone. White spots are formed on a brownish background for all compounds except p-methoxyphenol, which shows up as a pale orange spot. Since KMnO$_4$ reacts with impurities in the acetone to form MnO$_2$, which mars the detection, the reagent solution should be freshly prepared. Keeping in mind that checking the purity of HEMA is done by tlc of undiluted samples, one may conclude from the data in Table I that spraying with KMnO$_4$ allows the detection of impurities down to at least 0.1%.

Since the inhibitors are present in the various samples of HEMA, EDMA and methacrylic acid in concentrations of up to 0.02% only, they will escape detection with the above procedures. Therefore, two alternative methods of identification (12,13) have been tested: 1) spraying with a freshly prepared 0.5% solution of the diazo reagent Echtblausalz B in water, and either spraying with 0.1 N NaOH or, preferably, exposure to vapor of NH$_3$; 2) spraying with a mixture of 0.09 g sulphanilic acid dissolved in 10 ml 1.1 N HCl and 10 ml of an aqueous 4.5% NaNO$_2$ solution, kept at 0°C and diluted with an equal volume of a 10% Na$_2$CO$_3$ solution immediately before use. With both reagents, brownish spots show up on a nearly white background. As can be seen from the data in Table I, the use of even these reactions barely suffices for the identification of the inhibitors in methacrylic acid and its derivatives. Fortunately, improved results are obtained if sulphanilic acid is used to detect the phenolic compounds by a drop test procedure, in which 1 drop of the diazotised reagent is mixed with 1 drop of sample (dissolved in a minimum amount of ethanol, if necessary) and 1 drop of a 10% Na$_2$CO$_3$ solution. The detection limit now is approximately 0.01% for hydroquinone (green-to-yellow spots), and 0.001% for p - methoxyphenol (red-to-green spots).

Solvent Systems. After several trial runs of
the four main components with single-solvent systems
ranging from non-polar (n-hexane, cyclohexane) to
polar solvents (diethyl ether, methylisobutyl ketone
(MIBK), acetone), mixtures of n-hexane with diethyl
ether or (an)other component(s) were selected for
further research. Satisfactory results were obtained
with n-hexane - diethyl ether (1:1, v/v) and n-hexane-
MIBK - n-octanol   (9:2:1, v/v) (Fig. 2a and d). As a
drawback, however, it was observed that the spot due
to methacrylic acid tends to tail, especially in the
latter solvent system. Such tailing, which is particularly harmful in preparative-scale thin-layer or
column chromatography, can effectively be reduced by
adding an acidic component to the system. Two alter-
natives' have been investigated. 1) The solvent system
is saturated with an equal volume of 25% (v/v) HNO_3.
When a higher percentage of acid is used, decomposition of one or more of the compounds under investigation occurs during development, as evidenced by the
occurrence of a brown streaking zone on the chromato-
gram. 2) Development is done on a silica gel plate
soaked for 15 min with 1% H_2SO_4, and subsequently
dried overnight at 60°C. As is manifest from the data
in Fig. 2, with each of these techniques, the results
indeed improved.
      For qualitative analysis, development with a HNO_3-
saturated mobile phase should be preferred to tlc on
an acid-impregnated support, which is a more time -
consuming procedure. From Fig. 2 it is evident that
n-hexane - MIBK - n-octanol   (9:2:1, satd. with 25%
HNO_3) is a very powerful solvent system for the resolution of the four main components. Addition of nitric
acid to n-hexane - diethyl ether (1:1) also improves
the separation of HEMA from methacrylic acid; however,
the spots due to methacrylic acid and EDMA now show an
appreciable overlap. Therefore, as a tentative conclusion, we recommend the use of both n-hexane -
diethyl ether (1:1) and n-hexane - MIBK - n-octanol
(9:2:1, satd. with 25% HNO_3) for qualitative analysis.
      For preparative-scale tlc, chromatography on
H_2SO_4 -impregnated silica gel has been preferred to
development with a HNO_3-saturated mobile phase, because sulphuric acid, but not nitric acid, has a negli-
gibly small solubility in the solvent used to elute
HEMA from the support material (cf. below). n-hexane -
diethyl ether (1:1) has been selected as mobile phase
rather than the n-hexane - MIBK - n-octanol mixture,
because of the relatively large time of development of
the latter solvent system (approximately 4 vs. 2.5 h

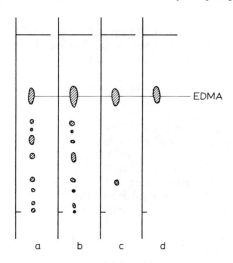

Figure 1. Thin-layer chromatograms of EDMA samples. (a) EDMA from BDH; (b) EDMA from Koch and Light; (c) redistilled EDMA (BDH); (d) EDMA (BDH) purified by preparative-scale tlc. System: silica gel/n–hexane–diethyl ether (1:1).

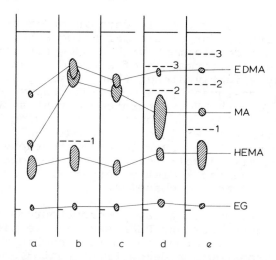

Figure 2. Tlc of HEMA, EDMA, methacrylic acid (MA), and ethylene glycol (EG) in the systems: (a) silica gel/n-hexane–diethyl ether (1:1); (b) silica gel/n-hexane–diethyl ether (1:1, satd. with 25% HNO₃); (c) H₂SO₄–impregnated silica gel/n-hexane–diethyl ether (1:1); (d) silica gel/n-hexane–MIBK–n-octanol (9:2:1); (e) silica gel/n-hexane–MIBK–n-octanol (9:2:1, satd. with 25% HNO₃). ---, acid (1), n-octanol (2) and MIBK (3) front.

for a 16-cm run). Due to the low volatility of
n-octanol, its complete removal from the thin layer is
not easily effected, and part of it will be eluted
together with HEMA.

Next, attention was paid to the selection of a
solvent suited to elute the separated component(s)
from the silica gel. At first sight, this choice does
not appear to be very critical, because HEMA dissolves
freely in solvents such as water, acetone, diethyl
ether, tetrachloromethane and, less so, n-hexane.
However, when using any of the first three highly polar
solvents, the acid used to impregnate the silica gel
is extracted along with HEMA. Moreover, when development is carried out on silica gel F_{254} plates, zinc -
originating from the manganese-doped zinc silicate
present as fluorescing indicator - as its sulphate
salt, is also extracted. The presence of both the acid
and its zinc salt in purified HEMA have been demonstrated by tlc on plain silica gel, with n-hexane -
diethyl ether (1:1) as developing solvent. Both compounds remain at the origin and are identified by
spraying with a suitable acid-base indicator, e.g.
thymol blue (pH range, 1.2-2.8), and 4-(2-pyridylazo)
resorcinol, respectively. As a consequence, the relatively non-polar tetrachloromethane is preferred as
eluent.

When examining the tlc data in order to evaluate
systems suitable for column chromatography, one should
consider that development with a HNO_3-saturated mobile
phase causes the formation of an acid front at R_F
values only slightly higher than those of HEMA (Fig.
2b and e). Consequently, in chromatography on columns
equilibrated with HNO_3-saturated solvent systems, the
separation of HEMA from contaminants having higher R_F
values, presumably will be rather marginal. Therefore,
preference should be given to chromatography on H_2SO_4-
impregnated silica gel; for the same reasons as mentioned above, n-hexane - diethyl ether (1:1) is again
recommended as mobile phase.

Application. Thin-layer chromatograms of all
reference substances and HEMA samples are pictured in
Fig. 3. Further information is recorded in Table II.

The chromatograms demonstrate that - apart from
the inhibitors - ethylene glycol and/or other polar
material, EDMA and methacrylic acid (BDH sample only)
are present in the HEMA samples from Merck and BDH.
Detectable amounts of methacrylic acid and EDMA are
absent from HEMA obtained from Hydro Med Sciences.

Table II. R_F data for HEMA and its principal contaminants in the system silica gel/n-hexane – diethyl ether (1:1)

Compound	Abreviations (cf. Fig. 3)	hR$_F$ of/in: Pure compounds	HEMA (BDH)	HEMA (Merck)	HEMA (Hydro)	Methacrylic acid
EDMA	EDMA	70	70	70	-	-
p-Methoxyphenol	pM	60	60	60	60	-
Methacrylic acid	MA	40	40	-	-	40
Hydroquinone	Hy	25	*	*	*	25
HEMA	HEMA	25	25	25	25	-
Diethyleneglycol methacrylate	DGMA	10	10	10	10	-
Ethylene glycol/ polar material	EG/P	00	00	00	00	-
Drop test			+	+	+	+

* Cannot be detected: cf. text.

Still, even this highly pure product contains at least two impurities. The spot with R_F 0.0 again may be attributed to ethylene glycol and/or other polar material. As for the second spot, a compound displaying R_F approximately 0.10 in both chromatographic solvent systems, has also been observed in the HEMA samples obtained from Merck and BDH. Milligram amounts have been collected by preparative-scale tlc and column chromatography. On the basis of the mass spectrum (m/e value of 144; presence of 3 oxygen atoms) the unknown compound was initially thought to be hydroxypropyl methacrylate. However, comparison of the tlc behavior of the unknown compound and a sample of pure 2-hydroxypropyl methacrylate demonstrated our hypothesis to be incorrect. Recently, from information obtained from Hydro Med Sciences, it has become evident that HEMA samples may contain several tenths of a per cent of diethyleneglycol methacrylate. The latter product has been synthesised in our laboratory (cf. above) and has been shown to be identical with the unknown compound by both tlc and mass spectrometry[*]. In addition, we note that a sharp separation of HEMA and diethyleneglycol methacrylate is obtained by carrying out development with pure diethyl ether instead of n-hexane - diethyl ether (1:1). The R_F values now are 0.80 and 0.55 respectively.

As regards the inhibitors, with n-hexane - diethyl ether (1:1), hydroquinone displays an R_F value approximately equal to that of HEMA and cannot be detected with either Echtblausalz B or diazotised sulphanilic acid. When using n-hexane - MIBK - n - octanol (9:2:1, satd. with 25% HNO_3) as mobile phase, hydroquinone has a slightly faster migration rate than has HEMA, its spot overlapping that of methacrylic acid. However, detection of small amounts of hydroquinone even now is not successful. In separate experiments, we have demonstrated that this is due to severe streaking of the hydroquinone spot, effected by the presence of the large excess of HEMA.

As a consequence, hydroquinone is detected in isolated instances (methacrylic acid!) only. No such difficulties are encountered with p-methoxyphenol; with this inhibitor, identification by tlc is successful.

––––– [*]As regards the discrepancy between the mass of diethyleneglycol methacrylate (174) and the value quoted above (144), it is a well-known fact that in mass spectrometry polyglycols easily lose a part of their molecule having a mass of 30; i.e., the m/e value of 144 should be attributed to $[m - CH_2O]^+$.

Still, with both compounds, in view of the rather
unfavorable detection limits quoted in Table I, it is
recommended to apply several microliters of sample
solution to the chromatoplate, using development over
a distance of 10-15 cm. Alternatively, preconcentra-
tion of the inhibitors by means of treatment of the
sample solution with Amberlyst A-27 resin (cf. below)
may be used. In conclusion, as is evident from the
data in Table II, the drop test procedure with sulphanilic acid is excellently suited to detect the presence, though not the type, of an inhibitor.

Comparison of the overall picture of the series
of chromatograms obtained with n-hexane - diethyl
ether (1:1), and the series obtained with n-hexane -
MIBK - n-octanol (9:2:1, satd. with 25% HNO_3) demonstrates that a smaller number of spots is observed
when developing with the latter solvent system. This
is mainly due to the fact that n-octanol has low volatility and cannot easily be removed from the chro-
matogram prior to identification. As a consequence,
upon spraying with a $KMnO_4$ solution, the color of the
background turns brown rather rapidly, thus obscuring
several of the smaller spots. In summary, n-hexane -
diethyl ether (1:1) should be preferred for qualitative analysis, both on account of its faster migration
rate, and the better detection achieved for actually
all components present in the HEMA and EDMA samples.

Preparative-scale tlc on H_2SO_4-impregnated silica
gel has successfully been carried out with HEMA samples from Hydro Med Sciences and BDH. (For all compounds studied, the R_F values in this system are equal
to, or slightly higher than those in the system silica
gel/n-hexane - diethyl ether, satd. with 25% HNO_3; cf.
Fig. 2). After development with n-hexane - diethyl
ether (1:1) and elution with tetrachloromethane, the
samples turn out to be chromatographically pure. If
hydroquinone has been added to the HEMA sample (Merck)
this inhibitor will still be present after chromatographic purification (cf. above). Generally, the
presence of an inhibitor is not harmful since it can
be removed quantitatively from the PHEMA gel by
prolonged washing with water. However, should quantitative removal of hydroquinone at an early stage be
imperative, then two consecutive treatments of the
sample to be purified with Amberlyst A-27 resin (Rohm
and Haas, Philadelphia, Pa. 19105, U.S.A.) suffice to
remove up to 0.2% of hydroquinone, and also other
polar compounds; these can subsequently be eluted with
methanol. The concentrated extract so obtained is
excellently suited to detect the presence of inhibi-

Table III. n_D^{20} values for HEMA and some of its contaminants

Compound	n_D^{20}	Ref.
HEMA	1.4525	[1], [2], [4]
Diethyleneglycol methacrylate	1.4568	[8]
EDMA	$\begin{cases} 1.4549 \\ 1.4549\text{-}1.4553 \end{cases}$	[2], [4] [1]
Ethylene glycol	1.4313	[5]
Methacrylic acid	1.4314	[4], [5]
Methyl methacrylate	1.4142	[5]

tor(s), as was referred to above.

A final word about two further criteria which are used to assess the purity of HEMA and EDMA, viz. the refractive index and the infrared spectrum. According to our experiences, n_D^{20} values, which are rather generally quoted in the literature, are not very reliable criterion, since the impurities normally present in HEMA exhibit n_D^{20} values which are both higher (EDMA and diethyleneglycol methacrylate) and lower (methacrylic acid and ethylene glycol) than are those of HEMA itself. As an illustration, a series of data is recorded in Table III. Besides, the refractive index of HEMA strongly decreases with increasing water content, as is demonstrated in Fig. 4. Infrared spectroscopy has been recommended to check the purity of EDMA. Our results indicate that compounds such as HEMA and ethylene glycol can be detected down to approximately 1 wt.% due to the strong absorption of their hydroxyl groups (broad band at 2800-3600 cm^{-1}). However, one should keep in mind that the infrared spectrum does not easily allow a conclusion regarding the nature of the contaminant(s) - e.g. HEMA, ethylene glycol or water. Besides, a compound such as methacrylic acid goes undetected at the 1% level.

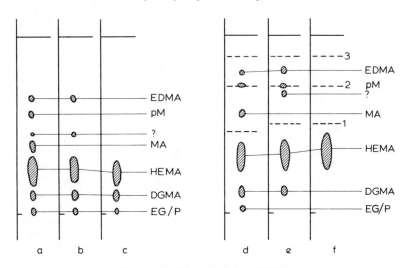

Figure 3. Tlc of three HEMA samples on silica gel, using n-hexane–diethyl ether (1:1) (a–c), and n-hexane–MIBK–n-octanol (9:2:1, satd. with 25% HNO₃) (d–f) as mobile phases. HEMA samples obtained from BDH (a,d), Merck (b,e), and Hydro Med Sciences (c,f). For abbreviations, see Table II. – – – –, acid (1), n-octanol (2), and MIBK (3) front.

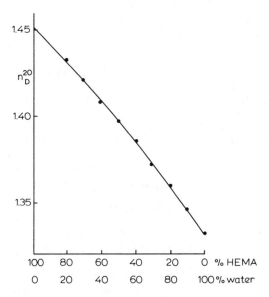

Figure 4. Dependence of n_D^{20} on water content of HEMA

Abstract

Poly(hydroxyethyl methacrylate) is generally accepted as a biocompatible, safe and stable hydrogel for medical use. The present paper describes the use of tlc for the analysis and small-scale preparation of the initial monomer, 2-hydroxyethyl methacrylate. Tlc on silica gel, with n-hexane - diethyl ether (1:1, v/v) and/or n-hexane - MIBK - n-octanol (9:2:1, v/v, satd. with 25% HNO_3) as mobile phase is recommended for qualitative analysis. Preparative-scale work is preferably carried out on H_2SO_4-impregnated silica gel, development being done with n-hexane - diethyl ether (1:1). Inhibitors are detected by means of tlc and of a drop test procedure with diazotised sulphanilic acid. The nature of the contaminants present in several commercially available HEMA samples is discussed. n^{20}_D values are reported for the system HEMA- water.

Literature Cited

1. Wichterle, O., and Lim, D., Nature (1960) **185**, 117.
2. Wichterle, O., and Lim, D., U.S. Patent, (1961) 2, 976, 576.
3. Kopecek, J., Jokl, J., and Lim, D., J. Polymer Sci., [C] (1968) **16**, 3877.
4. Refojo, M.F., J. Appl. Polymer Sci., (1965) **9**, 3161.
5. Goedberg, A.I., Fertig, J., and Stanley, H., U.S. Patent, (1962) 3, 059, 024.
6. Refojo, M.F., and Yasuda, H., J. Appl. Polymer Sci., (1965) **9**, 2425.
7. Hasa, J., and Janácek, J., J. Polymer Sci., [C] (1967) **16**, 317.
8. Kopecek, J., Thesis, Inst. Macromol. Chem., Prague, (1965) p. 35.
9. Mark, M.F., Gaylord, N.G., Bikales, N.M., (eds), "Encyclopedia of Polymer Science and Technology", Vol. 15, 1971, 273.
10. Zweig, G., and Sherma, J. (eds), "Handbook of Chromatography", CRC Press, Interscience, New York, 1971.
11. Crawford, J.W.C., U.S. Patent, (1939) 2, 143, 941.
12. Jatzkewitz, H., and Lenz, U., Hoppe-Seylers Z. Physiol. Chem., (1956) **305**, 53.
13. Jatzkewitz, H., Hoppe-Seylers Z. Physiol. Chem., (1953) **292**, 99.
14. Merck, E., Chemicalien, Reagenzien (1974), Darmstadt.

Rotational Viscometry Studies of the Polymerization of Hydrophilic Methacrylate Monomers

SHAO M. MA, DONALD E. GREGONIS, CHWEN M. CHEN,
and JOSEPH D. ANDRADE

Department of Materials Science and Engineering, University of Utah,
Salt Lake City, Utah 84112

There is considerable literature on the mechanical behavior
of poly(hydroxyethyl methacrylate) (PHEMA) gels (1). To our
knowledge, no study has been made on the flow properties of this
polymer during the course of polymerization. In this study we
attempted to use this easily obtained property to obtain knowledge
concerning (1) the relative rate of polymerization with various
initiators; (2) the effect of polymerization time on the flow
properties of PHEMA; (3) the relations between the flow curve,
molecular weight and the molecular weight distribution.

At a given temperature the viscosity (η) vs. shear rate ($\dot{s}$)
curve for concentrated polymer solutions and pure polymers depends
on the molecular weight and molecular weight distribution of the
system. However, an increase in viscosity of the polymerization
mixture indicates only that conversion is increasing. How viscosity changes during the course of polymerization from Newtonian
to non-Newtonian as a function of molecular weight or molecular
weight distribution is not totally understood.

Qualitatively, the polymerization mixture behaves as a Newtonian fluid during the early stage of polymerization. As the
degree of polymerization or the concentration of the polymeric
fraction increases, the viscosity also increases and the system
becomes non-Newtonian. Polymers with broad molecular weight
distributions show a higher viscosity dependency on shear rate
(2,3); non-Newtonian behavior starts to occur at lower shear
rates than similar polymers with narrow molecular weight distributions. Therefore, the effect of polymerization time at this
stage would depend on the polydispersity of the polymer produced.
Further polymerization encounters branching and gel formation.
Graessley (4) reported that long-chain branching affects the
viscosity-shear rate curve in a similar way as broadening of the
molecular weight distribution. However, the effects of branching
cannot be separated from the effect of polydispersity in viscometry measurements. The viscosity of a branched polymer may be
higher or lower than a linear polymer. This depends on whether
the molecular weight is higher or lower than the molecular weight

of tangling segments. Because of the complicated nature of the problem, no single theory can adequately describe the mechanical behavior of the polymeric system during the course of polymerization. However, , several theories (5-7) have worked well for linear polymers, which might be the case for PHEMA produced with very little crosslinking agent in a relatively poor solvent such as water.

The present investigation consists of: (1) flow curves obtained for hydroxyethyl methacrylate (HEMA)-water mixtures, polymerized at 60°C as functions of polymerization time and initiator; (2) three theories of non-Newtonian viscosity, the Bueche-Harding method (5), Ree-Eyring activated-state theory (6) and Bartenev's empirical method (7), are briefly described and the flow parameters of our systems are analyzed with these theories; (3) the relations between initiators and polymerization rate and between flow curve, molecular weight, molecular weight distribution and polymerization time are discussed.

Material and Methods

The monomer-solvent mixture used in this study consisted of six parts of HEMA and three parts of water, by volume. The initiators used include ammonium persulfate, azobisisobutyronitrile (AIBN), azobis(methyl isobutyrate), azobis(methoxyethyl isobutyrate), and azobis(methoxydiethoxyethyl isobutyrate). A concentration of 5.71 mmol/liter was used for ammonium persulfate and 5.21 mmol/liter for the others. The synthesis, purification and chemical characterization of the PHEMA are given elsewhere (8). It should be noted that although HEMA monomer is completely miscible with water, PHEMA is not water soluble.

A Haake Rotovisco rotational viscometer was used for viscosity measurement. Polymerization was carried out in a coaxial cylinder sensor system, with a MV cup and a MVI bob. Temperature was kept at 60°C with a Lauda K-2/R circulating bath. Flow curves were determined as functions of time at shear rates from 0 to 685 sec^{-1}. To avoid permanent mechanical breakdown, lower shear rates were used as the polymerization increased.

Non-Newtonian Viscosity

In non-Newtonian flow the chance in apparent viscosity (η) as a function of shear rate ($\dot{s}$) generally takes the form

$$\eta \, \eta_o = f(\tau\dot{s}), \qquad\qquad [1]$$

where η_0 is the viscosity at zero shear rate, $\dot{s}_0$, and τ is the characteristic relaxation time, which is molecular weight-dependent.

Many theories have been proposed to explain non-Newtonian behavior in condensed systems. Molecular theories based on the equivalence hypothesis (9), on the entanglement concept (10), and on the activated-state model (6) have gained considerable acceptance.

Bueche-Harding Standard Curve. Bueche introduced a shear-rate dependence to Rouse theory (11). According to his hypothesis, macromolecules in solution under dynamic deformation are assumed to behave similarly to those under steady shearing. The change in viscosity with shear rates is considered as due to the results of deforming and rotating of the coiling polymer molecules under a shearing force. Below a certain characteristic time, τ_b, (which equals approximately the reciprocal of zero shear rate, $\dot{s}_0$) the viscosity decreases rapidly with increasing shear rate, while above it the viscosity approaches its maximum value. This relaxation time can be calculated from the properties of the system at low shear rates.

Bueche has ignored the effect of chain entanglements on non-Newtonian behavior and assumed that the local properties of the system, such as the relaxation time distribution, are independent of its state of motion. His theory, therefore, does not correlate well with experimental data (4, 12-14).

Graessley (10) proposed a theory by assuming that the viscosity of a polymeric system is controlled by intermolecular chain entanglements and that an increase in shear induces changes in the network of entanglements and hence causes the viscosity to decrease. This entanglement approach has a sound theoretical basis. However, information concerning the polydispersity and the entanglement density are needed to carry out actual computation.

For a linear coiling polymer, Bueche-Harding, later, suggested a method for determining its absolute molecular weight from its flow curve (5). In this method they used a standard curve which follows the empirical equation

$$\eta_0/\eta = 1.00 + 0.60 \ (\tau \dot{s})^{3/4}, \tag{2}$$

to match their experimental data. The value of η_0 and s_0 are determined by superimposing the standard curve in the form of log (η/η_0) vs. log $(\dot{s}\tau)$ with an experimental curve in the form of log (η) vs. log $(\dot{s})$ while both curves were plotted on the same scale. The molecular weight of the polymer is then calculated from the expression

$$M = \pi^2 NckT/12\eta_0 \ \dot{s}_0 \ , \tag{3}$$

where N, c, k, and T are Avogadro's number, the concentration of polymer in g/cc, Boltzmann's constant, and the absolute temperature, respectively.

Bueche's relaxation time $(\tau_b = 1/\dot{s}_0)$, as pointed out by
Graessley (15), governs the magnitude of the shear rate when the
viscosity begins to decrease. It should be an important parameter for those properties in which the longer relaxation times
are determining factors. The Bueche-Harding standard curve,
which agreed moderately well with data on unfractionated polysty-
rene in benzene and poly(methyl methacrylate) (PMMA) in chloro-
form, should predict the non-Newtonian behavior of any linear
polymeric system with not too broad a molecular weight distribution (5). PHEMA and PMMA have the same backbone structure.
Since the shear rate used in this study is relatively low; and
the system is expected to have low entanglement density (water is
a poor solvent), low branching and crosslinking (the diester
concentration is low), the Bueche-Harding method should work well
for our system.

Ree-Eyrings's Activated-State Model. This model (6)
assumes that there exists i groups of flow units which differ
in relaxation time and in geometrical dimensions. Some of these
flow units are Newtonian, others are non-Newtonian. A Newtonian
flow unit is a molecule or a group of molecules isolated from
other units, while a non-Newtonian unit is a Newtonian unit
bonded (or entangled) with another Newtonian unit or units.
Thus, for flow of non-Newtonian units, this bond (or entanglement) must be broken (or disentangled).

Based on this concept, the generalized viscosity equation is

$$\eta = \Sigma_i \frac{x_i \beta_i}{\alpha_i} \frac{\sinh^{-1} \beta_i \dot{s}}{\beta_i \dot{s}} , \qquad [4]$$

where x_i is the fractional area occupied by the ith flow unit on
the shear surface, and

$$\alpha_i = (\lambda \lambda_2 \lambda_3)_i / 2kT ; \qquad [5]$$

$$\beta_i = [(\lambda/\lambda_1)2k']_i^{-1} ; \qquad [6]$$

α_i and β_i are the characteristic shear volume divided by kT

which is related to the inverse of the shear modulus and the

relaxation time, respectively. β_i/α_i is the Newtonian viscosity.

λ is the jumping distance. λ_1, λ_2, and λ_3 are the molecular

dimensions of a flow unit. k' is the jumping frequency of the
flow unit when there is no stress. According to the theory of
rate processes (16):

$$k' = \frac{kT}{h} \frac{G\ddagger}{G} \exp (-\varepsilon_0/kT), \qquad [7]$$

where h is Planck's constant; $G^{\ddagger}$ and G are the partition functions of the activated and the initial state, respectively; and ε_o is the activation energy.

Considering the HEMA-H_2O mixture, one can assume the presence of two types of flow units, the Newtonian type (unstructured water, HEMA monomer, and the lower members of HEMA polymers), and the non-Newtonian type (higher members of PHEMA with bonded water). For these considerations, Equation [4] takes the form

$$\eta = x_o \eta_s + \frac{x_1 \beta_1}{\alpha_1} + \frac{x_2 \beta_2}{\alpha_2} \frac{\sinh^{-1} \beta_2 \dot{s}}{\beta_2 s} \quad . \quad [8]$$

Here, η_s is the solvent viscosity. When the rate of shear is very small, one has

$$\eta_o = x_o \eta_s + \frac{x_1 \beta_1}{\alpha_1} + \frac{x_2 \beta_2}{\alpha_2} \quad . \quad [9]$$

This theory makes no assumptions as to the specific molecular conformations or configurations to the flow units. Also, it recognizes the discrete boundary between flow units and medium and between separate flow units.

To obtain a relationship between the relaxation time and molecular weight (17), equation [6] can be rewritten as

$$\frac{1}{\beta_i} \propto k_i^{\ddagger} = \frac{kT}{h} e^{-\Delta F_i^{\ddagger}/RT} \quad , \quad [10]$$

where ΔF_i is the standard free energy of activation per mole of ith flow unit. For a condensed phase this free energy approximately equals the Helmholtz free energy which is related to the partition function, f_i, by

$$A_i = -kT \ln f_i^N \quad . \quad [11]$$

Introducing equation [11] into [10] one has

$$\frac{1}{\beta_i} \propto f_i \quad . \quad [12]$$

According to Ma, Jhon and Erying (18), the partition function for linear polymers can be separated into two parts, a part dependent on chain entanglement and a part independent of chain entanglement. The first part can be written as

$$(\frac{M}{m_1})^{4/3} (\frac{M}{m_2})^2 \quad .$$

Here M, m_1 and m_2 are the molecular weights of the polymer, the kinetic segment and the tangling segment, respectively. In concentrated solutions and in pure polymeric systems, kinetic segments can move into neighboring vacancies independently except for the restriction that they remain tied together while all tangling segments except one are obliged to follow around the randomly located tangling points.

We expect that flow units, which contain molecules with M > m_2, are mostly non-Newtonian in behavior. For two systems with the same molecular weight it follows that

$$\frac{(\beta_2)_1}{(\beta_2)_2} = \frac{(f_2)_2}{(f_2)_1} = \frac{(m_2)_1^2}{(m_2)_2^2} = \frac{(\bar{\ell}_2)_1^2}{(\bar{\ell}_2)_2^2} . \quad [13]$$

The subscript 2 inside the parentheses refers to the non-Newtonian flow units and the subcripts 1 and 2 outside the parentheses refer to systems 1 and 2, respectively. Equation [13] indicates that the average length or the molecular weight of the tangling segments is proportional to the square root of the relaxation time of the flow units which contain such segments. This would approximately be the case when the same degree of polymerization was reached from different initial HEMA-water mixtures. On the other hand we can write (19)

$$\beta_p = \beta_s (M/m_2) (M/m_1)^{1/3} \quad \text{and}$$
$$\alpha_p = \frac{\alpha_s}{(M/m_1) (M/m_2)} , \quad [14]$$

where the subscripts p and s refer to properties for the polymer and unattached kinetic segments, respectively. If the degree of polymerization is sufficiently large, i.e., if M > m_2, the value of m_1 , m_2 should not be affected by further polymerization. It follows that

$$\frac{(\beta_p)t_1}{(\beta_p)t_2} = \frac{(M_{t_1})^{4/3}}{(M_{t_2})^{4/3}} \quad \text{and}$$
$$\frac{(\alpha_p)t_1}{(\alpha_p)t_2} = \frac{(M_{t_2})^2}{(M_{t_1})^2} , \quad [15]$$

where t_1 and t_2 refer to two different polymerization times.

Results and Discussion

The viscosity of HEMA-water mixtures at 60°C is plotted against the rate of shear at various polymerization times for five initiators (Figure 1). According to Nakajima ([20]), the steady-state flow curves for eight polyethylene samples have a similar shape to their corresponding cumulative fractionation curves. He assumed that ΣWi be the cumulative weight fraction and B_i be the relative chain length at $\Sigma Wi = H$ (a fractional number) and concluded that

$$\eta/\eta_0 = H = \Sigma Wi \qquad [16]$$

and

$$\dot{s}_H = KB_{iH}^{-\alpha} \quad . \qquad [17]$$

With three adjustable parameters, η_0, K, and α, he showed that with his data the flow curve and fractionation curve are equivalent for most of the range. The amount of crosslinker (diester) present in our system is less than 0.01%. Equations [16] and [17], which apply satisfactorily to linear polymers, are probably also applicable to the current system. To prove this, a fractionation curve for PHEMA has to be obtained independently through other measurements. Note that in the polymerization system employed herein that a diester concentration in excess of 0.035% will result in an insoluble, crosslinked polymer.

Flow parameters were calculated from the Bueche-Harding standard curve, from Equation [8] and from Bartenev's empirical equation, the latter taking the form

$$\log \eta/\eta_0 = -K' \tau_w , \qquad [18]$$

where τ_w is the shear stress.

The Bueche-Harding standard curve represents composite viscosity data for polydisperse polymer systems. Our data can be superimposed onto this standard curve and form a single master curve. Figure 2 is an example. The values of τ_b, $\dot{s}$ and η_0 obtained hereafter are given in Table I. The polymer concentration in mole/liter is calculated from Equation [3]. Because of the lack of data concerning the value of c as a function of polymerization time, the absolute values of molecular weight are not computed. However, two points can be concluded from our data: 1) our experimental curves correlate well with the Bueche-Harding standard curve, indicating polymers produced during the course of measurement are probably all linear polymers; and 2) assuming half of the initial HEMA monomer (0.7 g/cc) was converted into polymers at the end of our measurement, the viscosity average molecular weight reached a value of

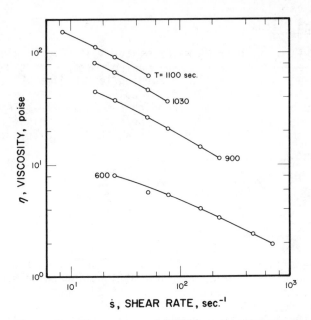

Figure 1a. Flow curves of HEMA–water mixtures at 60°C. Initiator: ammonium persulfate.

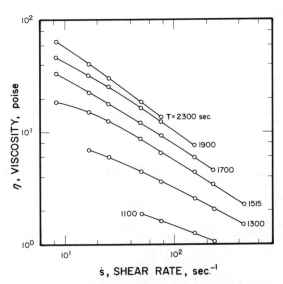

Figure 1b. Flow curves of HEMA–water mixtures at 60°C. Initiator: AIBN.

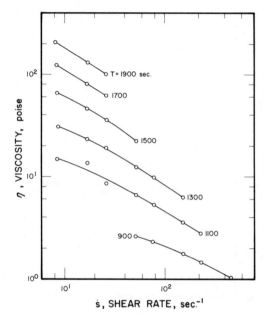

Figure 1c. Flow curves of HEMA–water mixtures at 60°C. Initiator: azobis(methyl isobutyrate).

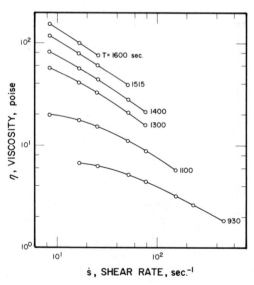

Figure 1d. Flow curves of HEMA–water mixtures at 60°C. Initiator: azobis(methoxyethyl isobutyrate).

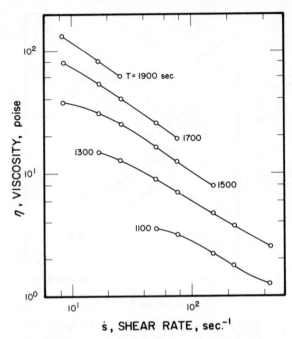

Figure 1e. Flow curves of HEMA–water mixtures at 60°C. Initiator: azobis(methoxydiethoxyethyl isobutyrate).

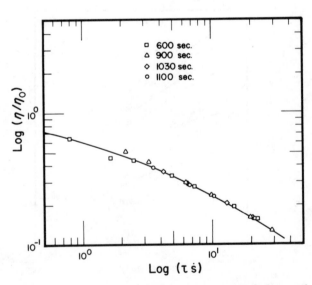

Figure 2. A superimposed plot of experimental data and Bueche-Harding standard curve. Initiator: ammonium persulfate.

TABLE I.

The Effect of Polymerization Time, t, on the Flow Parameters
(Bueche-Harding Method) for HEMA-water Mixtures

Solvent: Water (η_s = 4.665 x 10^{-3} poise at 60°C)
Concentration: 6 parts HEMA/3 parts water

t (sec)	τ_b (sec)	S_0 (sec^{-1})	η_0 (poise)	1000 C/M (mole/liter)
(1) Initiator: Ammonium persulfate				
600	.0364	27.5	13.03	1.57 x 10^{-5}
1030	.208	4.8	213.0	4.46 x 10^{-5}
1100	.323	3.1	357.3	4.9 x 10^{-5}
(2) Initiator: Azobisisobutyronitrile				
1100	.0186	53.7	3.0	7.07 x 10^{-6}
1515	.286	3.5	44.67	6.8 x 10^{-6}
1900	.588	1.7	134.9	1.03 x 10^{-5}
2300	1.111	0.9	281.8	1.1 x 10^{-5}
(3) Initiator: Azobis(methyl isobutyrate)				
900	0.0288	34.7	4.92	7.46 x 10^{-6}
1300	0.476	2.1	91.2	8.26 x 10^{-6}
1700	1.042	0.96	496.6	2.08 x 10^{-5}
1900	1.389	0.72	954.0	3.03 x 10^{-5}
(4) Initiator: Azobis(methoxyethyl isobutyrate)				
930	0.038	26.3	10.6	1.22 x 10^{-5}
1300	0.455	2.2	151.4	1.46 x 10^{-5}
1600	1.099	0.91	626.0	2.51 x 10^{-5}
(5) Initiator: Azobis(methoxydiethoxyethyl isobutyrate)				
1100	0.0364	27.5	7.2	8.77 x 10^{-6}
1500	0.333	3.0	97.7	1.29 x 10^{-5}
1900	1.887	0.53	758.5	1.75 x 10^{-5}

TABLE II.

The Effect of Polymerization Time, t, on the Flow Parameters
(Eyring's Theory) for HEMA-Water Mixtures

Solvent: Water ($\eta_s = 4.665 \times 10^{-3}$ poise at 60°C)

Concentration: 6 parts HEMA/3 parts water

t (sec)	$x_0\eta_s + \dfrac{x_1\beta_1}{\alpha_1}$	x_2/α_2 (dyne/cm^2)	β_2 (sec)	Correlation^2 Coefficient	η_0 (poise)
(1) Initiator: Ammonium persulfate					
600	1.31	1.2×10^2	0.077	0.981	10.56
900	3.14	4.66×10^2	0.124	0.999	60.91
1030	8.92	6.21×10^2	0.213	1.000	141.08
1100	13.34	7.37×10^2	0.294	1.000	230.24
(2) Initiator: Azobisisobutyronitrile					
900	0.09	5.0×10^1	0.0074	0.96	0.46
1100	0.28	6.8×10^1	0.0288	0.996	2.24
1300	0.55	9.53×10^1	0.0785	0.999	8.03
1515	0.64	1.46×10^2	0.151	1.000	22.69
1700	0.86	1.68×10^2	0.280	0.998	47.89
1900	0.999	2.19×10^2	0.340	1.000	77.4
2100	0.42	2.12×10^2	0.754	1.000	159.95
2300	1.28	2.05×10^2	1.03	1.000	212.15
(3) Initiator: Azobis(methyl isobutyrate)					
800	.13	1.58×10^2	0.0074	0.912	1.30
900	.28	1.02×10^2	0.0287	0.999	3.21
1100	1.33	1.84×10^2	0.0528	0.984	11.04
1300	1.06	1.91×10^2	0.214	1.000	41.93
1500	1.70	2.94×10^2	0.381	1.000	113.96
1700	3.54	4.14×10^2	0.736	1.000	308.38
1900	15.13	4.40×10^2	1.755	1.000	871.8

Table II. (cont.)

t (sec)	$x_0\eta_s + \dfrac{x_1\beta_1}{\alpha_1}$	x_2/α_2 (dyne/cm^2)	β_2 (sec)	Correlation^2 Coefficient	η_0 (poise)
(4) Initiator: Azobiz(methoxyethyl isobutyrate)					
600	0.32	2.28×10^1	0.0675	0.94	1.86
930	1.48	5.57×10^1	0.216	0.990	13.51
1100	0.55	2.43×10^2	0.0878	1.000	21.90
1300	1.13	3.09×10^2	0.258	1.000	80.72
1400	1.54	3.82×10^2	0.335	1.000	129.21
1515	4.65	4.23×10^2	0.586	1.000	252.96
1600	9.50	4.33×10^2	1.002	1.000	442.99
(5) Initiator: Azobis(methoxydiethoxyethyl isobutyrate)					
900	0.043	1.05×10^2	0.01	1.000	1.096
1100	0.28	1.25×10^2	0.0346	0.997	4.62
1300	1.02	1.49×10^2	0.140	1.000	21.87
1500	1.04	2.71×10^2	0.179	0.997	49.62
1700	1.71	3.09×10^2	0.526	1.000	164.15
1900	4.72	3.42×10^2	1.44	1.000	497.32

TABLE III.

The Effect of Polymerization Time, t, on the Flow Parameters
(Bartenev's Method) for HEMA-Water Mixtures

Solvent: Water (n_s = 4.665 x 10^{-3} poise at 60°C)

Concentration: 6 parts HEMA/3 parts water

t (sec)	K' x 10^3	n_0 (poise)	Correlation ^{2} Coefficient
(1) Initiator: Ammonium persulfate			
600	0.521	9.0	0.962
900	0.327	74.851	0.992
1030	0.214	138.38	0.957
1100	0.216	299.77	0.999
(2) Initiator: Azobisisobutyronitrile			
900	1.64	0.518	0.945
1100	1.47	2.47	0.978
1300	1.22	8.51	0.968
1515	1.17	27.98	0.986
1700	1.14	62.68	0.993
1900	1.06	126.34	1.000
2100	1.40	377.28	0.997
2300	1.30	419.89	0.994
(3) Initiator: Azobis(methyl isobutyrate)			
800	1.53	1.73	1.000
900	1.31	3.94	0.998
1100	0.581	13.48	0.913
1300	1.03	59.74	0.999
1500	0.816	200.34	0.996
1700	0.596	556.69	0.999
1900	0.394	1046.6	0.989
(4) Initiator: Azobis(methoxyethyl isobutyrate)			
600	1.07	0.99	0.771
930	0.917	9.37	0.938
1100	0.778	29.25	0.986
1300	0.755	133.75	0.998
1400	0.629	222.29	0.999
1515	0.510	389.16	0.995
1600	0.473	633.97	0.994

Table III. (cont.)

t (sec)	$K' \times 10^3$	η_0 (poise)	Correlation ^{2} Coefficient
(5) Initiator: Azobis(methoxydiethoxyethyl isobutyrate)			
900	1.64	1.47	0.998
1100	1.18	5.67	0.977
1300	0.877	22.67	0.979
1500	0.812	78.49	0.489
1700	0.804	297.38	0.999
1900	0.72	876.56	0.999

the order of 10^7. With such a high molecular weight one would
expect the viscosity of the system to be much higher. However,
since water is a poor solvent for PHEMA, a large portion of
the polymer molecule might remain unentangled and, as a consequence, result in a lower viscosity.

The parameters $x_0 \eta_s + x_1 \beta_1/\alpha_1$, x_2/α_2 and β_2 all increase
with an increase in polymerization time (Table II) at a given
temperature ($60°C$). The quantity $x_0 \eta_s + x_1 \beta_1/\alpha_1$ increases with
increasing polymerization time because x_1 is concentration dependent. The quantity x_2 should increase with both an increase
in polymer concentration and increased molecular weight. Thus,
the increase of x_2/α_2 with polymerization time seems to be
natural since α_2 is a characteristic property for flow unit 2.
The relaxation time, β_2, is a quantity which is independent of
concentration at low concentrations but increases with concentration at high concentrations. Our results show that β_2 increases with polymerization time. The early increase is
probably due to an increase in molecular weight alone, while
the later increase is due to changes in concentration and in
molecular weight. Originally, we applied a method developed by
Gabrysh et al. (21) to determine these parameters. However, we
were unable to determine $x_0 \eta_s + x_1 \beta_1/\alpha_1$ meaningful by such a
method. The parameters given in Table II were determined by
using a computer program to obtain a best straight line fit for
a η vs. $\sinh^{-1} \beta_2 \dot{s}/\beta_2 \dot{s}$ plot. The square of the correlation
coefficient for such a fit is greater than 0.98 with a few
exceptions, presumably due to greater experimental errors when
solutions are dilute.

In Bartenev's expression K' increases with increasing
breadth of molecular weight distribution and η_0 increases with
increasing molecular weight. Table III shows that the correlation between our data and Bartenev's expression is not very
satisfactory (ten curves have the square of correlation coefficient less than 0.98). However, the changes in K' and η_0
with polymerization time follows a reasonable trend.

Bueche-Harding's η_0 is greater than Eyring's. The ratio
between them is ranging form 1.23 to 2.18 (with one exception).
Bartenev's η_0 started with a value close to that of Eyring's,
increased steadily and finally passed both Bueche-Harding's
and Eyring's value. Bueche's relaxation time, τ_b, and Eyring's
relaxation time, β_2, are of the same order of magnitude, presumably both τ_b and β_2 measure the molecular time constant
controlling non-Newtonian behavior.

From Equation [3], one obtains

$$\frac{\tau_b}{\eta_0} \propto \frac{M}{CT} \qquad [19]$$

In a concentrated polymer solution, we can assume $\beta_2 \cong \beta_p$, and

$\eta_0 \cong \beta_s/\alpha_s$. Then, from equation [14],

$$\frac{\beta_2}{\eta_0} \cong \alpha_s (\frac{M}{m_2}) (\frac{M}{m_1})^{1/3} \frac{M^{4/3}}{T} \; . \tag{20}$$

Comparing Equation [19] with Equation [20], it follows that

$$\beta_2 \infty c \tau_b M^{1/3} \; . \tag{21}$$

Experimentally the molecular weight exponent is less than one and the concentration exponent is greater than one [22]. Graessley, et al. [13] suggested that the characteristic relaxation time be proportional to M/cT at low entanglement density and proportional to $1/c^2 T$ at high entanglement density. We expect that the entanglement density is relatively low in our system. Hence, both τ_b and β_2 should govern.

Figure 3 gives the plot of Eyring's zero shear viscosity against the polymerization time. The relationship between ln η_0 and ln t is linear for samples 2, 3 and 5 and can be fitted with two straight lines for sample 4. Because there are not enough data points, no such fitting was attempted for sample 1. At the same length of polymerization time, we can conclude that: (a) the polymerization initiated by ammonium persulfate is the fastest; and (b) the polymerizations initiated by the three AIBN derivatives have comparable rates which are slightly faster than the rate induced by AIBN itself. Similar conclusions can be drawn by using Bueche-Harding's or Bartenev's zero shear viscosity.

Conclusions

From viscometry data we can conclude the following:
(1) PHEMA produced from HEMA-water mixtures with less than 0.01% crosslinker is mostly linear polymer.
(2) The relative rate of polymerization initiated with the various initiators are of the following order: ammonium sulfate >> AIBN derivatives > AIBN;
(3) Both Ree-Eyring's theory of non-Newtonian flow and Bueche-Harding's method can describe the behavior of HEMA-water mixtures during the course of polymerization. Bartenev's empirical expression works less well presumably because the ratio η/η_0 for our system is still some function of molecular weight and cannot be assumed to depend on shear stress alone;
(4) The molecular weight of linear PHEMA can be obtained by using the Bueche-Harding method with independent information concerning concentration of the polymer. However, we have not carried out concentration measurements in this

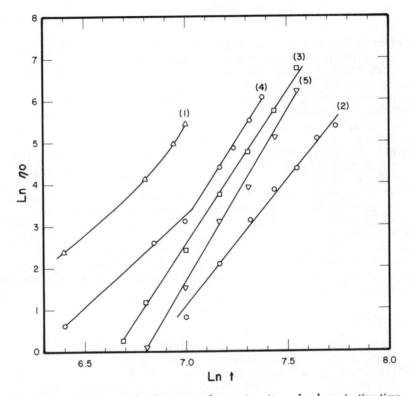

Figure 3. The relationship between polymer viscosity and polymerization time (Eyring's Model). Initiator: (1) ammonium persulfate; (2) AIBN; (3) azobis-(methyl isobutyrate); (4) azobis(methoxyethyl isobutyrate); (5) azobis(methoxy-diethoxyethyl isobutyrate).

study and only the ratio $^c/M$ is estimated. The relative molecular weight at different polymerization times can be obtained from Eyring's relaxation time;
(5) Nakagima's theory (20) correlates the flow curve with the cumulative molecular weight distribution curve of linear polymers. One can explore this theory further and see whether it is applicable to our systems.

The work presented here is only a preliminary study. The data obtained with a Haake Rotovisco viscometer are not a sufficient test. To further test the applicability of the theories requires that: (1) experimental data be obtained over a wide range of rates of shear, at several different temperatures and using various solvents; (2) concentration and polydispersity be determined side by side with viscometric measurement so that results from these measurements can be directly compared.

Abstract

Flow curves for hydroxyethyl methacrylate-water mixtures were determined as functions of polymerization time and initiator. The non-Newtonian behavior of these systems was analyzed by a Bueche-Harding standard curve, by the Ree-Eyring generalized viscosity equation and by Bartenev's empirical equation. The relative rate of polymerization initiated with AIBN and various AIBN esters and the relationships between flow curves, molecular weight, molecular weight distribution and polymerization time are discussed.

Literature Cited

1. Janacek, J., J. Macromol. Sci.-Revs. Macromol. Chem. (1973) C9 (1) 1-47.
2. Rudd, J. F., J. Poly. Sci. (1960) 44, 459.
3. Sabia, R., J. Appl. Poly. Sci. (1964) 8, 1053.
4. Graessley, W. W., and Prentice, J. S., J. Poly. Sci., A-2 (1968) 6, 1887.
5. Bueche, F., and Harding, S. W., J. Poly. Sci. (1958) 32, 177.
6. Ree, T., Eyring, H., J. Appl. Phys. (1955) 26, 793, 800.
7. Bartenev, G. M., Vysokomolekyluarnye Soedineniya (1964) 6, 2155.
8. Gregonis, D., Chen, C. M., and Andrade, J. D., this symposium.
9. Bueche, F., F. J. Chem. Phys. (1954) 22, 1570.
10. Graessley, W. W., J. Chem. Phys. (1965) 43, 2696.
11. Rouse, P. E., J. Chem. Phys. (1953) 21, 1272.
12. Ballman, R. L., and Simon, R. M., J. Poly. Sci. A-2 (1964) 2, 3557.
13. Graessley, W. W., Hazleton, R. L., and Lindeman, L. R., Trans. Soc. Rheology, (1967) 11, 267.

14. Shih, C. K., Tran. Soc. Rheology (1970) 14, 83.
15. Graessley, W. W., J. Chem. Phys. (1967) 47, 1942.
16. Glasstone, S., Laidler, K., and Eyring, H., "The Theory of Rate Processes," p. 483, McGraw Hill Book Company, New York, 1941.
17. Jhon, M. S., and Eyring, H., private communication.
18. Ma, S. M., Eyring, H., and Jhon, M. S., Proc. Natl. Acad. Sci. (USA) (1974) 71, 3096.
19. Eyring, H., Ree, T., and Hirai, N., Proc. Natl. Acad. Sci. (USA) (1958) 44, 1213.
20. Nakajima, N., Proc. 5th Intl. Cong. on Rheology, Ed. by Shigeharu Onogi, (1968) 4, 295.
21. Gabrysh, A. F., Eyring, H., Shimizu, M., and Asay, J., J. Appl. Phys. (1963) 34, 261.
22. DeWitt, T. W., Markovitz, H., Padden, F. J., and Zapas, L. J., J. Colloid Sci. (1955) 10, 174.

10

The Effect of Tacticity on the Thermal Behavior of Various Poly(methacrylate esters)

G. A. RUSSELL, D. E. GREGONIS, A. A. DeVISSER, and J. D. ANDRADE
Department of Materials Science and Engineering, University of Utah,
Salt Lake City, Utah 84112

D. K. DALLING
Department of Chemistry, University of Utah, Salt Lake City, Utah 84112

Tacticity, the stereochemical placement of pendant side chains along the polymer backbone, has long been recognized as an important factor in determining bulk polymer physical properties. Such properties as crystallizability, solubility, melting point, glass transition temperature, etc. have been correlated with the tacticity of a given polymer (1,2,3,4). Introduction of Ziegler-Natta catalysts and other catalyst systems capable of controlling tacticity during synthesis has opened up new areas for commerical development based on systematic variation of molecular configuration. Little work has been reported, however, correlating chain tacticity to the surface and interfacial properties of polymers. Recent work (5) suggests that reorientation of the chains in the surface zone of the polymer can affect wettability of the surface If that is indeed the case, the tacticity of the polymer chains may well have an influence on wettability, since the barriers to chain rotation are a function of tacticity. Also, the availability of hydrophilic and hydrophobic sites for interaction at the interface is influenced by the chain configuration and conformation. Consequently, it was felt that a study of the influence of tacticity on bulk physical and interfacial properties in the poly(hydroxyethyl methacrylate) (pHEMA)/water system might indicate some means of altering the interfacial properties of the polymer by control of its tacticity during synthesis. In this paper we report preliminary results of some experiments in which various methods of altering tacticity during synthesis have been examined and their effect correlated with changes in the thermal properties of the resulting polymers.

Synthesis of methyl and other alkyl methacrylates of high stereoregularity, either of high isotactic content or high syndio-tactic content, is well-known (6). Unfortunately, the organometallic initiators used for production of stereoregular methacrylates are highly reactive to the hydroxyl functionality in the side chain of HEMA. Consequently, some means of blocking the hydroxyl group during polymerization was required. Methoxyethyl methacrylate (MEMA) was chosen as a model for a HEMA-like monomer

with its hydroxyl group protected, even though the methyl ether
linkage is too stable to permit hydrolysis back to pHEMA. A
second protected HEMA derivative used was trimethylsilylethyl
methacrylate (TMSEMA), whose trimethylsilyl protection group can
be cleaved readily, forming HEMA. This allows polymerization of
TMSEMA using an anionic catalyst to produce a stereoregular
pTMSEMA, followed by hydrolysis to stereoregular pHEMA. Polymers
of HEMA, MEMA, and TMSEMA were synthesized by a variety of techniques, their tacticity was determined by ^{13}C-NMR spectrometry,
and the variation in tacticity was correlated with changes in
the DSC thermograms of each polymer.

Experimental

Monomer Preparation. All monomers used in this study except
TMSEMA were obtained commercially. Methyl methacrylate (MMA) and
methoxyethyl methacrylate (MEMA) were supplied by Aldrich Chemical Co.; 2-hydroxyethyl methacrylate (HEMA) was supplied by
Hydro-Med Sciences. TMSEMA was synthesized in our laboratory by a
procedure to be described elsewhere (7). The hydroquinone
inhibitor was removed from the commercial monomers by extraction
with aqueous NaOH. The monomer was then dried over MgSO₄ and
distilled from LiAlH₄ and CuCl. The distilled product was then
stored at 4°C over 5A molecular sieve until used. The structures
of the monomers and abbreviations used are shown in Figure 1.
The polymerization conditions used are summarized in Table I.

Anionic Polymerization. MMA, MEMA and TMSEMA were each
polymerized anionically in dry toluene at -78°C using
n-butyllithium as initiator. These conditions had been shown
previously to produce pMMA and other alkyl methacrylates containing a high percentage of isotactic triads (8). In all cases the
reaction was terminated by addition of methanol, and the polymer
precipitated by a non-solvent, petroleum ether for pMMA and
pMEMA, water for pTMSEMA. The crude products were then redissolved in benzene (pMMA and pMEMA) or some suitable solvent, and
centrifuged to remove cross-linked polymer and precipitated LiOH.
The resulting product was then dried overnight in vacuo at about
80°C.

Free Radical Polymerization. Free radical polymerization of
HEMA and MEMA were accomplished by adding 1.8 mg/ml of azobismethylisobutyrate (AMIB) to the degassed monomer. The monomer/
initiator solution was injected into a split polypropylene mold
and placed in an oven at 80°C for 20 hours. The resulting polymer sheet (3-4 mm thick) was then removed for further characterization. Since the pHEMA produced by bulk polymerization was
too highly cross-linked to be redissolved for determination of
its tacticity, the monomer, solvent (generally pyridine) and initiator were added directly to a 10 mm NMR tube along with 0.1-0.2
ml of p-dioxane as an internal standard and the resulting mixture
was polymerized in situ by placing the tube into the oven at 80°C

for 20 hours. TMSEMA was polymerized in dry toluene using AMIB
(TMSEMA/AMIB 1:100) as initiator. The polymer formed was preci-
pated in petroleum ether and dried overnight at 80°C in vacuo.

TABLE I.

MONOMERS USED AND POLYMERIZATION CONDITIONS*

Monomer	Initiator	Solvent	T (°C)	Approx. Yield
Methylmethacrylate (MMA)	n-BuLi	Toluene	-78°	80%
	n-BuLi	THF	-78°	20%
2-Hydroxyethylmethacrylate (HEMA)	AMIB	None	60°	>95%
	AMIB w/UV	MeOH	-60°	25%
trimethylsilylethylmeth- acrylate (TMSEMA)	AMIB	Toluene	60°C	>90%
	n-BuLi	Toluene	-78°	20%
methoxyethylmethacrylate	n-BuLi	Toluene	-78°	30%
	AMIB	None	60°	>95%

*n-BuLi : n-butyl lithium
AMIB : azobismethylisobutyrate
UV : ultraviolet radiation

One low temperature polymerization of HEMA was carried out
in an effort to produce a highly syndiotactic pHEMA. It was performed by dissolving HEMA and AMIB in methanol and exposing the
solution to a 254 nm UV source for seven hours at -60°C. The
resulting polymer was then precipitated in toluene and dried
overnight at 60°C in vacuo prior to subsequent characterization.
Tacticity Determination. Samples for tacticity determination were prepared by placing 0.8-1.0 g of the dried polymer into
a 10 mm NMR tube and adding 1.5-2 ml of an appropriate solvent
along with 0.1-0.2 ml of p-dioxane as an internal reference. The
tube was capped, and the solvent was allowed to swell or dissolve
the polymer in the tube. $CDCl_3$ was used for pMMA and pMEMA, and
pyridine for pHEMA and pTMSEMA. The proton decoupled 25,2 MHz
^{13}C spectrum was then obtained using a Varian XLFT-100 NMR
spectrometer in the Fourier Transform mode. Spectra obtained at
ambient temperature contained peaks which were too broad to
resolve. Consequently, the samples were heated as high as practical without causing refluxing of solvent. Probe temperatures
ranged from 50°C for MMA and pMEMA in $CDCl_3$ to 70°C for pHEMA and
pTMSEMA in pyridine. At elevated temperatures sharp spectra were
obtained in which each carbon absorption was resolved.

Assignment of the peaks in the pMMA spectra were based on published spectra (10). Assignment of peaks exhibiting shifts
due to tacticity for pMEMA, pHEMA, etc. were made by analogy to
pMMA.

    <u>Thermal Analysis.</u> Differential Scanning Calorimetry (DSC)
curves for each polymer were obtained over the range -50° to
$+250^\circ$C using a DuPont Model 990 Thermal Analysis System. Samples
were weighed and placed in covered aluminum sample pans, then
placed into the DSC. Each sample was annealed above its glass
transition temperature for approximately 5 minutes, then cooled
to the starting temperature for the thermogram. The sample was
then heated at 10°C/minute under a nitrogen atmosphere. An empty
aluminum sample pan was used as a reference mass for each run.

Results and Discussion

    <u>pMMA.</u> Figure 2 contrasts the ^{13}C-NMR spectra of two differ-
ent samples of pMMA, one produced by free radical initiation and
the other by low temperature anionic polymerization. Each peak
in the spectrum is assigned to the various carbons in the polymer,
and the chemical shift of each relative to the p-dioxane singlet
at 0.0 ppm is shown. Of particular interest are the three peaks
which correspond to the α-CH$_3$ carbon atoms which are in the
center of isotactic (i), heterotactic (h) and syndiotactic (s)
triads, respectively. Similar splitting is observed for the
carbonyl and quaternary carbon atoms, although the peaks are less
well-resolved than the α-CH$_3$ peaks. The area under each of the
peaks is proportional to the number of carbon atoms in each type
of triad in the sample. Hence, a ratio of the peak areas yields
the relative amounts of each type of triad. Since the α-CH$_3$
peaks are split relatively far apart, they have been used to
calculate tacticities in this study. The free radical pMMA
specimen contains 52%s / 41% h/ 7% i triads, based on the relative peak areas. The anionic pMMA contains 10% s/ 19% h/ 71% i
triads. The peak areas and chemical shifts of both pMMA samples,
as well as those for the other polymers studied, are listed in
Table II. Figure 3 contrasts the DSC thermograms of the same two
polymers. Clearly, the predominantly isotactic anionic polymer
exhibits different thermal behavior from that of the predominantly
syndiotactic free radical polymer. The isotactic polymer shows a
glass transition at 76°C, whereas the syndiotactic polymer has a
T_g of 110°C. The thermal behavior of the specimens can thus be
correlated with the tacticity observed by NMR.

    <u>pMEMA.</u> The behavior of pMMA described above corresponds
well with that reported previously (1). Consequently, it was
expected that polymers of MEMA produced by free radical and anionic methods would show a corresponding difference in tacticity.
The α-CH$_3$ portions of the ^{13}C-NMR spectra of two different pMEMA
samples are shown in Figure 4. The first specimen was produced
using AMIB as a free radical initiator, while the second was

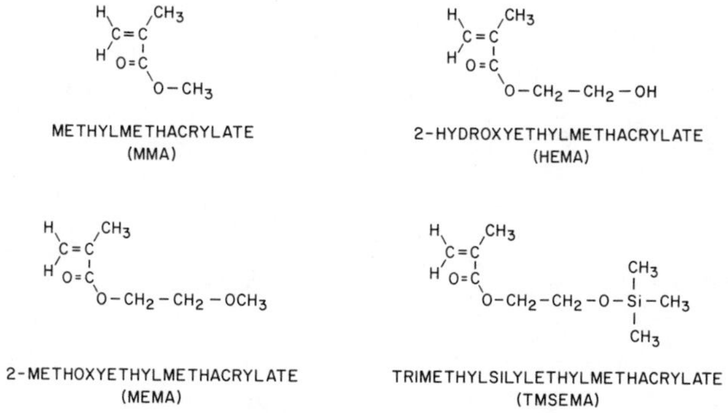

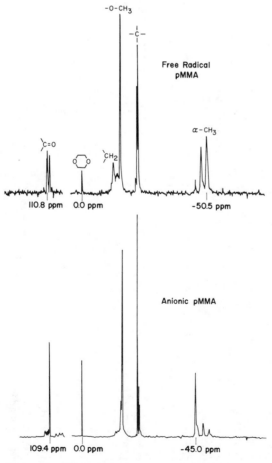

METHYLMETHACRYLATE
(MMA)

2-HYDROXYETHYLMETHACRYLATE
(HEMA)

2-METHOXYETHYLMETHACRYLATE
(MEMA)

TRIMETHYLSILYLETHYLMETHACRYLATE
(TMSEMA)

Figure 1. Monomer structure and nomenclature

-O-CH₃

$-\overset{|}{\underset{|}{C}}-$

Free Radical
pMMA

α-CH₃

$^{\backslash}C=O$

CH₂

110.8 ppm   0.0 ppm   -50.5 ppm

Anionic pMMA

109.4 ppm   0.0 ppm   -45.0 ppm

Figure 2. ¹³C NMR spectra of poly(methyl methacrylate)

TABLE II.
Chemical Shifts for Various α-CH₃ and Carbonyl Triads*

Polymer	δα-CH₃(ppm)**			Area: α-CH₃			δC=O(ppm)			Solvent	Temperature(°C)
	i	h	s	i	h	s	i	h	s		
pMMA (anionic)	-45.0	-45.6	-45.7	57.1 71%	15.6 19%	8.1 10%	109.4	109.7	110.4	CDCl₃	Ambient
pMMA (commercial)	-45.8	-48.2	-50.5	14 7%	88.5 42%	109.3 51%	--	109.9	110.8	CDCl₃	Ambient
pHEMA (free radical)	--	-47.7	-49.5	-- <1%	54.0 33%	107.9 66%	--	110.1	110.7	Pyridine	70°
pHEMA (free radical-from pTMSEMA)	-44.6	-47.6	-49.4	6 4%	57.3 38%	89.1 58%	--	110.2	110.8	Pyridine	70°
pHEMA (anionic-from pTMSEMA)	-44.5	-47.5	-49.4	137 30%	164.4 35%	165.0 35%	110.0	110.2	110.8	Pyridine	70°
pHEMA (UV-low temp)	-45.9	-47.6	-49.3	21.1 9%	31.2 13%	189.6 78%	--	--	110.8	Pyridine	70°
pMEMA (anionic)	--	-48.5	-50.2	-- <1%	28.8 33%	55.8 66%	--	109.4	110.3	CDCl₃	50°
pMEMA (free radical)	--	-48.5	-50.3	-- <1%	77.5 45%	95.0 54%	--	109.5	110.2	CDCl₃	50°
pMEMA (anionic from pMMA)				45.4 71%	11.9 18%	7.2 11%				CDCl₃	50°

* i = isotactic; h = heterotactic; s = syndiotactic
** Chemical shifts expressed as ppm from dioxane singlet at 0 ppm

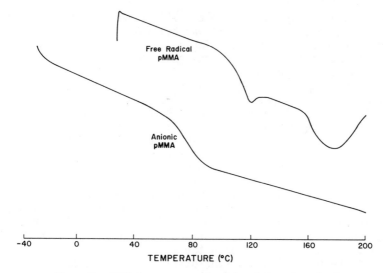

Figure 3.   DSC thermograms of poly(methyl methacrylate)

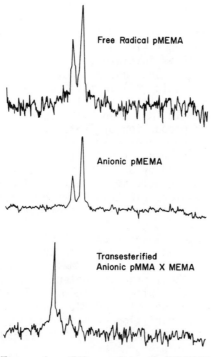

Figure 4.   α-CH₃ portions of ¹³C-NMR spectra of poly(methoxyethyl methacrylate)

produced anionically in toluene at -78°C. Both polymers contain
a preponderance of syndiotactic and heterotactic triads, with few
isotactic triads, as listed in Table II. In an effort to show
that no irregularities in chemical shift had caused the anionic
polymer to appear syndiotactic when it was actually isotactic, a
sample of the isotactic pMMA described above was transesterified
to pMEMA. The α-CH₃ portion of its ¹³C-NMR spectrum is shown in
Figure 4c. Since it shows the same chemical shifts and relative
peak areas as the parent pMMA, it must be concluded that the high
syndiotactic content of the anionic pMEMA is not an artifact.
Rather, it must be due to some interaction between the oxygen of
the methoxy side chain and the anionic growth center which
changes the local dielectric constant and alters the gegenion
separation, thus favoring syndiotactic placement of incoming monomer units. Similar effects have been noted previously (11) with
MEMA initiated by t-butyllithium at 20°C and -20°C. In that
study, fewer than 10% isotactic triads were found at either
temperature. The fact that both the anionic and free radical pMEMA
specimens are predominantly syndiotactic is reflected in their
DSC traces, shown in Figure 5. Both show an inflection at 25°C
with no melting apparent, and begin to decompose above 250°C.

pHEMA. Figure 6 shows the α-CH₃ portion of the ¹³C-NMR
spectra of pHEMA produced by four different methods. The relative tacticities calculated from the peak areas are given in
Table II. The data show that pHEMA produced at elevated temperatures with AMIB (Figure 6(a)) contains predominantly syndiotactic
(66%) and heterotactic triads (33%), with few isotactic triads
(< 1%). Figure 6(b) shows similar results for a polymer formed
from TMSEMA using AMIB as initiator, then subsequently hydrolyzed
to pHEMA. Syndiotactic (58%) and heterotactic (38%) triads
predominate over isotactic (4%) in this case as well. The third
polymer, shown in Figure 6(c), was polymerized anionically at
-78°C in toluene from TMSEMA, then hydrolyzed to pHEMA. Here
we see a significant proportion of isotactic triads (30%) for the
first time, with the remainder evenly divided between syndiotactic
(35%) and heterotactic triads (35%). The fourth pHEMA, shown in
Figure 6(d), was produced by UV initiation using AMIB at -78°C.
As the figure shows, a highly syndiotactic polymer has been produced, having 78% syndiotactic triads, with 9% isotactic triads
and 13% heterotactic triads. The DSC thermogram of each polymer
is shown in Figure 7. The glass transition of the pTMSEMA produced anionically is 13° lower than that of free radical pTMSEMA,
indicating that an increase in isotactic triads leads to a
decrease in T_g, just as it did with pMMA. Further work will be
required to increase the isotactic content of pHEMA polymers
in order to fill out the full range from highly syndiotactic
pHEMA to predominantly isotactic pHEMA. However, it now appears
likely that systematic variation of tacticity will be achievable
in the future and may provide a means of varying mechanical,

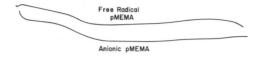

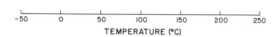

Figure 5. *DSC thermograms of poly(methoxyethyl methacrylate)*

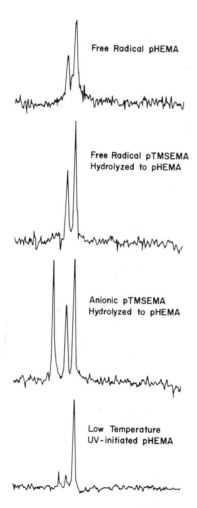

Figure 6. *α-CH₃ portions of ^{13}C-NMR spectra of poly(hydroxyethyl methacrylate)*

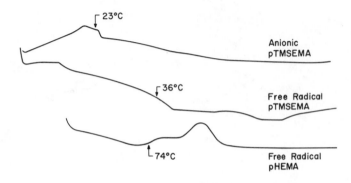

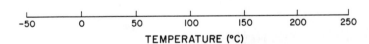

Figure 7. DSC thermograms of poly(hydroxyethyl methacrylate) and
derivatives

thermal and surface properties in the pHEMA/water system for future studies.

Conclusions

Based on the data presented above, we conclude:

1. ^{13}C-NMR provides a method for directly determining the tacticity of methacrylate polymers. This eliminates the need for lengthy hydrolysis to poly(methacrylic acid) and re-esterification to pMMA.
2. Methacrylate-based polymers display significant changes in their DSC thermograms as a result of changes in tacticity.
3. Methacrylate esters with heteroatoms in their side chains interact with anionic initiators, reducing the isotactic content of the resulting polymers. Whether the effect of the heteroatoms can be reduced at temperatures below -78°C or by sterically hindering the heteroatom during polymerization remains to be seen.
4. Techniques are now available for producing soluble pHEMA polymers having a variety of tacticities. The effect of that tacticity variation on interfacial and bulk properties will be the subject of future work.

Acknowledgements:

This work was supported by NIH Grant HL 16921-01. We thank Hydro Med Sciences, Inc., New Brunswick, N. J. for generous donations of pure HEMA monomer.

Abstract

In this study a series of hydroxyethyl methacrylate (HEMA) and methoxyethyl methacrylate (MEMA) polymers of varying tacticities was produced. Both free radical and anionic initiators were used and their effect on tacticity determined. Tacticity of each polymer was measured by ratioing the ^{13}C-NMR peak areas for the α-CH$_3$ carbon atoms in isotactic, syndiotactic, and heterotactic triads. Both the α-CH$_3$ and carbonyl carbon peaks exhibit splitting due to tacticity, but the α-CH$_3$ peak was better resolved. Chemical shifts for both atoms appear to be independent of the ester side chain for the polymers studied.

The thermal behavior of each polymer was observed over the range -50° to +250°C by differential scanning calorimetry (DSC). Differences in melting point and glass transition of the polymers were relatable to the tacticity measured by ^{13}C-NMR. This indicates potential for some control of the properties of pHEMA and other related hydrogels by control of tacticity during synthesis.

Literature Cited

1. Bovey, F. A. and Tiers, G. V. D., J. Poly. Sci., 44, (1960), 173.
2. Matsuzaki, K., Ishida, A., and Tateno, N., J. Poly. Sci., Part C, (1967), 16, 2111
3. Goode, N. E., Owens, F. H., Fellman, R. P., Snyder, W. H. and Moore, J. E., J. Poly. Sci., (1960), 46, 317
4. McCrum, N. G., Read, B. E. and Williams, G., "Anelastic and Dielectric Effects in Polymeric Solids", pp. 249-55, John Wiley & Sons, New York, 1967.
5. Holly, F. J. and Refojo, M. F., J. Biomed. Materials Res., (1975) 9, 315.
6. Lenz, R. W., "Organic Chemistry of Synthetic High Polymers", pp. 439-41, Interscience, New York, 1967.
7. DeVisser, A. C., Unpublished results.
8. Chlanda, F. P., and Donamura, L. G., J. Appl. Poly. Sci., (1971), 15, 1195.
9. Lando, J. B., Litt, M., Kumar, N. G., and Shimko, T. M., J. Poly. Sci., (1974), Symposium Series No. 44, 203.
10. Bovey, F. A., High Resolution NMR of Macromolecules, pp. 80-82, Academic Press, New York, 1972.
11. Trekoval, J., Vlcek, P., and Lim, D., Coll. Czech. Chem. Comm., (1971), 36, 3032.

11

Calcification and Bone Induction Studies in Heterogeneous Phosphorylated Hydrogels

J. G. N. SWART
Department of Oral Surgery, Free University, Amsterdam, The Netherlands

A. A. DRIESSEN and A. C. DeVISSER
Department of Materials Science, Free University, Amsterdam, The Netherlands

Development of poly(hydroxyethyl methacrylate), poly(HEMA) ,hydrogels for application in the medical field has up to date mainly been directed towards the replacement of soft-tissue structures and organs, e.g. the soft contact lens.

Because of its relatively poor mechanical properties the hydrogel as such can not be applied as substitute for hard tissue. However, based on the findings that calcification has occurred in heterogeneous poly(HEMA) hydrogels (1-6), Calnan et al. (3) suggested that the hydrogel possibly could promote the deposition of calcium salts in its matrix, thus being a "challenger" for calcification.

Some authors (2,4) reported bone formation following the occurrence of calcification. Calcification and bone induction appeared to be accelerated by the presence of methacrylic acid (MAA) groups in the gel (4). Sprincl et al. (6), however, reported that modification of a poly(HEMA) gel by incorporation of MAA (up to a mole ratio of MAA/HEMA 1:5) or dimethyl - aminoethyl methacrylate (DMAEMA) did not affect the calcification process, whereas Cernij et al. (7) found that incorporation of 4% MAA inhibited calcification. From the above results it is evident that the effect of MAA, incorporated in the hydrogel, on calcification or bone formation has not been unambiguously established yet. Possibly, other factors than chemical modification such as pore size of the gel, animal species and implantation site, play a more dominant role in the occurrence of these phenomena. A hydrogel that, in addition to calcification,could induce bone formation in its matrix would have great potential for restoration of large defects in bone tissue. Such a material could for example be applied to facilitate the healing process in the post extraction alveolus or bone in-

growth in large cysts. On the other hand, calcification
is often an undesirable phenomenon in a hydrogel serving as soft tissue substitute since it affects the
functional and esthetic properties of the material
adversely. In this case, inhibition of calcification
by chemical modification would be beneficial.

In view of the above considerations our objective
is to study calcification and bone formation in heterogeneous hydrogels as function of their chemical and
physical properties and to find means to control these
processes by suitable modification of the hydrogel.
The study described hereafter deals with an experiment
designed to determine the effect of incorporation of
the phosphate group in a heterogeneous poly(HEMA) gel
on its calcification-inducing ability.

The hydrogel should be heterogeneous because it
was found that a sufficiently large pore size (> 40μm)
(4) is a prerequisite for ingrowth of the surrounding
tissue which in turn can lead to calcification and
formation of bone. The phosphate group was selected to
be built in the gel because this group is one of the
building blocks of calcium hydroxyapatite the main inorganic constituent of bone and, although being present
as HEMA-phosphate, should have a high affinity towards
calcium ions.

Material and Methods

HEMA-phosphate was prepared by reacting HEMA
(Hydro Med Sciences, Inc., U.S.A., purity min. 99.2%)
with phosphorus pentoxide (Merck,Darmstadt, Germany,
purity min. 98%) in a 1:1 mole ratio in dichlorometha-
ne at 0 - 5°C. The reaction proceeds almost quantitatively. After removal of solids by centrifugation and
evaporation of the solvent in vacuo at 20°C and 1 mm
Hg, a mixture of mono- and di-ester was obtained in a
mole ratio of 2 to 1 assuming that the amount of tri-
ester formed is negligible.

Ethyleneglycol dimethacrylate (EGDMA) (Merck)
was purified by distillation at 83°C and 1.5 mm Hg.
HEMA was used as obtained from Hydro Med Sciences.

Preparation of the hydrogels was performed in sealed ampoules of 9 mm inner diameter. The various monomers were dissolved in a Tyrode solution containing
1% (w/w) of ammonium persulfate as initiator. Polymerization was carried out at 50°C during 24 hours.

Composition of the polymerization mixtures is as follows:

Hydrogel	HEMA	EGDMA (crosslinker)	HEMA− phosphate	Tyrode sol.
H 1	19.8	0.2	--	80% w/w
H 2	17.8	0.2	2.0	80% w/w

The gels were kept in distilled water for several days and then boiled twice in distilled water to remove low molecular weight substances. Then the gels were soaked in sterile salt solution and cut into discs about 2 mm thick and 9 mm in diameter.

These discs were inserted in muscle pockets in the back muscle of rats. In each rat we implanted 3 discs of the same hydrogel and one pocket was filled with gelatin foam (Spongostan^R, Ferrosan, Will-Pharma N.V., Holland) as a control. Excision followed after periods of 3 days up to 24 weeks. The skin was cut away and the implants removed with the surrounding muscle tissue. The removed tissue blocks were placed for 10 minutes between gauze strips saturated with isotonic saline to allow the contractile properties to become quiescent. Thereafter the excised tissue was fixed in 10% buffered formalin.

After fixation and dehydration in absolute alcohol the biopsies were embedded in paraffin and cut into sections of 6 μm. These sections were stained with the Haematoxylin and eosin stain. Calcium salts were visualized according to Von Kossa.

Pore size of the materials was determined by scanning electron microscopy.

Results and Discussion

Macroscopically all implants were accepted and well tolerated; no signs of severe inflammation or implant rejection could be observed. (figure 1).

Microscopic investigation revealed the following: 3 days after implantation a slight edema and invasion of granulocytes and mononuclear cells into all implants was observed. The phosphorylated hydrogels showed less cellular infiltration and stained basophilic probably because of the acid phosphate group. Some of the control gelatin foams (Spongostan^R) showed invasion of fibroblasts and mononuclear cells as well as hyperaemia around the implants as a sign of acute inflammatory response.

In the one week biopsies there was an increase
in cellular ingrowth of about 0.2 mm for the poly
(HEMA) implants. The phosphorylated implants hardly
showed any ingrowth or signs of inflammation.

After two weeks we saw a further increase of
ingrowth up to a maximum of 1 mm into the poly(HEMA)
specimen. The implants were slightly thicker (3 mm)
and for the first time basophilic clusters could be
seen as the earliest sign of calcification (figure 2).
The phosphorylated poly(HEMA) implants behaved quite
indifferently towards the tissue. There was practically no fibrous lining around the implant, it even seemed to lay directly upon the muscle fibres.

In the 4 week biopsies the "Spongostan^R" gelatin
foam implants showed no signs of acute inflammation:
tissue proliferation with fibroblasts, foreign-body,
giant cells and capillaries dominated the picture.
The gelatin foam gradually disappears as a result of
phagocytosis by macrophages.

In the poly(HEMA) hydrogels basophilic granules
are scattered all through the implant.

The phosphorylated implants were well tolerated,
and still showed practically no ingrowth or encapsu-
lation.

After 8 weeks the "Spongostan^R" gelatin foam has
completely disappeared, with only small strands of
fibrous scar tissue left. The poly(HEMA) implants
showed considerably more fibroblasts than 4 weeks before. The basophilic granules within these implants
formed larger colonies. Frequently, a thin fibrous
capsule surrounded the phosphorylated implants, which
still hardly showed any ingrowth.

At 16 and 20 weeks whole granular fields were
observed in the poly(HEMA) implants. These fields
particularly appeared in the edges of the implant.
Where these granular fields were present, tissue ingrowth did not take place or seemed to be prevented.
Some of the poly(HEMA) implants showed less basophilic
granules, but considerably more tissue ingrowth. Calcium determination by the Von Kossa stain revealed
that the socalled basophilic granules were calcified
(figure 3).

The 24 week biopsies, the last period evaluated,
showed some interesting features.

A röntgenogram (60 kv, 3 mA, 0.10 sec) of biopsies from the 5 rats out of this group showed a radio-
paque outline in the poly(HEMA) implants, phosphorylated poly(HEMA) implants did not show this phenomenon.

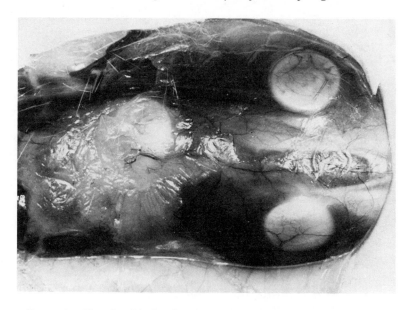

Figure 1. *Phosphorylated poly(HEMA) implants in the back of a rat 24 weeks after the implantation showing the excellent biocompatibility of this material*

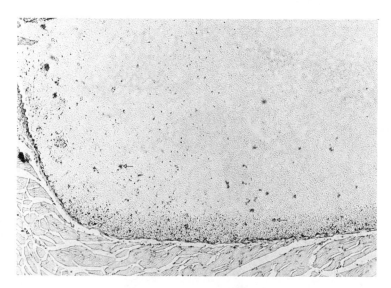

Figure 2. *Photomicrograph of a poly(HEMA) implant section, 14 days after implantation. For the first time calcified granules (arrows) appear within the poly(HEMA) implants. (Haematoxylin and eosin stain, magnif. ×39)*

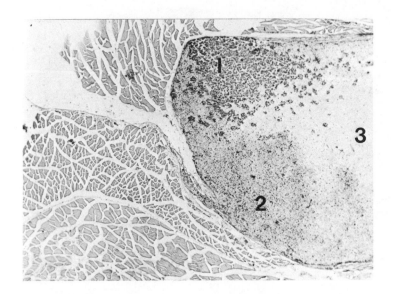

Figure 3a.   Photomicrograph of a poly(HEMA) implant 20 weeks after insertion showing areas of dense calcification (1) and areas of massive tissue ingrowth (2) with only sparse calcification. The center (3) of the implant shows only a slight ingrowth of cells into the pores of the implant material. (Haematoxylin and eosin stain, magnification ×20)

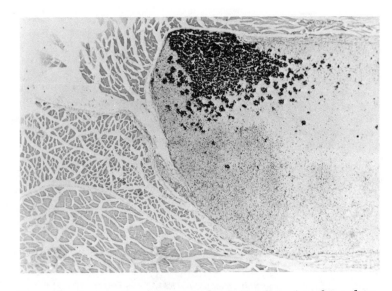

Figure 3b.   Same section stained according to Von Kossa for calcium determination. With this method calcified tissues stain black and uncalcified tissues red, which shows plain grey on this photomicrograph.

The Von Kossa stain proved the presence of calcium within the poly(HEMA) implants which showed radio-paque lines in the röntgenogram and the absence of calcium salt deposits in the non-radiopaque specimen. The phosphorylated implants showed only a thin fibrous lining and ingrowth, if any, predominantly in the outer margin (figure 4).

In one rat we observed a remarkable macroscopic difference between two poly(HEMA) implants that were chemically identical; one implant had a white aspect and a normal size; the other implant showed an in-crease in size and the same colour as the surrounding tissue. The microscopic picture revealed that the white implant was more calcified compared to the other implant and showed less tissue ingrowth. This differ-ence might be the result of a different blood supply (figure 5).

In order to relate the results with the physical structure of the gels, the pore size of the various implants was determined by scanning electron micros-copy. The poly(HEMA) implants had pores from 30 - 70 µm, while the phosphorylated hydrogels possessed pores from 70 to over 100 µm (figure 6).

The inhibition of tissue infiltration in the phosphate containing gels is surprising because one would expect that substantial ingrowth would occur in an implant having such a large pore size. Wether this phenomenon is due to the higher acidity in these gels, as compared to the poly(HEMA) gels, or the nature of the phosphate group itself is not known and needs further study.

Conclusions

Summarizing the results we can conclude that:
1. Incorporation of the phosphate group in heteroge-neous poly(HEMA) hydrogels when implanted intra-muscularly in rats, does not promote calcification or bone formation, on the contrary it seems to prevent calcification and tissue ingrowth. This may make the phosphorylated poly(HEMA) hydrogels a better material for soft tissue substitution than regular poly(HEMA), although it is to early yet to make definite conclusions and recommenda-tions.
2. None of the implants showed bone or cartilage for-mation. The calcification occuring within the poly-(HEMA) implants is most likely due to calcification of degenerate tissue (dystrophic calcification) as the result of insufficient blood supply.

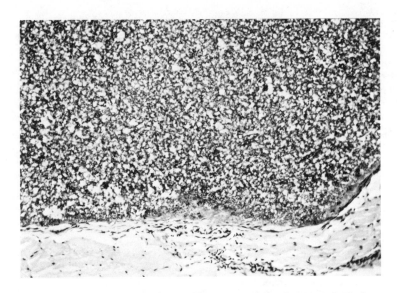

Figure 4a. Photomicrograph of a phosphorylated poly(HEMA) implant section 24 weeks after insertion showing the thin fibrous lining of two or three cell layers and rather superficial tissue ingrowth.

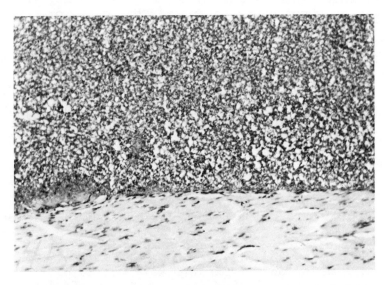

Figure 4b. Photomicrograph of a section of the same material which has been excised after 14 days shows almost the same picture. (Haematoxylin and eosin stain, magnification × 85)

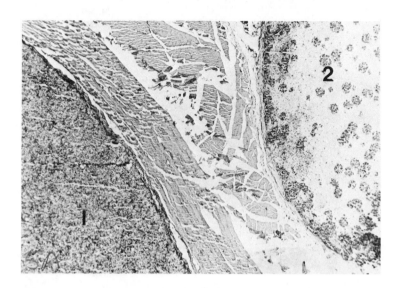

Figure 5a. *Poly(HEMA) implants out of the same rat 24 weeks after implantation. In one implant (1) there is a diffuse ingrowth of tissue practically without calcification. The other implant (2) shows colonies of calcification and only tissue ingrowth in the outer margin of the implant. (Haematoxylin and eosin stain, magnification × 21)*

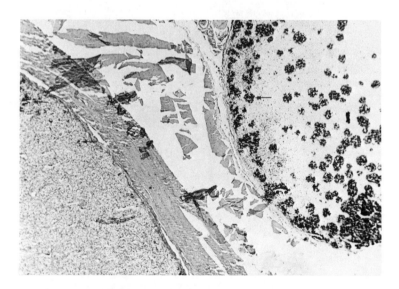

Figure 5b. *Comparable section, stained according to Von Kossa, proving the presence of calcium salts which stain black (arrows)*

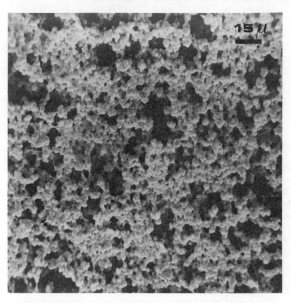

Figure 6a. Scanning electron micrograph of poly(HEMA) having pore sizes from 30–70 μ

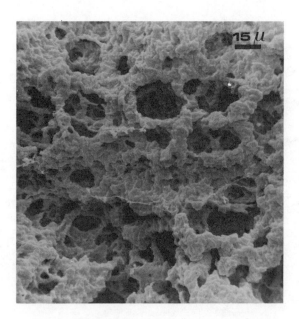

Figure 6b. Scanning electron micrograph of phosphorylated poly(HEMA) having pores from 70 to over 100 μ

3. No lymphocytes were seen in the implants which
indicates that poly (HEMA) and phosphorylated poly -
(HEMA) are immunologically inert materials.
4. A large variation in calcification within the
regular poly (HEMA) implants has been observed.
Variation in blood supply with implant site possibly plays an important role in this phenomenon.

Abstract

The effect of chemical modification of a heterogeneous poly (hydroxyethyl methacrylate) hydrogel on
calcification in its matrix has been studied by intra-
muscular implantation in Wistar rats. Ten percent
phosphorylated hydroxyethyl methacrylate, based on
total amount of monomer, was incorporated in hydrogels
containing 80% (w/w) water. These gels which exhibit
poor mechanical properties can not be used as direct
replacement for hard tissue, but may have potential
as calcification "challenger" for living tissue.
Scanning electron micrographs show that the gels
have pores with an average diameter 70 - 100 μm.
Histological examination of the implants after
3 days up to 24 weeks revealed that the poly (HEMA)
and the phosphorylated poly (HEMA) gels were very well
tolerated by the organism.
Tissue ingrowth and calcification occurred in
the poly (HEMA) gels, but bone formation was not found.
The phosphorylated gels were encapsulated by a thin
fibrous lining whereas ingrowth only sporadically and
to a slight extent was observed; calcification did
not take place.

Literature Cited

1. Kliment, K., Stol, M., Fahoun, K., Stockar, B.,
J. Biomed. Mater. Res. (1968), 2, 237.
2. Winter, G.D., and Simpson, B.J., Nature (1969),
223, 88.
3. Calnan, J.S., Pflug, J.J., Chhabra, A.S., Raghypat-
li, N., Brit. J. Plastic Surgery,(1971),24, 113.
4. Smahel, J.,Proserova, J.,Behounkova, E., Acta Chi-
rurgiae Plasticae (1971), 13, 193.
5. Sprincl, L.,Vacik, J.,Kopecek, J., J. Biomed. Mater.
Res. (1973), 7, 123.
6. Sprincl, L.,Kopecek, J.,Lim, D., Calc.Tiss.Res.
(1973), 13, 63.
7. Cernij, E., Chromeček, R., Opleta, A., Papousek, F.
Otoupalová, J., Scripta Medica (1970).

12

Continuous Monitoring of Blood Glucose Using Gel Entrapped Glucose Oxidase

K. F. O'DRISCOLL and A. KAPOULAS
Department of Chemical Engineering, University of Waterloo, Waterloo, Ontario, Canada

A. M. ALBISSER and R. GANDER
Medical Engineering Department, Hospital for Sick Children, Toronto, Ontario, Canada

The use of enzymes as efficient and highly specific catalysts in chemical analysis has been given considerable attention during the past twenty years. The development of effective procedures for the immobilization of enzymes by physical occlusion within a hydrophilic polymeric matrix offers new opportunities for the practical use of enzymes in chemical analysis (1,2).

Glucose oxidase is an intracellular enzyme exhibiting a high specificity for β-D-glucose. As the only sugar normally present in the bloodstream is D-glucose, immobilized glucose oxidase would seem ideally suited for determining blood sugar on a continuous basis during surgery or for diabetics.

This report describes a practical system for the analysis of glucose concentration in both simple and complex biological fluids, using glucose oxidase enzyme immobilized by occlusion within a polymeric, hydrophilic matrix. In designing a system for such a purpose, there were two basic criteria to be considered. First, results must be rapidly obtained and be reproducible. Second, flow through the system must be kept to a minimum since the ultimate use involves blood flow from a human body.

In the system we have devised the oxygen concentration of a sample solution is continuously measured, after passing through immobilized glucose oxidase, by means of a commercially available polarographic electrode. The amount of oxygen consumed by oxidation of glucose is related to the glucose concentration of the sample by means of a calibration graph. The linearity of the graph in the range of 10-150 ppm glucose, permits a single point calibration graph. The design of the electrode holder insures high linear velocity of liquid past the electrode membrane and results in a minimum pressure drop. Thus oxygen concentrations recorded are 98% or better of the true oxygen concentration of liquids.

Both the procedure and associated hardware are simple in design and operation and have demonstrated exceptional operational stability. Blood, continuously withdrawn from a dog and

and diluted 30-fold has been used, and the system has been satis-factorily operated for five hours continuously. With synthetic glucose solutions the system has been operated for up to 60 hours continuously without need for recalibration. One reactor has been used with the same gel in it for more than 200 hours over a three month period without any detectable loss of enzymatic activity.

The only instabilities of the system we have observed have been associated with the electronic equipment used and temperature variations rather than with the immobilized enzyme column. Work in progress is aimed at eliminating such problems so that this system may become the glucose sensing portion of an artificial pancreas.

The System

The prototype system, shown schematically in Figure 1, has three principal components in sequence:
1. An aeration coil, consisting of 0.23 cm I.D. glass tube, 45 cm long, followed by a debubbler.
2. A 0.5 cm I.D. fixed bed reactor, containing 1.2 gm (dry) of hydrated gel in which is the immobilized glucose oxidase.
3. The detector assembly, incorporating (in a specially designed electrode holder) a membrane covered polarographic electrode, (Beckman Instrument Company 39533 O_2 Sensor), for the measure-ment of dissolved oxygen in the effluent from the reactor.

Solution transport through the system is achieved by a Watson-Marlow (Buckinghamshire, England), Multichannel, Variable Speed Peristaltic Pump.

Procedure

The glucose oxidase immobilization in a poly(HEMA) gel has been described (3).

Calibration Graph.

Sodium acetate buffer solution (0.1 M, pH 5.6) containing 0.5% NaCl and 1.3 x 10^{-4}% KCN is pumped through the system for at least five minutes and the oxygen concentration of the outlet stream is recorded. Synthetic glucose samples of different con-centrations are prepared, allowed to equilibrate for twenty four hours and then pumped through the system at a fixed flow rate of ca. 1 ml/min. The effluent oxygen concentration of these solu-tions is continuously recorded, taking approximately 3 - 5 minutes to reach a steady state.

A graph of O_2 concentration at steady state, vs. glucose concentration, is linear and establishes the relation between total glucose concentration and percent conversion. Flow rate should be checked and maintained constant for obvious reasons. Figure 2 shows calibration curves for two reactors containing

slightly different activity levels of gel entrapped glucose
oxidase.

Serum Samples.

Individual serum samples diluted thirty-fold, thus having
glucose concentrations between 0.01 - 0.15 mg/ml may be analyzed
by the same method used to obtain the calibration graph. The
steady state oxygen concentration is related to initial total
glucose concentration by means of a calibration graph.

For the calibration graph, it is important to use glucose
solutions of the same ionic strength as the solution used for
dilution of serum samples because dissolved oxygen concentration
in liquids is a function of ionic strength and because ions such
as halides depress the activity of the immobilized enzyme.

Continuous Glucose Analysis of Blood.

Oxygen concentration of a blood stream is much lower than the
oxygen concentration of air saturated liquids. Moreover, red
cells continue to metabolize thus consuming oxygen and glucose.
Several chemicals exist which depress or stop the rate of metabolism of red cells. Some of these chemicals also deactivate the
immobilized enzyme, glucose oxidase. Thus a solution to be used
for the dilution of blood must not coagulate or hemolyze the
blood, must maintain a constant pH, and must not seriously inhibit
the enzyme activity and thereby appear to reduce the glucose conversion or oxygen concentration.

Recognizing these constraints, the blood stream, 0.05 ml/min.
is mixed with 0.05 ml/min. of heparin solution in a catheter and
drawn off continuously. This stream, 0.1 ml/min., is mixed at the
beginning of the aeration coil with 1.15 ml/min. of diluent solution containing the following chemicals: sodium acetate buffer
0.1 M, pH = 5.6, 0.5% sodium chloride, 0.2% sodium fluoride, 0.13
mg/100 ml potassium cyanide. This solution satisfies the above
criteria and has functioned satisfactorily as a diluent solution
for continuous blood glucose analysis. Potassium cyanide is added
to depress the activity of catalase, existing in the blood stream,
which catalyzes the decomposition of hydrogen peroxide to oxygen
and water, thus introducing a serious error in measurement of
steady state oxygen concentration of the reactant stream. Sodium
fluoride suppresses the red cell activity toward oxygen.

Diluent solution and the blood stream are mixed with 2.5 ml/
min. air and transported to the reactor through the aeration coil
and associated tubing. The reactor contained 1.3 gms (dry) gel
with an activity of 12 units/gm. Debubbling and stream splitting
takes place at the inlet of the immobilized glucose oxidase reactor. The reactant stream goes through the packed bed reactor
and then to the oxygen electrode flow cell. Its oxygen concentration is recorded continuously as a function of time. Glucose

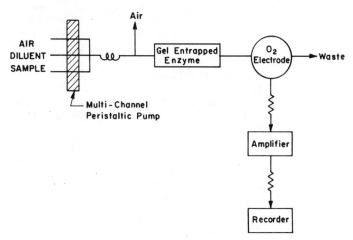

Figure 1. Schematic of glucose analyzer with gel entrapped glucose oxidase

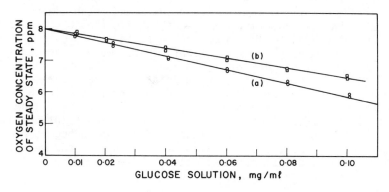

Figure 2. Calibration plots for (a) reactor No. 1 and (b) reactor No. 2

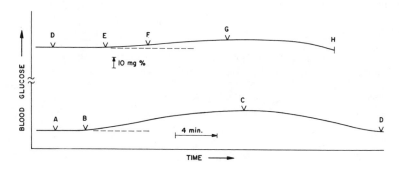

Figure 3. Monitoring of dog blood glucose

concentration of the reaction stream is obtained from the steady state oxygen concentration by means of the calibration graph. There is no apparent effect of the blood on either the oxygen electrode or the immobilized enzyme activity.

Flexible tubing from the reactor to the oxygen electrode flow cell must be kept as short as possible in order to reduce the oxygen diffusion from air to the reactant stream, thus resulting in a lower sensitivity of glucose measurements. Uncompensated errors may be introduced if the oxygen concentration of the reactant stream is less than air saturation levels. This can be checked by on line measurements of oxygen concentration of this stream after debubbling, using another oxygen electrode or by bypassing the stream from the glucose oxidase reactor after debubbling and recording its oxygen concentration.

Results and Discussion

Figure 3 shows the results of monitoring the blood glucose of a dog. Point A represents the stable level after approximately 3 hours of monitoring. At that time, the blood diluent being used was changed to include 1% glucose, a response being noted after a few minutes beginning at point B and stabilizing in 10 minutes. The diluent was changed back to the glucose free solution at C and steady state reached again at D. Over a time period of 5 minutes (E - F) 5 gms of glucose were injected into the dog and the heightened blood sugar was controlled by the dog's pancreas (note the maximum at point G). The experiment was terminated when the level had returned to normal at point H.

This experiment, and many others with blood sera, show the compatibility of the system with whole blood. The solutions used for diluent successfully compensate for the hemoglobin and catalase content of the blood without affecting the enzyme activity. No adverse "adsorption effects" as described by Guilbault (4) have been observed with these gels.

Acknowledgement

Support of this research by the National Research Council of Canada is appreciated.

Literature Cited

(1) Hicks, G.P., and Updike, S.J., Anal. Chem. (1966), 38, 726.
(2) Weetall, H.H., Anal. Chem. (1974), 46, 602A.
(3) Hinberg, I., Kapoulas, A., Korus, R., and O'Driscoll, K.F.,
      Biotech. & Bioeng. (1974), 16, 159.
(4) Guilbault, G.G., Chem. and Eng. News (October 2, 1972), 45.

13

The Slow Release of Hydrocortisone Sodium Succinate from Poly(2-hydroxyethyl methacrylate) Membranes

J. M. ANDERSON, T. KOINIS, T. NELSON, M. HORST, and D. S. LOVE

Department of Macromolecular Science and Department of Anatomy,
Case Western Reserve University, Cleveland Ohio 44106

Love and co-workers have suggested that the endocrine system plays a key role in regulating the pattern of growth and skeletal muscle differentiation in chick embryos (1,2). To determine the validity of this hypothesis, attempts to inject intravenously serial doses of specific hormonal agents into hypophysectomized chick embryos were made. This method of drug delivery was unsatisfactory as dosage levels were variable, serial injection into the chorioallantoic membrane was difficult and constant manipulation of the chick embryo resulted in a high rate of mortality. These difficulties were overcome by using hydrogel membranes which gave a slow, sustained release of drug to the chick embryo. This paper, the first of a series, will deal with the development and in vitro and in vivo release behavior of hydrocortisone sodium succinate from crosslinked poly(2-hydroxyethyl methacrylate) membranes.

In recent years, the concept of a sustained release system, which, upon implantation in the body, would release biologically active levels of drug for prolonged time periods has received increased attention. Sustained release systems have been developed and examined for their potential application in the treatment of glaucoma (3), trichoma (4), narcotic addiction (5), malaria (6), diabetes (7), cancer (8) and contraception (9, 10, 11). Most of these systems have used silicone rubber as the matrix material in which the drug was either embedded or contained. In addition, silicone rubber has been extensively studied in regards to its transport properties and those factors which control or alter the transport of drugs through the matrix (12, 13, 14).

Poly(2-hydroxyethyl methacrylate) and other hydrogel type polymers have also been examined for potential use in sustained delivery systems (15). As the diffusivity of drugs from the polymer membranes is probably the most important criteria for the polymer, aside from its biological compatibility, the hydrogel materials have excellent potential as their physical

167

characteristics (degree of hydration, crosslink density,
porosity, etc.) can be easily altered and controlled to vary
the rate of drug diffusion. Through controlled alteration of
the amount of crosslinking agent, monomer to water ratio, and
the conditions of polymerization, hydrogel structures ranging
from compact gels to cellular sponges with varying physical
properties can be obtained (16, 17).

Levowitz and co-workers have shown that sutures and
catheters coated with an ethyleneglycomethacrylate gel in which
cephalothin was incorporated decreased the rate of induced
infection when used in animal studies (18). In vitro release
rates were shown for norethandrolone and 5-fluorouracil and
reference was made to the sustained release of antibiotics,
local anesthetics and cortisol. Human studies using Hydron ®-
cephalothin urethral catheters have shown a delay in the onset
of bacteriuria that was statistically superior to the uncoated
controls. In addition to serving as a vehicle for the delivery
of a wide variety of antibiotics such as neomycin, ampicillin,
tetracycline, etc., Hydron ® coated catheters in which
cephalothin was absorbed were efficacious against susceptible
bacteria after being stored sterilly for up to six months (19).

The use of soft contact lenses for prolonging the action
of drugs administered drop-wise or for sustained release of
drugs has been investigated. Kaufman and co-workers have shown
that soft contact lenses can prolong the therapeutic action of
polymixin B, α-phenylephrine, and pilocarpine when these drugs
are administered drop-wise to human and animal eyes. In addition,
these studies point out the fact that near-toxic drug levels
do not have to be administered to provide therapeutic efficacy
when soft contact lenses are used to mediate and prolong the
delivery of drugs (20). The above authors were also able to
show that corneal ulcers in rabbit eyes, previously infected
with McKrae herpes virus, healed faster with 5-iodo-2-deoxyuridine
drops mediated by a soft contact lens than with 5-iodo-2-
deoxyuridine drops alone.

Podos et al. have investigated the administration of
pilocarpine, to decrease ocular hypertension and relieve miosis
and other glaucoma symptoms, by the use of soft contact lenses
pre-soaked in pilocarpine (21). Their results are encouraging
and point to the need for further work in retarding the rate of
drug release from presoaked lenses. Pre-soaking the lenses,
while providing for easy preparation, usually results in too
rapid release of the drug.

Abrahams and Ronel (22) using poly(2-hydroxyethyl methacrylate) hydrogels, have examined the effect that copolymer
composition and ionogenic groups on the hydrogel chain have on
the in vitro release behavior of cyclazocine, a narcotic
antagonist, from such varied forms as capsules, barrier-film
coated tablets and bulk polymerized rods. Using capsular devices
the release of cyclazocine could be increased markedly by

adjusting the ionogenic group content (methacrylic acid) of the hydrogel from 0.3% to 2.4%. Homogeneous bulk polymerized rods to which a coating of a copolymer of 80% ethoxyethyl methacrylate and 20% hydroxyethyl methacrylate has been applied, provided zero-order release in excess of one month whereas uncoated rods exhibited a first-order release pattern with the total amount of cyclazocine being released in less than 10 days.

Other hydrogel systems such as polyacrylamide and poly-vinylpyrrolidone have also been examined for their potential in providing sustained release (23, 24, 25). These hydrogels have been shown to release such biologically active substances as prostaglandins, immunoglobulins, luteinizing hormone, albumin, insulin, and ethinyloestradiol.

In summary, hydrophilic polymers such as poly(2-hydroxyethyl methacrylate) have potential as matrix materials for drug delivery systems. The distinct advantages offered by hydrogel systems for sustained or slow release of biologically active agents are the ease by which the hydrogels can be chemically modified to control their physical properties, i.e., their release behavior, the minimal formation of fibrous capsule which may impede drug release in vivo, and the ability to release polar and large molecular weight compounds such as peptide hormones. However, further characterization of the respective hydrogels is necessary and much remains to be done in investi-gating those factors such as hydration, crosslink density, porosity, drug-polymer interactions, etc. which will control the release of any drug.

Materials and Methods

Films containing hydrocortisone succinate were prepared by film casting polymer solutions containing hydrocortisone sodium succinate, ^{3}H-hydrocortisone sodium succinate, Hydron ® polymer-Type N, and, when desired, known amounts of photosensitive polymer crosslinking agent, ammonium dichromate. For each film, nanograms of hydrocortisone sodium succinate per milligram dry film and milligrams of polymer crosslinker per milligram of dry film were calculated. The films were cast on glass plates and when films containing the photosensitive crosslinking agent were prepared, film casting was carried out in a darkroom using a Wratten OA filter. Crosslinked polymer films were prepared by irradiating with ultraviolet light those films which con-tained the photosensitive crosslinking agent for 18 hours at 26°C.

Hydron ® polymer - Type N and the photosensitive cross-linking agent, ammonium dichromate, were graciously provided by Dr. Sam Ronel of the Hydron Laboratories, Inc., New Brunswick, New Jersey. Hydrocortisone sodium succinate was obtained as Solu-Cortef from the Upjohn Company, Kalamazoo, Michigan, and ^{3}H-hydrocortisone sodium succinate was purchased from New England Nuclear, Boston, Massachusetts.

Discs for the diffusion studies were prepared by reswelling the polymer films slightly with ethanol (to make the films pliable) and the discs were then cut from the film. The disc diameter was 0.61 cm (240 mils) and the dry thickness varied from 0.023 to 0.028 cm (9-11 mils). Wet thicknesses were measured immediately following the completion of each diffusion study and varied from 0.033 cm to 0.041 cm (13-16 mils). Table 1 summarizes the data on the in vitro hydrocortisone sodium succinate discs and includes nanograms of hydrocortisone sodium succinate per milligram dry disc, weight percent crosslinker in the respective discs, and total nanograms of hydrocortisone sodium succinate per disc.

To most closely approximate the in vivo diffusion system in which a disc of the polymer film containing the desired drug is placed on a small cut in the chorioallantoic membrane of the embryo, the following in vitro diffusion system was used. A twenty-five ml Erlenmeyer flask with a ground flat top and side arm for sample removal was used. A cellulose millipore filter (0.65 μm pore diameter) was sealed with acetone to the top of the Erlenmeyer to function as a porous support for the polymer disc and the side arm was capped with a rubber stopper. The Erlenmeyer was filled with Ringer's solution and the polymer disc was placed on the millipore filter. A glass cap, to provide a constant humidity and temperature environment was placed over the millipore filter and polymer disc. The in vitro diffusion apparati were then placed in a 38°C incubator and removed only for sampling purposes. One ml aliquots were removed from the Erlenmeyers at each time point and the Erlenmeyers refilled with Ringer's solution to bring the solution levels up to the millipore filter. All in vitro diffusion experiments were run in triplicate.

Diffusion of the tritiated hydrocortisone succinate was determined as follows: 1 ml aliquots of the desorbing Howard Ringers solution were withdrawn from the in vitro systems at appropriate time intervals and mixed with 5 ml of Triton X-100 toluene scintillation fluor in a glass scintillation vial. Counting was carried out in a Searle Isocap 300 preprogrammed for tritium using samples channels ratio.

Results and Discussion

The diffusion of hydrocortisone sodium succinate from hydrogel discs vary in their crosslink density is shown in Figure 1. In general, the diffusion curves show a rapid release of hydrocortisone sodium succinate in the early part of the experiment, 0 to 24 hours. This rapid release is decreased by increasing the crosslink density and at 24 hours, the 0% crosslinked discs have released ca. 75% of the steroid, the 5.1% crosslinked discs have released ca. 30% of the steroid, and the 9.9% crosslinked discs have released ca. 16% of the steroid. It is apparent that the 0% crosslinked discs have a first order release pattern in which the rate of steroid diffusion is dependent upon the amount of steroid still contained in the disc. In addition, the 0% crosslinked discs have released ca. 90% of the steroid within 48 hours.

Figure 2 shows eight day release patterns for the crosslinked discs with a twelve day release pattern shown for the 9.9% crosslinked discs. These data are an extension of those shown in Figure 1 and support the suggestion that the 9.9% crosslinked discs exhibit a nearly constant rate of steroid release. The curve for the 9.9% crosslinked discs is not strictly linear but does shown linearity from day 3 to day 12 of the experiment. Since the in vivo experiments require steroid release for a period of eight days, day 10 to day 18 of the chick embryo gestation period, it appears that the 9.9% crosslinked discs can successfully be used to deliver a relatively constant amount of steroid per day for this time period.

To determine the effect of the initial concentration of steroid in the disc on the steroid release pattern, 5% crosslinked discs containing 50,500 nanograms of steroid per milligram of disc, 2,140 nanograms of steroid per milligram of disc and 1,090 nanograms of steroid per milligram of disc were prepared and their respective release patterns examined over a five day period. The initial concentration of steroid appeared to have little effect on the percent released per time interval.

In an effort to place the experimental results on a quantifiable basis for the purpose of determining diffusion coefficients, a quasi-steady state approach was taken. This method was initially placed on a firm scientific basis by Barnes (26) and recently Garrett and Chemburkor (27) have applied this technique to drug diffusion across polymer membranes. Applied to our in vitro system, this approach requires that (1) the concentrations of the diffusing disc and the desorbing solutions are not held constant; (2) the concentration of the diffusing disc, initially C_0, decreases as the concentration, C_t, of the desorbing solution increases; and (3) the concentration values of both the disc and the desorbing solution approach the same equilibrium value with time. In the application of this method, Barnes stressed that the ratio of the membrane volume to desorbing solution volume should be less than 0.1. Since our membrane volume/desorbing solution volume ratio is 0.003, the method should be applicable and in the absence of anomalous partitioning effects all the drug in the membrane will eventually diffuse into the desorbing solution.

The following equation for the quasi-steady-state diffusion was used where V_1 is the volume of the desorbing solution, ca. 30 mls, V_2 is the volume of the diffusion disc, ca. 0.01 ml, C_0 is the initial concentration of steroid in the diffusing disc, C_t is the concentration of steroid in the desorbing solution at time t, X is the thickness of the

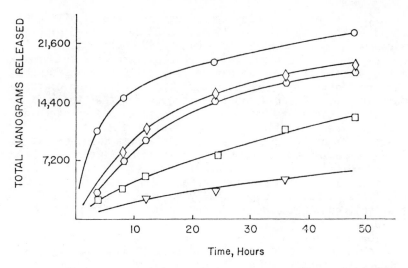

Figure 1.   The effect of crosslinking on hydrocortisone succinate in vitro *release. Percent by weight crosslinker:* ○, 0%; ◇, 0.5%; ⬠, 1.0%; ☐, 5.1%; *and* ▽, 9.9%.

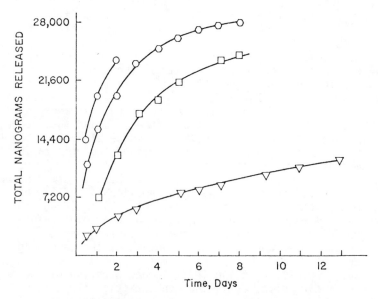

Figure 2.   The effect of crosslinking on hydrocortisone succinate in vitro release.   Percent by weight crosslinker: ○, 0%; ⬠, 1.0%; ☐, 5.1%; *and* ▽, 9.9%.

swollen disc, S is the surface area of the disc and D is the apparent diffusion constant:

$$\ln \frac{C_0}{\dfrac{C_0 - V_1 C_t}{V_2}} = \frac{D\,S}{X\,V_2}\,t$$

Figure 3 shows the plots of log $C_0 V_2/C_0 V_2 - V_1 C_t$ versus time in days and the linearity for each given cross-link density is consistent with the expectations of the above equation for quasi-steady state diffusion. Apparent diffusion constants, D, were calculated from the slopes in Figure 3 and are found in Table 1.

TABLE 1:   IN VITRO HYDROCORTISONE-SUCCINATE DISCS

Nanograms per Milligram Disc	Total Nanograms per Disc	Weight% Crosslinker	$D \times 10^{+13}$ L/cm sec.
1,090	7,700	5.0	51.0
2,140	9,800	5.0	42.0
5,000	24,000	0	-
5,200	28,100	0.1	213.0
5,350	25,300	0.5	-
4,750	30,200	1.0	90.0
4,990	28,300	5.0	44.0
4,640	20,900	9.9	8.3
50,500	293,000	4.9	43.0

Yasuda, Lamaze, and Ikenberry (28) have shown that diffusion through hydrophilic polymers, such as poly(2-hydroxyethyl methacrylate), can be described by the free volume theory of diffusion. In the free volume theory, the diffusion coefficient can be expressed by $D \propto \exp - (V^*/V_f)$ where V_f is the free volume in the sample and V^* is a characteristic volume required to accomodate the diffusing steroid molecules. Assuming a linear variation of the free volume with the hydration, H, and an inverse linear variation between the hydration and the crosslink density, or in our case the weight percent crosslinker, W, the following equation results:

$$\log D = \log D_0^- - K \cdot W + K'$$

where K is a constant. The validity of the above equation with
its accompanying assumptions is supported by the linearity of
the plot log D versus W using the apparent diffusion constants
found in Table 1 for the 5,000 ng per mg dry film membranes
(Figure 4).

As Love and coworkers (1, 2) have previously shown, the
fetal endocrine system plays an important role in the development of skeletal muscle in chick embryos by regulating the rate
of myoblast proliferation, the fusion of myoblasts to form
myotubes, and the maturation of myotubes to form mature muscle
fibers. We have chosen to examine this conversion from myoblast
to myotube to muscle by measuring the effect that the slow released hydrocortisone succinate has on the muscle enzymes:
phosphorylase, phosphoglucomutase, and glucose-6-phosphate
dehydrogenase. In general, the activity of glucose-6-phosphate
dehydrogenase in the developing embryonic muscle is indicative
of the rate of DNA synthesis and thus roughly measures the
myoblast replication or proliferation. On the other hand,
phosphorylase and phosphoglucomutase activities indicate the
level of anaerobic metabolism in the developed muscle. Thus,
development and maturation of the embryonic chick muscle can
be shown by increasing activity of phosphorylase and phosphoglucomutase and decreasing activity of glucose-6-phosphate
dehydrogenase. In order to test the ability of the hydrocortisone succinate-HEMA discs to function biologically in an
in vivo system, representative discs were implanted on
hypophysectomized chick embryo chorioallantoic membranes at day
10 of the gestation period. Comparison with three different
types of embryos were made: normal embryos, embryos in which
the pituitary glands were removed (hypophysectomized) at day
2 of the gestation period, and embryos which were hypophysectomized at day 2 and at day 14 a pituitary gland was trans-
planted to the embryo. All embryos were sacrificed at day 18
of the gestation period, and the enzymatic activities of
phosphorylase, phosphoglucomutase, and glucose-6-phosphate
dehydrogenase were determined.

Data on the effect of hydrocortisone-succinate poly (2-
hydroxyethyl methacrylate) implants on enzymatic activity in
developing muscle are found in Table 2. As the concentration
of hydrocortisone succinate in the implanted disc is increased,
the activities of phosphorylase and phosphoglucomutase also
increase suggesting that the effect of hypophysectomy is
reversed by these slow-release implants. At the same time,
the activity of glucose-6-phosphate dehydrogenase is decreased.
This trend, as discussed earlier, is indicative of a decreased
rate of DNA synthesis or of myoblast proliferation. Thus, the
hydrocortisone succinate reverses the effect of hypophysectomy.

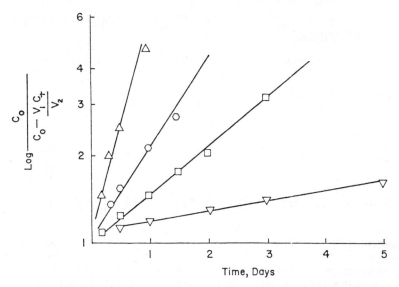

Figure 3. Quasi-steady state diffusion. Effect of crosslinking on hydrocortisone succinate in vitro release. Percent by weight crosslinker: △, 0.1%; ⬡, 1.0%; □, 5.1%; and ▽, 9.9%.

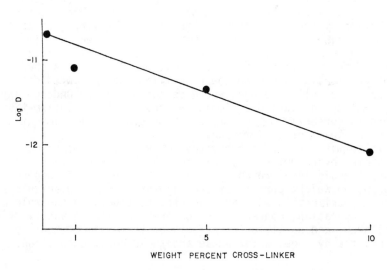

Figure 4. Free volume diffusion—variation of the apparent diffusion constant, D, with the weight percent crosslinker

TABLE 2:   EFFECT OF HYDROCORTISONE SUCCINATE HEMA IMPLANTS ON
ENZYMATIC ACTIVITY IN DEVELOPING MUSCLE

	Phosphoryl-lase b*	Phospho-glucomutase*	Glucose-6-PO$_4$* Dehydrogenase**
Normal	273	2200	12
Hydrocortisone Succinate			
41000	390	1970	20
20000	170	1330	31
2000	101	610	60
Hypophysectomized	82	420	63
Pituitary Transplant	152	1000	30

*   Nanomoles per minute per milligram DNA at day 18.
**   Total nanograms in polymer implant, 0% crosslinking.

Finally, to test the validity of our in vitro system and its
ability to imitate the in vivo system, diffusion experiments were
carried out using 2,000 ng hydrocortisone succinate per mg dry
film discs described in Table 1. The in vitro discs contained
5% crosslinker and the in vivo discs contained 0% crosslinker.
It is easily seen in Figure 5 that the in vivo discs release the
drug at a slower rate then do the in vitro discs even though
the in vivo discs contain no crosslinker and would be expected
to release the hydrocortisone succinate at a faster rate than
the in vitro discs which contain 5% crosslinker. Microscopic
examination of the implanted disc and the underlying chorioallantoic membrane during the experiment and up to ten days implantation reveals no fibrous capsule formation which might
decrease the rate of release. In addition, the vascular bed
in the chorioallantoic membrane underlying the implanted disc
appears to be well developed and does not appear to undergo any
change during the period of implantation. Further confirmation
of this is currently being obtained by histological examination
of the chorioallantoic membrane.
In summary, poly(2-hydroxyethyl methacrylate) membranes
containing hydrocortisone succinate have been shown to slowly
release steroid in vitro and in vivo. This release is dependent upon the weight percent of crosslinker of the implanted disc.
Further experiments are being conducted to elucidate the role of
membrane hydration, concentration of drug,type and quantity of
crosslinker, and partitioning effects of the drug between the
polymer, the hydrogel water phase and the desorbing solution.

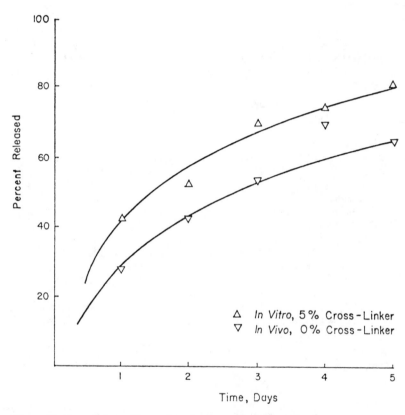

Figure 5. Comparison of in vitro *and* in vivo *release of hydrocortisone succinate (2,000 ng per mg dry film)*

ABSTRACT

   Poly(2-hydroxyethyl methacrylate)membranes containing
hydrocortisone succinate have been shown to exhibit slow -
relase behavior in vitro and in vivo. Apparent diffusion
constants, obtained from a quasi-steady-state approach,
correlate well with weight percent crosslinker suggesting a
free volume type of diffusion. In vivo experiments designed
to show the effect of replacement therapy in hypophysectomized
chick embryos were carried out. Using muscle enzymes, it has
been shown that implanted discs of poly(2-hydroxyethyl
methacrylate) containing hydrocortisone succinate can reverse the
effects of hypophysectomy (pituitary gland removal).

Literature Cited

 1. Love, D. S., Stoddard, F. J., and Grasso, J. A.,
    Developmental Biology, (1969), 20, pp. 563-582.
 2. Love, D. S. and Eisenberg, J., to be published.
 3. Leaders, F. E., Hecht, G., Van Hoose, M., and Kellogg,
    M., Annals of Opthalmology, (1973), 5(5), pp. 513-522.
 4. Dohlman, C. H., Langston-Pvan, and Rose, J., Annals of
    Opthalmology, (1972), pp. 823-832.
 5. Yolles, S. in "Controlled Release of Biologically Active
    Agents", A. C. Tanquary and R. E. Lacey, Ed., Plenum Press,
    New York, New York, 1974, pp. 177-193.
 6. Fu, J. C., Kale, A. K., and Moyer, D. L., Journal of
    Biomedical Materials Research, (1973), 7, pp. 71-78.
 7. Fu, J. C., Kale, A. K. and Moyer, D. L., Biomaterials
    Medical Devices and Artificial Organs, (1973), 1(4).
 8. Schmidt, V., Zapol, W., Prensky, W., Wonders, T.,
    Wodinsky, J., and Kitz, R., Transactions of the American
    Society for Artificial Internal Organs, (1972), 18, pp.
    45-52.
 9. Tatum, H. J., Contraception, (1970), 1(4), pp. 253-263.
10. Kincl, F. A., and Rudel, H. W., Acta Endocrinologica,
    (1971), 66, Accompanies Supplementum 151, pp. 5-30.
11. Dzuik, P. J., Cook, B., Niswender, G. D., Kaltenback, C. C.,
    and Doane, B. B., American Journal of Veterinary Research,
    (1968), 29(12), pp. 2415-2417.
12. Most, C. F. Jr., Journal of Biomedical Materials Research,
    (1972), 6, pp. 3-14.
13. Most, C. F., Journal of Applied Polymer Science, (1970),
    14, pp. 1019-1024.
14. Roseman, T. J., Journal of Pharmaceutical Sciences, (1972),
    61, No. 1., pp. 46-50.
15. Abrahams, R. A., Ronel, S. H., and Gould, F. E., Abstract,
    Eighth Annual Scientific Session of the Association for
    the Advancement of Medical Instrumentation, Medical
    Instrumentation, (1973), 7(1), p. 41.

16. Refojo, M. F., Journal of Applied Polymer Science, (1965), 9, p. 3161.
17. Sprincl, L., Kopecek, J., and Lim, D., Calcified Tissue Research (1973), 13, pp. 63-72.
18. Levowitz, B. S., LaGuerre, J. N., Calem, W. S., Gould, F. E., Francis, E., Scherrer, J., and Schoenfeld, H., Transactions of American Society for Artificial Internal Organs, (1968), 14, p. 82.
19. Lazarus, S. M., LaGuerre, J. N., Kay, H., Weinberg, S., and Levowtiz, B. S., Journal of Biomedical Materials Research, (1971), 5, p. 129.
20. Kaufman, H. E., Votila, M. H., Cassett, A. R., Wood, T. O., and Varnell, E. D., "Medical Uses of Soft Contact Lenses", Chap. 22, Ed. by A. R. Gassett and H. E. Kaufman, Soft Contact Lens, C. V. Mosley, Co., St. Louis, Mo., (1972).
21. Podos, S. M., Becker, B., Asseff, C., and Hartstein, J., American Journal of Opthalmology, (1972), 73(3), p. 336.
22. Abrahams, R. A. and Ronel, S. H., Journal of Biomedical Materials Research, (1975), 9, pp. 355-366.
23. Davis, B. K., Noske, I. and Chang, M. C., Acta Endocrinologica, (1972), 70, pp. 385-395.
24. Davis, B. K., Proceedings of the National Academy of Sciences U. S. A., (1974), 71(8), pp. 3120-3123.
25. Balin, H., Halpern, B. D., Davis, R. H., Akkapeddi, M. K., and Kyriazis, G. A., Journal of Reproductive Medicine, (1974), 13(6), pp. 208-212.
26. Barnes, C., Physics, (1934), 5(1), pp. 4-8.
27. Garrett, E. R., and Chemburkar, P. B., Journal of Pharmaceutical Sciences, (1968), Vol. 57, No. 6, p. 949.
28. Yasuda, H., Lamaze, C. E. and Ikenberry, L. D., Die Macromolekulare Chemie, (1968), 118, pp. 19-35; (1969), 125, pp. 108-118.

14

Controlled Release of Fluoride from Hydrogels for Dental Applications

D. R. COWSAR, O. R. TARWATER, and A. C. TANQUARY

Southern Research Institute, Birmingham, Ala. 35205

A century ago Erhardt(1) recommended oral administration of potassium fluoride to women during gestation
and to children during dentition to protect against
dental caries by hardening tooth structure. Numerous
studies since then have demonstrated conclusively the
effectiveness of fluoride prophylaxis for reducing the
incidence of carious lesions in teeth. Methods for
administering fluoride now include the fluoridation of
drinking water, the ingestion of fluoride tablets, the
incorporation of fluoride into mouthwashes and denti-
frices, and the topical application of fluoride solutions and gels.(2,3) When fluoride is present in
drinking water at the optimum level, which is about 1
ppm of fluoride added as sodium fluoride or sodium
fluorosilicate, the incidence of carious lesions is
reduced by 50 to 60%. However, a large portion of the
population does not have access to public water sup-
plies, and since many communities have elected not to
employ controlled fluoridation of water, a major segment of the population must rely on alternative means,
which are less efficient, to obtain the anticaries
benefits of fluoride.

During the past several years considerable progress
has been made in optimizing the therapeutic effectiveness (potency) of pharmaceuticals and other biologi-
cally-active agents by simply improving their delivery
to the target organs or organisms. Controlled-release
formulations that deliver small amounts of drugs at
constant rates for long times are usually much more
effective than conventionally-administered medicaments.
While not all drugs are amenable to controlled release,
the controlled delivery of fluoride appears practical
as well as highly desirable. An intraoral, controlled-
release device could provide continuous topical application of fluoride to tooth surfaces for six months to

a year or more depending upon design. Moreover, a device that can be applied or attached in the mouth by simple procedures would permit an economical anti-caries treatment applicable to virtually all segments of the population.

This paper describes research to develop a con-trolled-release delivery system that will release inorganic fluoride into the oral cavity at pre-determined rates of 1.0, 0.5, 0.2, and 0.02 mg/day for six months without maintenance or adjustment. Once the system is developed, the optimum fluoride release rate and the efficacy of this method of treat-ment can be determined clinically.

Our approach has been to fabricate reservoir con-trolled-release devices(4) that consist of a core, which contains a supply of sodium fluoride particles, and a coating, which serves as a membrane to control diffusion (release) of fluoride. The rate of release of fluoride is determined by the geometry (area and thickness) of the membranes and by the permeability parameters of the specific acrylic copolymers. The following equation is a simple expression of Fick's law that describes the steady-state rate of release of fluoride from such reservoir devices.

$$\frac{dM}{dt} = \frac{A\ D_{coat}}{h_{coat}}\ (C_S - K_f C_f)$$

where:

$\frac{dM}{dt}$ = rate of release of fluoride

A = area of the device

h_{coat} = thickness of the coating

D_{coat} = diffusion coefficient of fluoride in the polymeric coating

C_S = saturation solubility of fluoride in the coating

C_f = concentration of fluoride in the receiving fluid (saliva)

K_f = partition coefficient of fluoride between the polymeric coating and the receiving fluid

In order to design devices that release fluoride at the specified rates of 1.0, 0.5, 0.2, and 0.02 mg/24 hr, we needed to use, as rate-controlling coatings, polymers that had appropriate values of D, C_S, and K_f for sodium fluoride. Several years ago, Yasuda and co-workers(5) determined the diffusional parameters for sodium chloride in a series of hydrated acrylic copolymers and hydrogels. Various copolymers of methyl

methacrylate (MMA), glyceryl methacrylate (GMA),
hydroxyethyl methacrylate (HEMA), hydroxypropyl methacrylate (HPMA), and glycidyl methacrylate (GdMA) provided a series of hydrated membrane materials having
equilibrium water contents ranging from 10 to 70%.
Yasuda found that the diffusion coefficients of NaCl
in these membranes varied over five orders of magnitude with changes in the equilibrium levels of
hydration, and that the variance in diffusivity was
virtually independent of the specific chemical iden-
tities (structures) of the copolymers.

On the basis of Yasuda's findings, we anticipated
similar variance in the diffusivity of sodium fluoride
in hydrated acrylic copolymers. Specifically, we
anticipated that copolymers made by polymerizing HEMA
and MMA in ratios ranging from 80/20 to 40/60
(HEMA/MMA) would provide a series of core and coating
materials with values of D, C_S, and K_f suitable for
fluoride-releasing systems. And since MMA and HEMA
homopolymers are known to be tissue compatible and
toxicologically safe in the oral environment, we
anticipated no problems in demonstrating the biocompatibility and safety of devices based on HEMA/MMA
copolymers.

Materials and Methods

Acrylic Copolymers. Random copolymers of HEMA and
MMA were prepared in aqueous ethanol solutions by mixing HEMA and MMA monomers in various molar ratios and
initiating their polymerization with a redox catalyst.
All copolymerizations were carried out essentially as
follows: To a 1-liter, glass-stoppered bottle were
added 950 ml of a 3-to-2 mixture of ethanol and water
and 50 g of a mixture of purified hydroxyethyl methacrylate (Hydron Laboratories, New Brunswick, New
Jersey) and methyl methacrylate (Polysciences, Inc.,
Warrington, Pa.). Nitrogen was then bubbled through
the mixture for 50 min to purge it free of oxygen.
Polymerization was initiated by adding 0.125 g of
$K_2S_2O_5$ and 0.25 g of $Na_2S_2O_5$ to the mixture, and the
bottle was sealed. After ten days at room temperature,
the copolymer was precipitated by pouring the viscous
polymerizate into an excess (3000 ml) of water. The
white resin was washed thoroughly with water, isolated
by filtration, and dried at 50°C in vacuo.

This procedure was used to prepare HEMA/MMA copolymers having molar ratios of 20/80, 30/70, 40/60, 50/50,
60/40, and 75/25. Polymer yields ranged from 41 to
100% of theory. Inherent viscosities were determined

for each copolymer. These values ranged from 0.39 to 2.2 dl/g when measured at a concentration of 0.5% in a 60:40 (v/v) mixture of acetone and p-dioxane at 30°C. Most of our evaluations of the copolymers were carried out on film specimens that we prepared by spin casting.(6) An example of the general procedure for fabricating thin films of the acrylic copolymers is as follows: Exactly 3 g of copolymer was dissolved in 25 ml of a 60:40 (v/v) mixture of acetone and p-dixoane. This solution was poured into a 3-in.-diameter by 1-in.-deep, Teflon-lined, spin-casting cup which rotated at 3500 rpm until the solvent had evaporated. The cylindrical film was then lifted from the cup and cut transversely to give a highly uniform, rectangular film 15-mil thick, 1-in. wide, and 9.4-in. long. In some cases, the thickness of the films was varied by adding more or less solution to the spin-casting cup.

Permeability Parameters. The diffusion coefficients for sodium fluoride in the hydrated acrylic copolymers were determined by a modification of the desorption method of Yasuda et al.(5) In principle, the desorption method involves first equilibrating a thin membrane of the test material in a solution of the test solute (fluoride in this case), and then immersing the solute-loaded membrane in a fresh aliquot of receiving fluid and measuring the rate of increase in concentration of the solute in the fluid (or the rate of desorption of the solute from the membrane). A plot of the concentration of the solute in the receiving fluid vs the square root of time is prepared from the data, and the diffusion coefficient, D, is calculated from the linear portion of the curve by the following formula.

$$D = \frac{\pi}{16} K^2$$

where: $$K = \frac{d(M_t/M_\infty)}{d(t^{1/2}/h)}$$

and
$$M_t = \text{mass of solute released at time } t$$
$$M_\infty = \text{mass of solute released at time } \infty$$
$$h = \text{thickness of the membrane}$$
$$t = \text{time}$$

Film specimens of the 30/70, 40/60, and 50/50 HEMA/MMA copolymers were equilibrated at 37°C in a saturated aqueous solution of sodium fluoride. The equilibrating solution was maintained at saturation

by desorption of sodium fluoride from pellets of HEMA
polymer to which solid sodium fluoride had been added
before it was polymerized in bulk. This procedure is
a simple way of maintaining saturation of the solution
without contaminating the surfaces of film specimens
with particles of sodium fluoride.(7) After equilibration had been achieved (1 or 2 days), the film samples
were removed and blotted to remove surface solution.
The films were then placed in a polyethylene bottle
containing 50 ml of water which had been equilibrated
at 37°C. The fluoride specific-ion electrode (Orion
Model 96-09) was inserted in the solution, and the
bottle was shaken vigorously. The millivolt output of
the specific-ion electrode was recorded continuously
on a strip-chart recorder. Concentration vs $t^{\frac{1}{2}}$ plots
were made from the data on the strip chart. The film
samples were then equilibrated in several additional
50-ml volumes of water to allow extraction of all of
the fluoride. The desorbing solutions and the extracting solutions were analyzed to give the total mass, M_∞,
of fluoride released. The films were then weighed wet
and dried in vacuo.
    The diffusion coefficients, D, were calculated by
the desorption equation. The saturation concentration,
C_S, of sodium fluoride in the hydrated copolymers was
calculated by dividing the M_∞ value by the volume,
A x h, of the equilibrated sample. The partition coefficient, K_f, of fluoride between the polymer and the
receiving fluid was calculated from C_S and C_f, the
saturation concentration of sodium fluoride in the
fluid (determined by analyzing the saturated equilibrating solution).

$$K_f = \frac{C_S}{C_f}$$

The equilibrium water content (degree of hydration) was
calculated from the wet and dry weights of the films.

$$\text{Water Content} = \frac{\text{Wet Weight} - \text{Dry Weight}}{\text{Wet Weight}} \times 100$$

    Other Methods. Methods for fabricating devices
and for determining their fluoride-release rates in
vitro and in vivo are described in other sections. The
synthetic saliva for in vitro evaluations was a dilute
solution of electrolytes: 0.6 mM $CaCl_2$; 1.8 mM
NaH_2PO_4; 12 mM $NaHCO_3$; and 1.0 mM HCl. Human saliva
was collected from volunteers.

Device Design

To release fluoride at a constant rate for a prolonged period, a device must contain a reservoir of fluoride salt separated from the receiving fluid (saliva) by a rate-controlling membrane that is permeable to fluoride ions. To be permeable to fluoride ions, a membrane must become hydrated when immersed in aqueous media; and to control fluoride release, its degree of hydration must be easily reproduced. The acrylic hydrogels represent a class of polymers that can be formulated to provide membrane materials having a wide range of degrees of hydration. For our fluoride-releasing devices, we chose to use copolymers of hydroxyethyl methacrylate (HEMA) and methyl methacrylate (MMA).

In Figure 1 we show that the equilibrium water content of HEMA/MMA copolymer membranes varies directly with the mole % of the hydrophilic monomer (HEMA) but to different degrees above and below the equimolar ratios. Several values for HEMA/MMA copolymers that were reported by Yasuda, et al.(5) are also shown in Figure 1. Copolymers with 50 mole % or less of HEMA are preferred for both the core matrix and the rate-controlling membranes of our devices.

Two different copolymers were required for the devices. The one for the core had to be highly permeable to sodium fluoride, and the one for the rate-controlling coating had to be less permeable. Values of D, C_S, and K_f were determined for three copolymer compositions: 30/70, 40/60, and 50/50 HEMA/MMA. Table I shows the diffusion coefficient, D, the saturation solubility, C_S, the partition coefficient, K_f, and the calculated permeability, P, for sodium fluoride when water was the equilibrating (receiving) fluid. The values were essentially the same when synthetic saliva and human saliva were used as receiving fluids. On the basis of these data we chose the 50/50 HEMA/MMA copolymer for the core matrix material and the 30/70 HEMA/MMA copolymer for the coating materials.

Since the fluoride-releasing devices are expected to be worn by children as well as adults, we designed the final devices so that they would be as small as possible. The main factor restricting the overall size was the amount of fluoride salt that was required to sustain the desired daily rates of release for six months. From our work with design prototypes, we determined that the rate of release of fluoride from trilaminate devices remains constant until approxi-

Table I. Experimentally-Determined Values

Copolymer, ratio of HEMA/MMA	Degree of hydration, % water	D, cm^2sec^{-1}	Cs, gF/cm^3	K	$P = DK$ cm^2sec^{-1}
30/70	11.1	8×10^{-8}	2.5×10^{-4}	0.012	9.6×10^{-10}
40/60	13.8	26×10^{-8}	3.1×10^{-4}	0.015	39×10^{-10}
50/50	20.3	63×10^{-8}	3.3×10^{-4}	0.017	107×10^{-10}

mately 80% of the fluoride in the core is exhausted.
With this in mind, we designed the final devices to
contain a 20% excess of fluoride salt.

The core of the device was designed to consist of
powdered sodium fluoride dispersed in a matrix of
HEMA/MMA copolymer. The copolymer matrix serves two
functions. First, it provides a medium of fixed
geometry and water content in which the fluoride salt
dissolves prior to diffusing through the rate-controlling membranes. This design ensures that the concentration of dissolved fluoride at the interior surfaces
of the rate-controlling membrane will remain constant
throughout the lifetime of the device. And second,
the copolymer matrix provides a degree of safety to
prevent catastrophic release of the contents of the device if the rate-controlling membrane becomes accident-
ly punctured, torn, or worn through. Since the core is
actually a monolithic controlled-release device, (4)
stress failure of the rate-controlling membrane will
result in a high rate of fluoride release for a short
period until the fluoride salt in the immediate vicinity of the fault is depleted. Then, the core matrix
material will become rate controlling, and the release
of fluoride from the device will occur at a low,
declining rate.

Cores for trilaminate devices that are prepared
from the 50/50 HEMA/MMA copolymer to contain 80% (w/w)
of sodium fluoride have a loading density of 1.9 g of
sodium fluoride per cm^3 of core. Cores prepared to
contain 62% (w/w) of sodium fluoride in 50/50 HEMA/MMA
copolymer have a loading density of 1.53 g of sodium
fluoride per cm^3 of core. These density factors were
used to calculate the minimum core volumes required
for the four devices. From the minimum core volumes
we calculated "arbitrary" linear dimensions for the
cores, and from these we calculated the membrane surface areas. Then, we used the values of D, C_s, and K_f
that were determined for the 30/70 HEMA/MMA copolymer
and the Fick's Law equation to calculate the thickness
of the rate-controlling membranes required to give the
desired rates of release of fluoride. Two alternative
design specifications for each of the four final devices are given in Table II. Similar values can be
calculated for cylindrical, spherical, or irregular
devices. Since the rate-controlling membranes are
very thin, the device dimensions are very nearly the
core dimensions.

Table II. DESIGN SPECIFICATIONS FOR FLUORIDE-RELEASING DEVICES

Fluoride Release Rate	Core Specifications							Membrane Specifications		
	Core Copolymer	Wt NaF Contained, mg	Loading, % w/w	Volume, cm³	Length, cm	Width, cm	Thickness, cm	Membrane Copolymer	Area, cm²	Thickness, cm
1 mg/day	50/50	475	62	0.31	4.8	0.8	0.08	30/70	7.7	0.013
1 mg/day	50/50	475	80	0.25	3.9	0.8	0.08	30/70	6.2	0.011
0.5 mg/day	50/50	240	62	0.16	2.5	0.8	0.08	30/70	4.0	0.014
0.5 mg/day	50/50	240	80	0.12	1.9	0.8	0.08	30/70	3.0	0.011
0.2 mg/day	50/50	95	62	0.063	2.5	0.5	0.05	30/70	2.5	0.022
0.2 mg/day	50/50	95	80	0.049	2.0	0.5	0.05	30/70	2.0	0.017
0.02 mg/day	50/50	9.5	62	0.006	0.4	0.4	0.037	30/70	0.32	0.028
0.02 mg/day	50/50	9.5	80	0.005	0.3	0.3	0.05	30/70	0.18	0.016

Fabrication of Devices

The key to fabricating trilaminate devices having precise, predetermined rates of release is in accurately preparing the fluoride-containing cores to specified dimensions. To do this, we had stainless-steel replicas of the cores specified in Table II prepared in our machine shop. Using these replicas, we prepared multi-cavity silicone rubber molds for the cores. Molding mixtures were prepared by thoroughly mixing powdered sodium fluoride in appropriate amounts into viscous solutions of 50/50 HEMA/MMA copolymer in 60:40 acetone-dioxane.

Cores were molded by applying the fluoride mixtures by spatula into the cavities of the mold and allowing the solvent to evaporate overnight. When we removed the cores from the mold, they still contained a small amount of residual dioxane and were pliable. Two stainless-steel wires were attached to each core to provide a means for later attaching the device to an appliance in a dog's mouth. The large ribbon-shaped cores were also curved slightly while they were still pliable, and then they were allowed to dry. A sketch of a curved core with the wires attached is shown as Figure 2.

The cores were then coated with rate-controlling membranes by dipping them repeatedly into a 12% (by weight) solution of 30/70 HEMA/MMA copolymer in 60:40 acetone-dioxane. After each dip, the coating was allowed to dry in air for approximately 20 min. Each dipping increased the coating thickness by approximately 0.001 cm. Thus, approximately 14 dips were required to produce a coating thickness of 0.014 cm. The actual thickness was measured with a micrometer. After the last dipping, the devices were dried in air for 1 hr and then placed in a vacuum oven at 60°C overnight.

In Vitro Evaluation of Devices

A constant-temperature flow system was used for evaluating the release rates of fluoride from the devices. The devices were suspended by a stainless-steel wire in a 30-ml-capacity, polyethylene flow cell having inlet and outlet ports for the receiving fluid. Initially, high-purity deionized water was pumped at 0.85 ml/min (1220 ml/day) from a thermostated (37°C) reservoir, through the flow cell (also held at 37°C), and into a collection vessel. Later, devices were

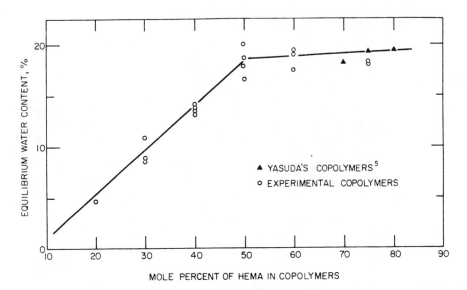

Figure 1.   *Equilibrium water content for HEMA/MMA copolymers*

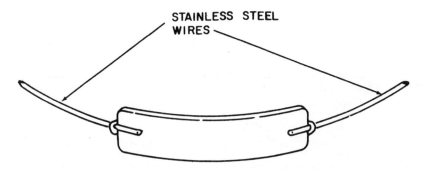

Figure 2.   *Core for fluoride-releasing device*

evaluated with synthetic and natural salivas as the
receiving fluids. At regular intervals the receiving
fluid in the collection vessel was assayed for fluoride
via a calibrated specific-ion electrode (Orion Model
96-09), and the volume of fluid collected during the
time interval was recorded. Figure 3 depicts the flow
system. Although only one flow cell is shown in Figure
3, we usually operated six cells simultaneously.

The fluoride-releasing devices all have similar
polymeric cores loaded with powdered sodium fluoride.
The copolymers for the core matrix were designed to be
highly permeable to fluoride. The rate of release of
fluoride from the devices is controlled by the membranes which surround the cores. To demonstrate the
validity of this design concept, we placed a disk-
shaped (1.18-cm-diameter x 0.05-cm-thick) core made of
50/50 HEMA/MMA copolymer and containing 40 mg of
fluoride in the flow cell and determined the rate of
release of fluoride into deionized water at 37°C.
Most of the fluoride was released during the first day,
and virtually all of the fluoride was released during
four days. When 0.023-cm-thick, rate-controlling membranes of 30/70 HEMA/MMA copolymer were pressure
laminated on both sides of a similar core, and the
edges of the laminate were sealed with a coating of
30/70 copolymer, a device that released fluoride at a
nearly-constant rate of approximately 0.8 mg/day was
obtained. Figure 4 shows the cumulative release of
fluoride from the core alone and from the complete
trilaminate device.

Long-term, in vitro evaluations of two of the final
devices were initiated soon after they were fabricated.
The devices were suspended in the flow cells and eluted
continuously at 37°C with synthetic saliva flowing at
approximately 0.85 ml/min. The release-rate data that
have been obtained so far are shown in Figures 5 and 6.

Device 7935-29-C (Figure 5) was designed to release
fluoride at a rate of 0.2 mg/day. The core was a
2.5 x 0.5 x 0.05-cm ribbon loaded with 62% of sodium
fluoride. The coating thickness is 0.022 cm. Over a
test period of 59 days the average rate of release of
fluoride has been 0.19 mg/day.

Device 7935-29-G (Figure 6) was designed to release
fluoride at a rate of 0.5 mg/day. It has a 1.9 x 0.8
x 0.08-cm core loaded with sodium fluoride at a level
of 80%. The coating thickness if 0.011 cm. Over a
test period of 63 days the device has released fluoride at an average rate of 0.49 mg/day.

The rate of release of fluoride from these devices
has been shown to be essentially independent of the

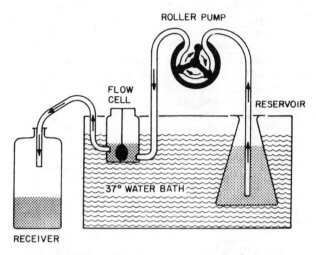

Figure 3. Diagram of constant-temperature flow system

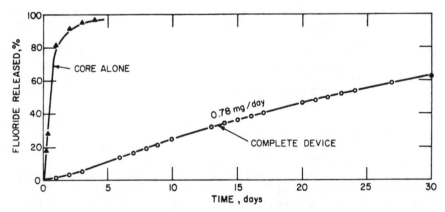

Figure 4. Cumulative release of fluoride from a core alone and from a pressure-laminated device

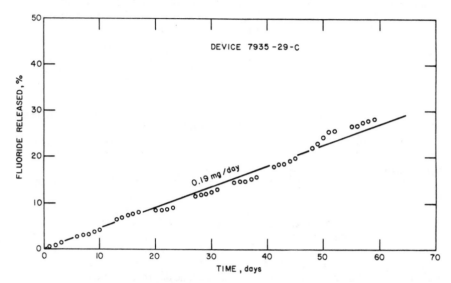

Figure 5.   Device designed to release fluoride at 0.2 mg/day

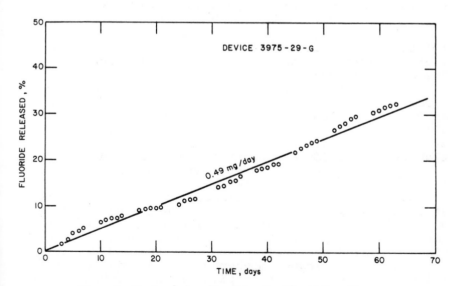

Figure 6.   Device designed to release fluoride at 0.5 mg/day

composition and pH of the receiving fluid. In tests
with synthetic saliva, human saliva, and water having
values of pH ranging from 4.8 to 7.8, the fluoride release rate varied by less than ±0.03 mg/day for a device with a daily mean release rate of 0.47 mg/day.

In Vivo Evaluation of Devices

Two fluoride-releasing devices identical to those
evaluated in vitro were also evaluated in vivo in
beagle dogs. The devices were attached via a special-
ly-designed intraoral appliance that was attached to
the upper canine teeth.

Fabrication of a fixed intraoral appliance which
would allow for attachment and removal of the con-
trolled-release device began with the taking of dental
impressions using an alginate impression material.
Models made of plaster of paris were poured and orthodontic bands for the upper canines were constructed
from 0.25-in.-wide, 0.002-in.-thick stainless-steel
ribbon. The bands were hand fitted and electrically
spot welded. A section of surgical arch bar was
adapted to the casts and soldered to the bands. The
labial bar was fitted with space left for the occluding
lower canines. The lugs were removed in the anterior
part of the arch bar to allow seating of the fluoride-
releasing device which was attached to the fixed
appliance by means of stainless-steel wires hooked
to the posterior lugs of the arch bar (Figure 7).

Seating and zinc oxyphosphate cementing of the
completed appliance were accomplished with the dogs
lightly anesthetized with Nembutal. Figure 8 is a
sketch showing the completed appliance attached in
the beagle's mouth.

Prior to attachment of the fluoride-releasing devices, saliva was collected from two beagle dogs and
analyzed for fluoride. The dogs were given 0.07 ml of
a solution of pilocarpine (75 mg/ml) s.c. in the nape
of the neck to stimulate saliva flow. Saliva collection by aspiration was begun immediately and continued
until a 2-ml sample had been collected. Average
saliva fluoride levels were found to be 0.015 ppm.

Daily saliva samples were collected from the dogs
after the devices were in place. Saliva was collected
with and without pilocarpine stimulation. The in vivo
data (Table III) show that the fluoride-releasing devices elevated the saliva fluoride levels significantly. The higher levels measured without pilocarpine
stimulation indicated that pilocarpine stimulation in-

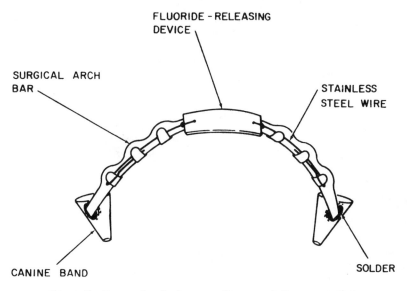

Figure 7. *Intraoral orthodontic appliance with device attached*

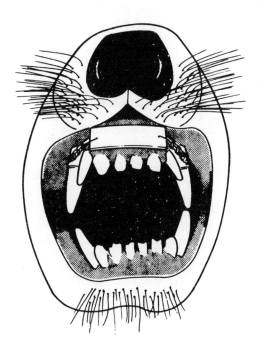

Figure 8. *Fluoride-releasing device in place in
a beagle's mouth*

creases saliva flow rate but not the rate of release
of fluoride from the devices.

Table III. Saliva Fluoride Levels of Beagle Dogs
Fitted with Intraoral Fluoride-
Releasing Devices

Saliva collected	Fluoride concentration, ppm, in saliva of dog with device releasing at	
	0.2 mg F/day	0.5 mg F/day
Before device	0.018	0.013
Without pilocarpine stimulation	0.26	0.76
With pilocarpine stimulation	0.15	0.25

Abstract

The continuous topical administration of fluoride
to teeth from an intraoral controlled-release device
is an untried method for reducing the occurrence of
dental caries. Pharmacokinetic dose-response data can
be obtained only after a delivery system has been
developed. In the present study we are developing biocompatible, fluoride-containing devices that release
fluoride in the oral environment at constant pre-
determined linear rates of 0.02 to 1.0 mg/day for at
least six months without maintenance or adjustment.
The trilaminate devices comprise a core of inorganic
fluoride salt, which is dispersed in a copolymer
(hydrogel) of hydroxyethyl methacrylate and methyl
methacrylate, and outer layers (or coatings) of semipermeable, rate-controlling membranes, which are made
from similar hydrated acrylic copolymers. Fluoride
from the core diffuses through the membranes into the
oral environment at a rate precisely controlled by the
thickness, area, and permeability of the materials.
The permeability parameters, which include the solubility of the fluoride salt in the polymeric core and
in the outer membranes, the diffusion coefficients of
fluoride in both materials, and the partition coefficient for fluoride between the outer membrane and
saliva, have been quantitated; and prototype devices
have been evaluated _in vitro_ and _in vivo_ in dogs.

Acknowledgments

This work is sponsored by the National Institute of Dental Research under Contract NO1-DE-42446. The authors express their thanks to Mrs. Melody D. Hamilton, Mrs. Brenda H. Perkins, Mr. Joe M. Finkel, Dr. Danny H. Lewis, and Dr. Lewis Menaker who contributed significantly to this work.

Literature Cited

1. Erhardt, Monatsschr. ration. Aerzte (Heil bronn) (1874), 19, p. 359.
2. Horowitz, H. S., Community Dent. Oral Epidemiol. (1973), 1, p. 104.
3. Carlos, J. P., Ed., "Prevention and Oral Health," Fogarty International Center Series on Preventive Medicine, Vol. 1, DHEW Publication No. (NIH) 74-707, Washington, D. C., 1974.
4. Baker, R. W., and Lonsdale, H. K., in "Controlled Release of Biologically Active Agents," pp. 15-71, A. C. Tanquary and R. E. Lacey, Ed., Plenum Press, New York, 1974.
5. Yasuda, H., Lemage, C. E., and Ikenberry, L. D., Die Makromolecular Chemie (1968), 118, p. 19.
6. Kaelble, D. H., J. Appl. Polym. Sci., (1965), 9, p. 1209
7. Lacey, R. E., and Cowsar, D. R., in "Controlled Release of Biologically Active Agents," pp. 117-144, A. C. Tanquary and R. E. Lacey, Ed., Plenum Press, New York, 1974.

15

The Hydrogel–Water Interface

A. SILBERBERG

The Weizmann Institute of Science, Rehovot, Israel

There are very few investigations which deal with the gel-
swelling medium interface specifically. The distribution of
solute components according to size or affinity is considered
in treatments of gel chromatography.   Contact angle measurements have been undertaken routinely (1), but the structure and
the statistical mechanics of this type of interphase has received
only a preliminary treatment (2).
     In selecting the hydrogel-water interface for special con-
sideration, it will have to be realized that it does not possess
features which are not characteristic of the gel-solvent interface in general and that the measure of our ignorance is probably about the same. This is provided, of course, that we put
aside - as we shall do here - such special potential aspects as
charged ionic surface groups and long range electrostatic interactions.
     Still, even without charge interactions, water is a rather
special solvent and the uncharged polymeric species which are
soluble, or swellable, are so, generally, because they possess
groups which hydrate well, i.e. will form a hydrogen bond with
water. Such groups are almost always also capable of forming
hydrogen bonds with each other and solubility, in general, implies
that the hydrogen bond with water is not too weak as compared with
the bond dimerizing two groups. While in water both the solution
and dimerization bond energies tend to be quite large, they are
also about equal and their difference remains small. Solubility,
i.e. swellability, of the material will be assured if the interaction between the polymer segments in the aqueous environment
is only mildly attractive, i.e. does not overfavor dimerization
over hydration (see Figure 1). Similar considerations will apply
to gels in general. Nevertheless, some special aspects govern
aqueous gels which could cause them to behave differently to
other gels.

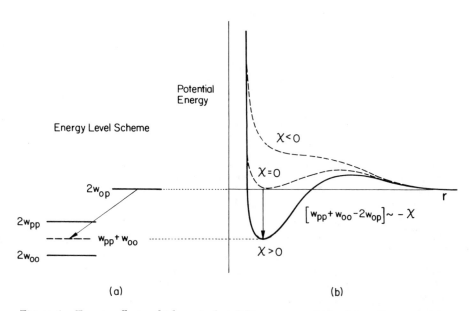

Figure 1. Energy effects which control stability. w_{op}, *energy level for polymer segment bound to solvent;* w_{pp}, *energy level for polymer segment-polymer segment interaction;* w_{oo}, *energy level for solvent–solvent interaction. According to the Flory-Huggins theory of polymer solutions,* $\chi = [2w_{op} - w_{oo} - w_{pp}]/2kT \leq 0.5$ *for solubility. The parameter* χ *is generally found to be in the range between 0 and 0.5. This implies, as the potential diagram (b) shows, that the "pp"-bond is probably stronger than the "op"-bond. If, however, it is too strong, then* $\chi > 0.5$, *and the system is potentially unstable. Hence, the hydrogen bond between gel substance groups can be somewhat stronger than the hydrogen bond of this group with water but not very much stronger.*

Features which Distinguish Hydrogels

Whereas in most non-polar solvents the mixing energies are
due to van der Waals interactions only (both the individual
energy level changes and their differences are small), in aqueous
environments we are dealing with relatively large energy effects.
Hence the same percentage change can produce much larger effects
on solubility in the aqueous medium. Conformational changes and
cooperative interactions between polymer and polymer and polymer
and solvent are thus much more dramatic and much more readily
precipitated in aqueous media than in other environments. A
hydrogel can be regarded as a potentially unstable system whose
state is easily influenced by slight variations in temperature,or
pressure, or by minor chemical modifications of the aqueous environment (e.g. addition of small amounts of such solutes as
neutral salts or alcohols).

To the extent that gradients exist in the gel, particularly
in the surface region, we might thus be faced with the possibility
of large changes in conformation (or degree of interaction between the segments of the gel substance) in going through the
interfacial region between bulk homogeneous gel and bulk water.

The Gel-Solvent Interface

Gels are characterized by the possession of crosslinks.
Generally speaking these are chemical units to which a number
of chains, three or more, are linked in some fashion either by
permanent bonds or by bonds of a sufficiently long lifetime so
that changes in structure do not occur, or occur only very slowly
under stress. These crosslinks, however, need not be defineable
chemical links between separate chains, but could, for example,
be more extensively organized regions containing a large amount
of gel substance between, which a small number of long coiling
polymeric chains establish the connections and hold the gel
together, topologically. These organized regions thus act as
effective crosslinks and the chain segments between them as
the effective network segments. We are comparing the actual gel
with the classical homogeneous ideal, three-dimensional network
model and try to establish correlations. These correlations will
depend upon the experiment considered and it is not to be expected that the division into effective crosslinks and effective
network segments need necessarily be unique. The kind of division which will account for the elastic modulus may not correspond to the division best suited to permeation studies. The
main points to remember are that gel structure is not necessarily
homogeneous,that not each chemical crosslink is an effective
crosslink,but that a meaningful separation into structural
elements is nevertheless possible.

If the presence of the crosslinks defines the presence of
the gel, the zone where the crosslink density (number of

crosslinks per unit volume) goes from its value in the gel
interior to its value, zero, in the water phase, represents the
gel surface, i.e. the interface with the solvent bulk phase.
Ideally we could picture this transition as geometrically abrupt,
but there is, in fact, little reality to an interfacial zone
more sharply located than the mean distance between crosslinks
in the bulk. The macromolecular fuzz, i.e. the region involving
the freely coiling chain segments between crosslinks will also
terminate at the interface, and in part, smooth out the coarse
grain of the crosslink density distribution. Even then it is
most unlikely that the gel surface can be much smoother than
the displacements characteristic of the intercrosslink distance.

Swelling Equilibrium

Since we are considering the open hydrogel–water interface,
swelling equilibrium has been established. This means essentially that the concentration of water in the gel has been
adjusted, by suitably contracting or expanding the distance
between crosslinks, by, in other words, coiling or uncoiling
the macromolecular network segment chains, so that solvent
chemical potential is the same inside and out.

If this experiment were done using a solution containing a
finite number of unlinked macromolecules in place of the gel,
the system would go to infinite dilution. If, however, the
volume of polymer solution were confined inside a closed membrane, which does not permit the polymer component to move out,
water will enter, or will tend to enter, the solution space
raising the pressure until the water chemical potential is
matched. Mechanically, this build-up of pressure is possible
because the hydrostatic pressure can be compensated by a stress
in the membrane. If the membrane is rigid (i.e. possesses
infinite elastic modulus) compensation of forces can be achieved
without volume change. In more practical situations the membrane
has a finite elastic modulus, will tend to distend slightly and
some swelling will take place. The point to note is that a real
pressure difference is established between solution and swelling
medium and that the water molecules in the solution phase are
"aware" of it by being forced closer together, i.e. being somewhat more compressed than in the pure water phase at ambient
pressure.

In the case of the equilibrium swelling of a gel, the gel
material would act as its own membrane. The question arises,
therefore, whether a hydrostatic pressure difference exists
between the interior of the gel and the water phase and whether
a pressure gradient accompanies the concentration transition
through the interface.

This is a more difficult question to answer than it should
be. First of all the literature is full of terms which suggest

that such a pressure arises. Moreover, mechanical work can be
done as a result of swelling. Thus swelling pressure seems to
be an appropriate term to apply when one wishes to characterize
the force which would have to be exerted on the gel in order to
prevent it from swelling in an experiment analogous to the
membrane-polymer solution experiment just discussed. In the
latter case, too, one speaks of a pressure, the osmotic pressure,
which can be translated into a real pressure effect when a
membrane is employed to resist a change in volume of the solution
phase, i.e. to prevent dilution. Osmotic pressure is, however,
not an inherent mechanical feature of the solution, but only a
method of expressing its concentration. To achieve a real pressure effect we have to perform an experiment which will depend
upon the presence of a mechanical device, the membrane, and
another phase, the pure solvent. Only then can a pressure
difference be generated, measured and maintained.

     Now exactly the same is true when a gel is studied under
similar conditions. A porous rigid support for the gel is required. Mechanically it is the equivalent of the membrane in
the osmotic pressure experiment. Through it the gel is put into
contact with the water phase, and by its device the volume of
the gel is confined mechanically. Precisely as in the osmotic
pressure measurement system, water will tend to enter, compression
of the water molecules will occur, the hydrostatic pressure will
rise, and the chemical potential of water inside the compartment
will rise to match that of the water outside. As in the case of
osmotic equilibrium, solvent chemical potential is balanced by a
real pressure rise and the mechanical interaction between the gel
and its porous rigid support provides the means by which water
compression is achieved. There is only one difference. The
pressure effect is not called osmotic pressure but swelling
pressure.

     The situation of swelling pressure measurement is, however,
not the situation of swelling equilibrium. In that case expansion is unresisted and will stop only when conformational changes
in the network segments have so distorted the statistical mechanical partition function of the system and so diluted the
system that the effect of the gel substance solute on the solvent
chemical potential is reduced to zero. Solvent chemical potential
is balanced by changes in solute chemical potential alone and
the need for a pressure rise on the solution is obviated. Indeed,
it would be mechanically impossible to have such a pressure arise
without a device which could compensate this effect mechanically.
True enough the network is distended but there is no force in
the network segments at swelling equilibrium. We can cut such
a gel at any point and no volume change will arise. Only then
in contact with pure solvent will forces arise when the network
is mechanically distorted out of the configuration in which
it is in thermodynamic equilibrium. Only then will

its segments carry force. In contact with a pure solvent phase,
therefore, the configuration of the network in its swelling
equilibrium situation is the reference unstressed nondistorted
configuration. Out of contact with pure solvent, the gel will
have been synthesized, or formed in other ways, at a concentration
above swelling equilibrium. For volume preserving deformations
from this state the system will tend to return to it as its then
reference state, but only if out of contact with solvent. The
reference state of the system is thus the equilibrium state of
the system under the environmental constraints imposed. It has
nothing to do with whether or not the degree of coiling of the
network segments by chance corresponds to the conformational
distribution of that polymer species in free solution.
    We can, moreover, change the reference state by changing
the outside medium. If, for example, as outside medium, an
aqueous solution were used where solute molecules are polymers
too big to enter the gel, a different swelling equilibrium would
be reached and the reference state would correspond to that
situation. Hence any state can be a reference state, but the
outside medium may have to possess rather special properties in
each case.
    It is the approach used by Flory in his treatment of the
isolated macromolecule (3). Here the macromolecule is considered
as a microscopic gel particle and swelling to equilibrium is
allowed to occur. The swollen state of the macromolecule and
not its Gaussian distribution state is now the normal conformational distribution of the segments in that solvent medium.

Segment Distribution in the Interface

    The swelling of the macroscopic gel can thus be considered
analogously to the Flory approach. The principal difference
derives from the fact that the segment distribution is rather
different to start with. Swelling will occur in most cases of
good solvents and the θ-conditions of ideal insolubility will
not be different since the θ-point corresponds, as before, to
the vanishing of the second virial coefficient i.e. to the
transition to critically attractive polymer-polymer interactions. While we can assume that network segment distribution
inside the gel is isotropic this distribution goes to zero over
some interfacial region of a slab thickness corresponding
approximately to the intercrosslink distance.
    A number of problems are, however, here encountered. The
distribution of conformations of a network segment between fixed
crosslinks, allowing for the exclusion of conformations due to
the simultaneous presence of other such network segments, is
not adequately known. Similarly, the distribution of dangling
chain ends from the outermost layer of crosslink points (taking
into account that here too there are network segments going

from these crosslinks to interior crosslink points and will inhibit conformational freedom) has not yet been derived. It may
be assumed that this distribution has some similarity to the
distribution of chains terminally attached to a rigid plane
surface (4). Account will, however, have to be taken of the
fact that the outermost layer of crosslinks does not lie in one
plane and that the dangling chain ends are not uniform in
length.

   We may expect, therefore, that the outermost layer of a
gel surface is much thicker than one intercrosslink distance
and that the surface phase will thus be much more diffuse than
the interior situation of the gel.

   It should be noted that using θ-conditions for the swelling
will give cases where chain statistics most closely approach the
ideal Gaussian random walk.

Relationship between Bulk and Surface Structure

   Laser scattering results from gels (5,6) and some recent
permeability studies (7) both confirm the idea that the internal
structure of gels does not necessarily conform to the simple concept that each crosslink molecule incorporated into the network
structure produces one independent crosslink. Many such links
apparently are wasted simply producing larger chains. Others
occur too close to each other and may only serve to extend the
region which mechanically will act as a single crosslink, all
chemical crosslinks in that region acting as one. Hence many
gels react as though they had far fewer effective crosslinks
than apparently incorporated and as though they had much wider
openings in them than would be suspected from the amount of
gel substance present.

   Effects like these, a coarsening of the grain of the gel,
may also be expected to affect the surface zone commensurately.

Hydrodynamic Effects

   Macromolecules, physically adsorbed or chemically attached
(at one or two points) to solid support surfaces produce a
diffuse macromolecular zone. In attempts to measure the thickness of such layers by the change in dimensions they produce,
it is found that the thicknesses so measured are effectively
much larger than those determined by other methods (9,10).
There is thus a hydrodynamic effect which produces an extra
energy dissipation in the fluid layers adjoining the macromolecular surface.

   Confirmation that flow past a gel,or gel-like,surface
is associated with an extra energy dissipation was obtained
from studies of the resistance to flow through cylindrical

channels in a gel (11). Compared with channels of the same
dimensions with rigid walls, the cylinder in the gel allows
only a reduced throughput. The effect could be correlated with
the elastic modulus of the gel and the thickness of the gel
layer. It seems to be due to oscillations excited in the
interface (12).

It is to be hoped that the interesting properties of the
hydrogel surface which have led to an increasing number of
applications in a variety of fields will also stimulate more
systematic work on their physico-chemical and mechanical
attributes.

Literature Cited:

1. Holly F.J. and Refojo M.F.   Polymer Preprints (1975) 16(2)
    426.
2. Silberberg A. Polymer Preprints (1970) 11, 1289.
3. Flory P.J. "Principles of Polymer Chemistry" (1953)
    Cornell Univ.Press, Ithaca, p.519.
4. Hesselink F.Th. J.Phys.Chem. (1971) 75, 65.
5. Dusek K. and Prins W. Adv.Polymer Sci. (1969) 6, 1-102.
6. Pines E. and Prins W. Macromolecules (1973) 6, 888.
7. Weiss N. and Silberberg A. Biorheology (1975) 12, 107.
8. Weiss N. and Silberberg A. This book.
9. Priel Z. and Silberberg A. Polymer Preprints (1970) 11, 1405.
10. Silberberg A. "Colloques Internationaux du C.N.R.S.
    No. 233 (1975) p.81.
11. Lahav J., Eliezer N. and Silberberg A. Biorheology
    (1973) 10, 595.
12. Hansen R.J. and Hunston D.L. J.Sound and Vibration (1974)
    34, 297.

16

Probing the Hydrogel/Water Interface

JOSEPH D. ANDRADE, ROBERT N. KING, and DONALD E. GREGONIS

Department of Materials Science and Engineering, University of Utah,
Salt Lake City, Utah 84112

The interfacial properties of gel/water interfaces are important in the biomedical applications of hydrogels, particularly
in the areas of blood compatibility, tissue compatibility, and
cell adhesion. The gel/water interface may also be important in
the interfacial chemistry of cell membranes (1).
Very little information is available on gel/water interfaces.
Many of the classical surface chemical methods involving solid/
liquid interfaces are inapplicable to the gel/liquid interface,
because the "solid" is highly deformable and, in general, the
liquid is highly diffusible in the solid. Gels, by definition,
are not sols so classical liquid-liquid interface techniques are
not generally applicable.
Gel/liquid interfaces are thus experimentally and theoretically somewhat frustrating. Nevertheless, they are of such great
practical and scientific importance that they merit careful study.
Silberberg has considered the gel/solvent interface, largely
in a theoretical sense (2,3).
The objective of this paper is to consider a variety of
experimental methods which can provide information on gel/water
interfaces. Much of the discussion will be brief and perhaps
shallow - but that is largely the present state of affairs with
regard to gel/liquid interfaces.

Topography - Structure

The first "measurement" which should be made on a surface
is gross examination, followed by microscopic examination. Is
the surface rough? Does it have an apparent structure or morphology? Is there any apparent orientation? Such questions
must be answered before any subsequent surface characterization
can be meaningful as virtually all surface characterization
techniques are surface roughness or topography dependent. Although a variety of techniques are available for measuring the
topography of hard solid surfaces (4), they are largely inapplicable to gel surfaces.

206

Optical microscopy is very useful (5) and can be used to study in situ surface topography using phase or interference contrast or with water immersion optics. Often, however, one requires higher magnification and must resort to the scanning electron microscope (SEM) or to the transmission electron microscope (TEM).

Simple air drying of the gel followed by metal coating often shows substantial differences in gel topography. Figure 1 shows a series of radiation-grafted poly(hydroxyethyl methacrylate) (PHEMA) coatings on polypropylene (6). The gross differences in topography are quite evident and strongly affect the surface or interfacial characterizations.

Air drying is seldom adequate for studying gel surfaces, particularly for less rigid gels. Most gels exhibit gross changes in structure and shape during drying and transformation to the xerogel state. Though a variety of techniques are available for examining delicate biological structures by SEM and TEM, most generally involve fixation or cross-linking steps, followed by dehydration, and critical point drying from Freon or liquid CO_2 (7). We would prefer to avoid such procedures with gels.

The most accepted and common method of observing highly deformable gels is by freeze-etching. The sample is rapidly frozen, often in liquid Freon, and fractured under liquid nitrogen. The fracture surface can then be directly observed in a cold stage or cold stub-equipped SEM or it can be replicated cold and the replica examined by SEM or TEM (7,8).

Cluthe has examined radiation cross-linked poly(ethylene oxide) (PEO) gels containing 1 to 10% polymer (9) by freeze etch replication. He showed that the structures observed from embedded and sectioned gels were similar to those of the freeze-etch replicated samples. He observed and discussed "cellular" structures for the PEO gels. He also noted that the "fibrillar" structures observed in the TEM by others may result from air drying artifacts.

Blank and Reimschuessel (10) recently reviewed various gel structure theories and presented photographs of polyacrylamide, poly(ethylene oxide), and poly(vinyl alcohol) gels. They considered the liquid phase in the gel in terms of free liquid, capillary or pore retained liquid, and polymer adsorbed liquid, i.e., a three-layer model. This is similar to the "X, Y, Z" water hypothesis of gel water (11). The solid phase was considered in terms of a cellular or micellar theory and a fibrillar theory (also called the "brush-heap" theory (9)). Their photographs clearly document roughness on the 10 micron level in most of the gels examined. High resolution studies on the structure of gelatin gels are also available (12,13). Geymayer has reviewed the problems involved in high resolution studies of freeze etched gels, including cooling rates, segregation and aggregation of diffusible solutes, and instrumentation (14).

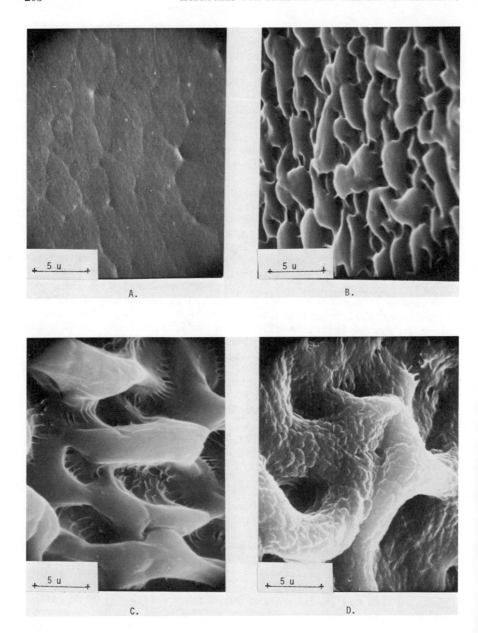

Figure 1.   SEM photos of HEMA grafts on polypropylene (original magnification, 3150 ×). (A) Control; (B) 15% HEMA, 0.25 Mrad; (C) 15% HEMA, 1 Mrad; (D) 20% HEMA, 0.5 Mrad.

We have carried out some very preliminary studies of PHEMA gels using freeze-etch SEM techniques. These studies are too premature to make any conclusions as to gel surface or bulk structure, but they do permit some conclusions about experimental methodology.

One can freeze dry a sample and observe it directly in the SEM at ambient temperatures (Figure 2a). This is generally unsatisfactory due to charging. One can freeze-dry, coat the sample at ambient temperature, and observe at ambient temperatures (Figure 2b). One can freeze the sample and observe it at liquid nitrogen temperatures (charging is not a serious problem as ice is a fair conductor); the sample can be slowly warmed and the ice sublimes away revealing the structure (Figure 2c). The sample of Figure 2c was mounted as indicated in Figure 3.

It is clear, even for a fairly rigid, highly cross-linked PHEMA gel, of relatively low water content, that air drying or freeze drying and ambient temperature examination may not reveal the 'real' structure. Freeze etching certainly has artifacts also, but at this stage it appears to be the method of choice for synthetic aqueous gels (14).

A more reliable procedure than outlined in Figure 3 is to place the sample onto a temperature controlled stage. Such stages are available for most SEM's. The samples can be observed frozen, freeze-etched, micromanipulated, and, if desired, coated under controlled temperature conditions.

Figure 4 is a sequence of photographs of a 70% H_2O - 30% PHEMA opaque gel, mounted and fractured under liquid nitrogen and examined in the ETEC Bio-SEM (ETEC, Inc., Hayward, California). The topography observed may be related to the rate of freeze etching.

We have not examined transparent gels - studies are in progress. Matas, et.al., however, reported some SEM studies of hydrophilic contact lenses (15). Surface scratches and roughness could be readily seen - no apparent bulk structure or porosity could be seen.

The surface roughness of the gels can be determined by stereo pair SEM photographs and stereophotogrammetric analysis (16).

Contact Angle Methods

Contact angle methods are widely used for measuring surface tensions or free energies of liquids and, less rigorously, of solids. Interfacial tensions can often be obtained by contact angle methods.

The traditional techniques for measurement of contact angles rely on determination of the static liquid profile encountered at the three-phase line (TPL) of the solid, a liquid, and a second fluid which may be liquid or vapor. These include (17) (a) the tilting plate method, (b) the sessile drop method,

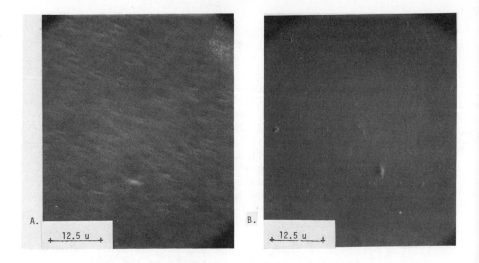

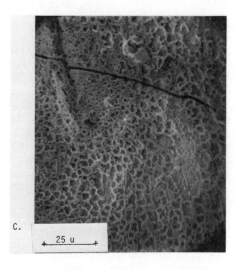

*Figure 2.   A PHEMA gel containing 7.5% ethylene glycol dimethacrylate crosslinker.
The gel is opaque and contains 38.1% wt water. (A) Freeze-dried, stored ambient, and
examined ambient; (B) freeze-dried, coated ambient, and examined ambient; (C) frozen;
freeze etched; examined cold.*

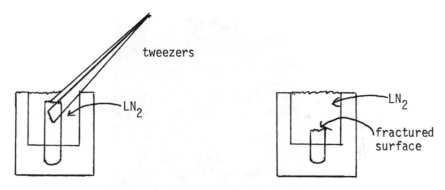

Figure 3.   *Cold stub and method of mounting and fracturing gel samples for freeze etch SEM examination*

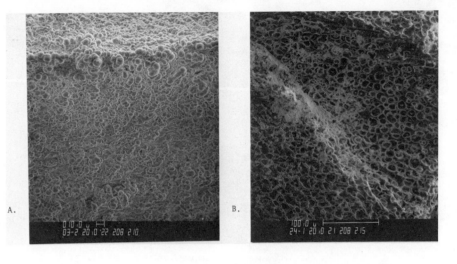

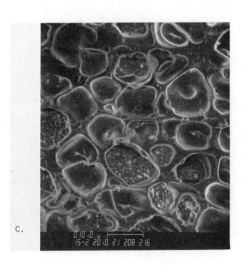

Figure 4. 30% PHEMA–70% water opaque gel freeze fractured in LN$_2$ and observed in the ETEC BioSEM before and during freeze etching. (A) Initial fracture surface and mold surface (top) showing ice crystals due to improper sample handling during fracture and admission to the SEM. (B) A groove produced by the micromanipulator. A cylindrical or cellular morphology is evident in the newly fractured zone. A similar but more random or disturbed morphology is evident outside of the micromanipulated zone. (C) Higher magnification of the micromanipulated zone showing "cells" which appear to be rupturing due to freeze etching.

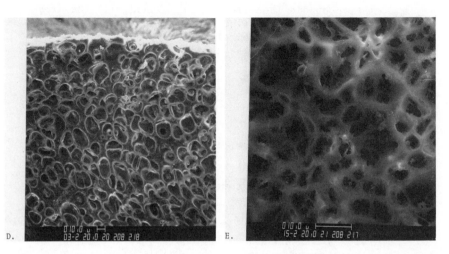

Figure 4. (D) A view of original fracture surface after extensive freeze etching (same areas as (A)). Note similarity with B and C. (E) A view prior to complete freeze etching. Note similarity to those regions of (B) outside the micromanipulated zone. All photos courtesy of ETEC, Inc., Hayward, Calif. All samples uncoated.

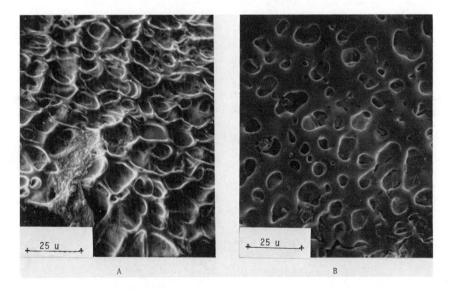

Figure 5. The same as in Figure 4. (A) Examined via the method of Figure 3. (B) Freeze-dried, coated ambient, and examined ambient.

(c) the captive bubble (sessile bubble) method, (d) the drop dimension method and (e) the Wilhelmy plate method. These methods have been adequately described elsewhere (17-20) and have been extensively used to investigate a large number of three-phase systems.

All of the above methods essentially rely on the classical contact angle or Young-Dupree equation:

$$\gamma_{SV} = \gamma_{SL} + \gamma_{LV} \cos\theta_e \quad , \tag{1}$$

where: γ_{SV} = solid/vapor interfacial tension

γ_{SL} = solid/liquid interfacial tension

γ_{LV} = liquid/vapor interfacial tension;

θ_e = equilibrium contact angle.

Several basic assumptions are inherent in this equation. The liquid must be considered incompressible, fluid, and coherent (25). Fluidity means that the application of a shear stress to the liquid must be accompanied by a shear strain, provided the stress is maintained beyond the molecular relaxation time. Coherence implies that the liquid resists a tensile stress until the stress magnitude is sufficient to cause rupture. The surface tension of a liquid is macroscopic evidence of coherence.

The solid surface in contact with the liquid must be considered smooth, homogeneous, isotropic and, most importantly, non-deformable, i.e., rigid in the sense that stress gradients at the TPL are insufficient to deform it significantly.

Equation 1 assumes that thermodynamic equilibria, at all three interfaces (solid/liquid, liquid/vapor, and solid/vapor), have been attained (21-24).

The kinetic interpretation relies on the measurement of dynamic contact angles and observation of the resultant contact angle hysteresis. The kinetics of surface wetting (or non-wetting) make use of the fact that the liquid/vapor interface will change its shape and total area as needed under the influence of the underlying surface in order to maintain a constant curvature at equilibrium. This directly results from the Laplace equation of capillarity (17):

$$\Delta P = \gamma_{LV} \left(\frac{1}{R_1} + \frac{1}{R_2}\right) \quad , \tag{2}$$

where: ΔP = pressure differential between the concave and convex sides of the interface;

γ_{LV} = liquid/vapor interfacial tension; and

R_1, R_2 = principle radii of curvature.

Differences in ΔP from region to region over the interface correspond to gradients in hydrostatic'pressure within the underlying bulk liquid, and therefore tend to cause motion in directions which will diminish these gradients. Such motion or yielding of the interface to meet the Laplacian ΔP requirement is nearly always accompanied by a displacement of the TPL which causes an instantaneous change in the shape and/or area of the liquid/vapor interface. The dynamic contact angle is an instantaneous measure of the angle between the liquid/vapor interface and the solid/liquid interface while the TPL is moving.

Dynamic contact angles may be measured by either spontaneous spreading or forced spreading of a liquid on a surface. In the spontaneous spreading method, the contact angle is observed as a function of time and of distance traversed by the TPL as a drop of liquid spontaneously spreads and approaches the equilibrium contact angle. In this case the driving force is the extent to which the γ_{SV}, γ_{SL}, and γ_{LV} cos $\theta_{inst.}$ are imbalanced. In forced spreading, the liquid/vapor interface is moved relative to the solid surface at a series of constant velocities by application of an external force. At each velocity the contact angle reaches a steady state value and a relationship between the TPL velocity and contact angle is thus established. In this case the driving force in which the TPL undergoes displacement is the result of a pressure-curvature imbalance (25).

The importance of dynamic contact angle measurements lies in the phenomenon of contact angle hysteresis, i.e., the ability to change the observed contact angle of a liquid on a surface without subsequent displacement of the TPL. This may be exhibited on many surfaces by measuring the instantaneous contact angle while, for example, increasing and decreasing the size of the sessile drop. Comparison of the resulting curves of advancing and receding contact angles in the case of most surfaces will show that a hysteresis loop is formed. The most common types of non-ideality are either a homogeneous but geometrically rough surface or a geometrically smooth but heterogeneous surface (19). One may also have both heterogeneities and surface roughness. Practically speaking then, both surface roughness and/or surface heterogeneity, both of which give rise to contact angle variations, can be the major undetected cause of contact angle hysteresis. Swelling, absorption, and molecular reorientation at the interface can also lead to apparent hysteresis (26).

The gel surface may be capable of molecular reorientation during dynamic contact angle measurements, as has been suggested (28).

The gel/liquid interface poses still further deviations from ideality which have yet to be considered. In addition to possibilities of surface roughness and surface heterogeneity, the typical gel surface is extremely deformable. Thus in the case of contact angle characterization of the gel surface, the vertical components of surface tension should cause appreciable distortion in the gel surface at the TPL and could cause gross misrepresentations of the angular measurements. This effect would thus significantly influence dynamic measurements and thus contact angle hysteresis. The contact angles of liquids at deformable solid surfaces have been theoretically treated by several investigators (23,27), but have yet to be experimentally applied to the surfaces of gel systems.

Interface Potentials

Ideally we would like to be able to probe the electrical double layer at gel/solution interfaces. Perhaps the most straightforward way is to use electrokinetic methods (29,30) generally electrophoresis (31), streaming potential (32), or electroosmosis (29). Such measurements allow one to calculate the potential at the shear plane - the zeta potential - given a number of assumptions.

Perhaps the biggest problem is the assumption involving the nature of the fluid dynamic boundary layer in such studies and the position of the shear plane. Even if the gel surface is perfectly smooth, we still have the problem of defining an interface position for a gel consisting of highly mobile segments and chains at the interface (2,3). The shear plane could be outside the interfacial zone, within the zone and free draining, or within the interfacial zone and non-free draining, as discussed by Brooks (33), and others (34). These same problems are present in viscometric or rheologic characterization of gel/liquid interfaces.

One must be particularly careful with streaming potential and electroosmosis measurements with respect to other fluid dynamic assumptions, particularly entrance effects and the establishment of parabolic flow profiles (35).

One can also obtain surface potential information using gel/air measurements, such as with a vibrating reed electrostatic millivoltmeter (36). These methods are commonly used to characterize monomolecular films at the liquid/air interface (37), including synthetic polymers (38). Such methods cannot be easily applied to the gel/solution interface, however.

The interpretation of gel/solution electrokinetic data is far from straightforward. One must of course consider a classical treatment in terms of fixed charges on the polymer "surface" and double layer counterions. In addition, the gel will absorb and partition ions from solution, even if the gel is largely "uncharged" in terms of fixed polymer charges, (this will be discussed later). If ions are partitioned between the gel and the solution, interface potentials will result which will, of course, influence electrokinetic measurements (39).

Adsorption of Polymers

One can study the adsorption of polymers at gel/water interfaces. Much of the work in this area has involved plasma protein adsorption (40-42). We have discussed the protein adsorption behavior of gels previously, in terms of interfacial free energies and water structure considerations (1).

One can obtain information on the nature of gel/water interfaces by adsorbing gel molecules on other substrates and then characterizing the adsorbed polymer/water interface. A large literature on polymer adsorption is available, including the study of water-soluble polymers (43).

The theories of colloidal particle stabilization by adsorbed non-ionic polymers via entropic and enthalpic repulsion (44,45) may be useful in understanding gel/protein and gel/cell interactions.

Effects of stereoregularity on interfacial behavior have been observed in poly(methyl methacrylate), poly(isopropyl acrylate), poly(2-vinyl pyridine 1-oxide) and other polymers (46,67).

The adsorbed layers can then be characterized by the instrumental techniques discussed in this paper.

Partitioning

Gels are very subtle probes of their environment. They will partition ions and other solutes and swell or deswell in response to their solution environments. Of particular importance to their surface properties is ion partitioning in the gels, which may significantly influence interfacial potential and interfacial tension studies.

Ion partitioning in gels has been observed for cellulose (47), cross-linked dextrans (48,49), poly(hydroxyethyl methacrylate) gels (50), and others. Of particular interest to us is the ion concentration profile of the gel/solution interface. Ideally we would like to know the concentrations in the electrical double layer, in the gel/water interfacial region, and in the sub-interfacial zone, perhaps to 1 or 2 microns below the surface. Such measurements are difficult to make by conventional techniques. One approach is to rapidly freeze the gel/water interface, fracture it, and perform an electron microprobe analysis or energy dispersive analysis of x-rays (EDAX) in the SEM using a cold stage to maintain the sample below -130°C to avoid ice crystallization and consequent ion segregation (51). Unfortunately the spatial resolution is limited to 0.1 to 1.0 micron or larger, making a high resolution interfacial region profile very difficult.

Microautoradiography of frozen samples is also possible, but is plagued by technical difficulties. More speculative methods of measuring concentration profiles will be discussed later.

Probing the Outermost Zone

There are very few methods with which to probe the outermost part of the gel/solution interface. We have already discussed the problems with contact angle measurements. Most optical spectroscopy methods, particularly IR, probe a zone of the order of microns in depth (next section). Although ellipsometry can ideally measure a film thickness of a few Angstroms (52), it may be difficult to apply to gels because of the lack of significant refractive index differences between the gel and the surrounding solution.

A variety of techniques are available for directly examining the nature of the surface itself. Only a limited number of these can be discussed here - see Reference 53 for the others. The two techniques most promising for polymeric and biological samples are electron spectroscopy (often called ESCA - electron spectroscopy for chemical analysis) and secondary ion mass spectroscopy (SIMS). Both techniques involve high vacuum environments. The sample must be frozen in liquid nitrogen, generally at T < -130°C. Experience with such methods on aqueous and biological samples is very limited at present.

ESCA was developed largely by the efforts of Siegbahn and his collaborators in Sweden (54). ESCA is based on the precise measurement of the kinetic energy of electrons ejected from the sample by the action of incident radiation, usually x-ray or UV. The binding energy of the electron prior to ejection (E_B) is obtained from the measured kinetic energy (E_k):

$$E_B = E_{h\nu} - E_k - C \ ,$$

where $E_{h\nu}$ is the energy of the monoenergetic excitation radiation (commonly Mg $K\alpha$, 1254 eV), and C is an instrument constant, which is readily determined experimentally. The high precision of the E_k measurement allows one to not only identify the elements present in the surface but also to identify their oxidation state. The photoelectron spectra can be shifted by up to 10 eV, depending on the oxidation state of the element (54,55).

Modern ESCA instruments sample an area of the order of several mm^2. The volume sampled depends on the photoelectron escape depth for the sample. The escape depth is of the order of 5 to 15 A over the energy range of interest for most metals (58). Data for polymers and low density solids are not readily available, though escape depths of the order of 50 to 100 A are generally accepted (58). The sampling depth can be decreased by decreasing the angle the escaping electrons make with the sample surface. This procedure is commonly called the "glancing angle" technique (59). A nondestructive depth profile over the 0-50 A range in polymers can be obtained by intensity ratios of emitted electrons of different kinetic energies (58).

The SIMS technique utilizes a focused ion beam which is rastered or scanned across the surface, sputtering off the outer 0 to 15 or so Angstrons of the surface and analyzing the sputtered ions by a very sensitive mass spectrometer. All elements and their individual isotopes can be detected. An elemental or isotopic image of the surface can be obtained with better than one micron resolution. The interface can be progressively ion-etched away and reanalyzed, providing a compositional analysis into the sample with about 100 A depth resolution. SIMS is in reality a destructive technique, as the ion beam continually sputters away the outer surface.

The SIMS method has been extensively applied to inorganic samples (56). Very limited application to biological samples is also underway (57).

The qualitative interpretation of ESCA spectra is relatively straightforward. Depth profiling of organics, using ESCA or SIMS, however, is speculative at this time and requires a great deal of empirical calibration work before it can be very useful.

Probing the Subsurface Zone

The most common method is the use of multiple internal re-flection infrared spectroscopy (60). Some depth dependence is available using different angles of incidence, though the distance probed is in the micron range. Careful studies have given monolayer sensitivities of known monolayers deposited directly on the internal reflection elements (IRE's).

The recent development of Fourier transform infrared spectroscopy (FT-IR) (61) overcomes many of the intensity and sensitivity limitations of conventional dispersive infrared spectroscopy, particularly for aqueous solutions (62,63). FT-IR can also be used in the multiple internal reflection mode (64). To our knowledge FT-IR has not yet been applied to gel/solution interface studies, however.

Raman spectroscopy should be very useful in characterizing gel/solution interfaces directly, as water is an ideal solvent for Raman studies (65).

Elemental analysis of the subsurface zone can be accomplished by energy dispersive analysis of x-rays in the SEM, as previously mentioned, or by the more accurate wavelength dispersive analysis of most electron microprobes (51). Subsurface penetration is in the micron region and can be adjusted somewhat by varying the incident electron beam energy. Again the sample must be rapidly frozen to liquid nitrogen temperatures and maintained below -130°C for analysis (51).

Further Discussion

Other techniques are available for study of the gel/water or gel/solution interface. Most solid surface characterization

techniques (53) can, in principle, be applied to the frozen gel
surface. Practically all of the microscopic and histologic
techniques used for the study of cell surfaces can also be applied, including selective fixatives, stains, etc. (7). Some of
the methods used in the study of liquid/liquid interfaces (17,66)
can be applied to gel interfaces. We have also ignored many
techniques which are generally applicable only for high surface
area systems.

We have discussed primarily those techniques with which we
are familiar or which we are considering to apply to the study of
gel/solution interfaces.

Conclusions

A variety of methods are available for the study of gel/
solution interfaces.
Direct in situ methods include:
    1. rheologic or viscometric analysis
    2. ellipsometry
    3. contact angles
    4. electrokinetics
    5. infrared spectroscopy
    6. Raman spectroscopy
    7. optical microscopy
Dry gel/air interface methods include:
    1. infrared spectroscopy
    2. scanning, transmission, and optical microscopy
    3. surface potential
    4. contact angles
    5. ESCA
    6. SIMS
Frozen gel surfaces can be studied by:
    1. scanning, transmission, or optical microscopy
    2. electron microprobe or EDAX
    3. ESCA
    4. SIMS
and many of the other methods.
These methods permit one to probe the gel/solution interface
with respect to:
    1. interface energetics
    2. interface potentials
    3. interface chemical groups and orientations
    4. interface structure or morphology
    5. interface elemental composition
The subsurface zone can be analyzed for:
    1. interface chemical groups and orientations
    2. interface elemental composition
In addition, soluble gel molecules can be studied as adsorbed films at solid/liquid interfaces or liquid/air interfaces.

Acknowledgements

Stimulating discussions with S. W. Kim, M. S. Jhon, Y. K. Sung and D. L. Coleman have been most helpful. The personnel of ETEC, Inc., Hayward, California were most generous with Bio-SEM time with which our first freeze-etch studies were done. The technical assistance of R. Middaugh, G. Iwamoto, and D. Kramer is appreciated. This work has been supported by NIH Grant HL 16921-01.

Abstract

The experimental characterization of gel/water interfaces is briefly discussed. Interfacial characteristics discussed include topography/morphology, interfacial tension or free energy, interface potential, adsorption, partitioning, the chemical nature of the gel surface, and the nature of the subsurface region. Techniques briefly discussed include microscopy, contact angle methods, electrokinetic methods, surface potentials, infrared spectroscopy - including Fourier transform and Raman, x-ray photoelectron spectroscopy (ESCA), and secondary ion mass analysis (SIMS). Most of these techniques are discussed in suggestive and speculative terms as so few have been applied to the gel/water interface. A variety of techniques are available for studying the gel/water interface either <u>in situ</u> or in the frozen state.

Literature Cited

1. Andrade, J. D., Lee, H. B., Jhon, M. S., Kim, S. W., and Hibbs, Jr., J. B., <u>Trans. Amer. Soc. Artif. Internal Organ.</u>, (1973) <u>19</u>, 1.
2. Silberberg, A., <u>Polymer Preprints</u>, (1970) <u>11</u>, 1289.
3. Silberberg, A., <u>This Symposium</u>.
4. Whitehouse, D. J., in P. F. Kane and G. B. Larrabee, eds., "Characterization of Solid Surfaces," Plenum, 1974.
5. McCrone, S. C., in P. F. Kane and G. B. Larrabee, eds., "Characterization of Solid Surfaces," Plenum, 1974.
6. Lee, H. B., Shim, H. S. and Andrade, J. D., <u>Polymer Preprints</u>, (1972) <u>13</u>, 729.
7. Koehler, J. D., ed., "Advanced Techniques in Biological Electron Microscopy," Springer-Verlag, 1973.
8. "Freeze Etch Replication," E. M. Ventions, Rockville, Maryland.
9. Cluthe, C. E., "Microscopical Study of the Structure of Radiation Cross-linked Aqueous Poly-(ethylene oxide) Gels and Frozen Cryoprotective Agents," Ph.D. Thesis, Cornell University, May 1972.
10. Blank, Z. and Reimschuessel, A. C., <u>J. Materials Science</u>, (1974) <u>9</u>, 1815.

11. Jhon, M. S. and Andrade, J. D., J. Biomed. Materials Res., (1973) 7, 509.
12. Belavtseva, Y. M., Titova, Y. F., Braudo, Y. Y. and Tolstoguzov, V. B., Biophysics, (1974) 19, #1, 15. (English translation of Biofizika: 19: No. 1 (1974) 19
13. Bohoneck, J., Coll. and Polymer Sci., (1974) 252, 417.
14. Geymayer, W. F., J. Polymer Sci. Symposium, (1974) 44, 25.
15. Matas, B. R., Spencer, W. H., and Hayes, T. L., Arch. Ophthal., (1972) 88, 287.
16. Johari, O. and Samudra, A. V., in P. F. Kane and G. B. Larrabee, eds., "Characterization of Solid Surfaces," Plenum, 1974.
17. Adamson, A. W., "Physical Chemistry of Surfaces," 2nd ed., Wiley, 1967.
18. Bikerman, J. J., "Physical Surfaces," Academic Press, New York, 1970.
19. Neumann, A. W., Adv. Coll. Interface Sci., (1974) 4, 105.
20. Wu, S., J. Macromol. Sci.-Revs. Macromol. Chem., (1974) C10, 1, 1.
21. Bikerman, J. J., Proc. 2nd Intern. Congress on Surface Activity III, (1957) 125.
22. Johnson, R. E., J. Phys. Chem., (1959) 63, 1655.
23. Lester, G. R., J. Coll. Sci., (1961) 16, 315.
24. Hansen, R. J. and Toong, T. Y., J. Coll. Interf. Sci., (1971) 37, 196.
25. Schwartz, A. M., Adv. in Coll. and Interf. Sci., (1975) 4, 349.
26. Timmons, C. O. and Zisman, W. A., J. Coll. Interf. Sci., (1966) 22, 165.
27. Braudo, Y. Y., Michailow, E. N. and Tolstoguzov, V. B., Phys. Chemie, Leipzig, (1973) 253, 369.
28. Holly, F. J. and Refojo, M. F., J. Biomed. Materials Res., (1975) 9, 315-326.
29. Davies, J. T. and Rideal, E. K., "Interfacial Phenomena," 2nd ed. Academic Press, 1963.
30. Shaw, D. J., "Electrophoresis" Academic Press, 1969.
31. Nordt, F. Knox, R. J., and Seaman, G. V. F., This Symposium.
32. Ma, S. M., Gregonis, D. E., Van Wagenen, R. and Andrade, J. D., This Symposium.
33. Brooks, D. E., J. Coll. Interf. Sci., (1973) 43 687.
34. Kavanagh, A. M., Posner, A. M., and Quirk, J. P., Faraday Disc., No. 59, (1975).
35. Van Wagenen, R., Andrade, J. D. and Hibbs, Jr., J. B., submitted to the J. Electrochem. Soc.
36. Instruction Manual for Model 162 Electrostatic Millivolt Meter, Monroe Electronics, Inc., Middleport, New York.
37. Gaines, G. L., "Insoluble Monolayers at Liquid-Gas Interfaces," Interscience, 1966.
38. Beredick, N., In. B. Ke, ed., "Newer Methods of Polymer Characterization," Wiley, 1964.

39. Johansson, G., Biochem. Biophys. Acta, (1970) 221, 387.
40. Holly, F. J. and Refojo, M. F., This Symposium.
41. Horbett, T. A. and Hoffman, A. S., Adv. in Chem. Series #145, "Applied Chem. at Protein Interfaces," R. E. Baier, ed., ACS 1975, 230.
42. Kim, S. W., unpublished work.
43. Lipatov, Yu, S. and Sergeeva, L. M., "Adsorption of Polymers," J. Wiley and Sons, 1974. (Translated from Russian by R. Kondor).
44. Bagchi, P., J. Coll. Interfac. Sci., (1974) 47, 86.
45. Hesselink, F. Th., Vrij, A. and Overbeek, J. Th. G., J. Phys. Chem., (1971) 75, 2094.
46. Botham, R. and Thies, C., J. Coll. Interf. Sci., (1969) 31, 1.
47. McGregor, R. and Ezuddin, K. H., J. Appl. Polymer Sci., (1974) 18, 629.
48. Kalasz, J., Nagy, J., and Knoll, J., J. Anal. Chem., (1974) 272, 22.
49. Marsden, N. V. B., Naturwissenschaften, (1973) 60, 257.
50. Adamcova, Z., Coll. Czech. Chem. Comm., (1968) 33, 336.
51. Hall, T., Echlin, P., and Kaufmann, R., eds., "Microprobe Analysis as Applied to Cells and Tissues," Academic Press, 1974.
52. Muller, H., Adv. Electrochem. and Electrochem. Engr., (1973) 9, 167.
53. Kane, P. F. and Larrabee, G. B., eds., "Characterization of Solid Surfaces," Plenum, 1974.
54. Siegbahn, K., et.al., "ESCA--Atomic, Molecular and Solid State Structure Studied by Means of Electron Spectroscopy," Almquist and Wiksells, Publ., Uppsala, Sweden, 1967.
55. Dekeyser, W., Fiermans, L., Vanderkelen, G., and Vennik, J., "Electron Emmission Spectroscopy," D. Reidel Publ. Co., 1973.
56. Werner, H. W., Surface Science, (1975) 47, 201.
57. Galle, P., in Reference 51.
58. Clark, D. T., Feast, W. J., Musgrove, W. K. R., Ritchie, I., J. Polymer Sci., Polymer Chem., (1975) 13, 857.
59. Fadley, C. S., Baird, R. J., Siekhaus, W., Novakov, T., and Bergstrom, S. A. L., J. Electron Spect. Related Phen., (1974) 4, 93.
60. Harrick, N. J., "Internal Reflection Spectroscopy," Interscience, 1967.
61. Koenig, J. L. and Tabb, D. L., Canad. Res. and Dev., (Sept-Oct, 1974) 25.
62. Low, M. J. D. and Yang, R. T., Spectrochem. Acta, (1973) 29A, 1761.
63. Low, M. J. D. and Yang, R. T., Spectroscopy Letters, (1973) 6, 299.
64. Yang, R. T., Low, M. J. D., Haller, G. L. and Fenn, J., J. Coll. Interf. Sci., (1973) 44, 249.

65. Buechler, E. and Turkevich, J., J. Phys. Chem. (1972) 76, 2325.
66. Albertsson, P. A., "Partition of Cell Particles and Macromolecules," 2nd ed., Wiley, 1971.
67. Dobreva, M., Dancheva, N. and Holt, P. F., Brit. J. Ind. Med., (1975) 32, 224.

Elimination of Electroosmotic Flow in Analytical Particle Electrophoresis

F. J. NORDT, R. J. KNOX, and G. V. F. SEAMAN

Department of Neurology, University of Oregon Health Sciences Center, Portland, Oreg. 97201

Interest in surface coatings which will markedly reduce or eliminate the zeta potential at a chamber wall stems from the practical issue of eliminating electroosmotic flow during electrophoresis. In a closed cylindrical glass electrophoresis chamber containing an electrolyte the negative charge at the glass wall results in an increase in concentration of cations close to this surface. Application of an external electrical field results in movement of fluid near the wall (electroosmosis) toward the cathode and a concurrent forced return flow through the center of the tube. It can be shown from hydrodynamics that there is a cylindrical envelope (stationary level) in the chamber where no net flow of fluid occurs during electrophoresis. Figure 1 illustrates the general features of laminar electroosmotic fluid flow for a closed cylindrical tube including the parabolic fluid flow profile, regions of electroosmotic flow, return fluid flow and location of the stationary level.

In analytical particle electrophoresis true electrophoretic velocities of particles may be measured at the stationary level while velocities determined elsewhere in the chamber will be comprised of contributions from both electrophoresis and electroosmosis. In preparative applications of electrophoresis the boundary of a concentrated suspension (sample) becomes paraboloidal in contour as a result of electroosmosis of the suspending medium in the chamber. The non-planar sample distribution introduces difficulties in separating or resolving particle populations which differ in electrophoretic mobility. In analytical particle electrophoresis the presence of electroosmotic flow requires that measurements be carried out at the stationary level. Since this level is infinitely thin electroosmotic flow will always contribute to experimental error.

The wall charge in electrophoresis chambers arises from either the ionization of surface charge groups or as a consequence of redistribution of ions from the suspending medium (adsorption or desorption). The wall charge may be reduced or eliminated by: a) use of adherent or adhesive films (1).

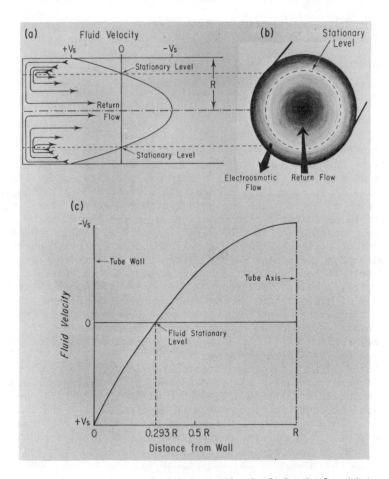

Figure 1. *Electroosmotic fluid flow in a closed cylindrical tube.* (a) A longitudinal crossectional view of the fluid velocity profile and fluid streamlines for a tube with radius, R (fluid velocity is plotted in terms of V_s, the fluid flow at the tube wall). (b) A transverse crossection where maximum electroosmotic fluid flow is shown by dense shading, and the unshaded area shows the region of the stationary level where fluid flow tends to zero. (c) Fluid flow parabola which is a plot of fluid velocity vs. distance from the tube wall.

b) covalent bonding materials (2).

c) physical adsorption of substances (3).

Electrophoretic testing of the stability and completeness of different surface treatments or coatings may be carried out in various ways:

i) After coating the inside of the electrophoresis chamber the electroosmotic flow is calculated from experimental measurements of the electrophoretic velocities of standard particles at various distances from the tube wall.

ii) The electrophoretic mobilities of coated or modified particles made from the same materials as the electrophoresis chamber are measured by standard analytical electrophoresis.

iii) The zeta potential of coated tubes may be also determined from electroosmotic flow or streaming potential measurements.

The first approach is the more desirable test of any coating procedure but for screening purposes the second test approach will significantly reduce the time needed for the initial survey or testing of coatings or modification procedures.

Theoretical

The electrophoretic mobility, u, is defined as the electrophoretic velocity, v, of a particle per unit field strength, χ:

$$u = \frac{v}{\chi} \tag{i}$$

The relationship between electrophoretic mobility, u, and zeta potential, ζ, for nonconducting particles whose radius of curvature, a, is large in comparison to the effective thickness of the electrical double layer, $1/\kappa$, is usually described accurately for $\kappa a > 300$ by the Helmholtz-Smoluchowski equation (4):

$$u = \frac{\zeta \varepsilon}{4 \pi \eta} \tag{ii}$$

where ε and η are the dielectric constant and viscosity, respectively, within the electrical double layer which are assumed to be the same as the bulk values of the suspending medium. Experimental measurements of v result in an observed electrophoretic velocity, v_e, which is subject to error as is the experimentally derived electrophoretic mobility, u_e.

Application of an electric field to a suspension of charged particles contained in a closed cylindrical chamber with a charged wall results in electrophoresis of the particles and electroosmotic flow of the suspending medium. The observed velocity, v_o, of a particle in the tube is thus the sum of its electrophoretic velocity, v_e, and the velocity of the suspending medium, v_w:

$$v_o = v_e + v_w \qquad \text{(iii)}$$

It has been shown by Bangham et al. (5) that the observed velocity of the particle is related to its distance, r, from the axis of the tube by the expression:

$$v_o = v_e + v_s[\frac{2r^2}{R^2} - 1] \qquad \text{(iv)}$$

where v_s is the fluid velocity adjacent to the tube wall and R is the radius of the tube. Solution of the flow equation (iv) shows that at a distance r = 0.707R from the axis there is no net flow of fluid, i.e., a stationary level.

In theory electrophoretic mobility measurements which are made at the stationary level are not subject to error as a result of fluid flow. However, in practice errors arise as a result of:

a) the finite size of the particles which cannot be contained in an infinitely thin stationary level;

b) focusing errors at the stationary level (requirement for appropriate optical corrections to rectify the effects of refraction and aberration, depth of focus, and shape and size of focal field relative to the radius of curvature of the stationary level);

c) heterogeneous distribution of charge on the tube wall resulting in a shift in location of the stationary level;

d) Brownian motion and sedimentation of the particle; and

e) thermal convection arising from Joule heating.

The magnitude of errors in the electrophoretic velocity as a result of fluid flow may be estimated from a differentiated form of equation (iv):

$$dv_o = \frac{4v_s r}{R^2} dr \qquad \text{(v)}$$

Division of equation (v) by the field gradient, χ, gives an expression for the change in the experimentally observed electrophoretic mobility at small radial increments from the stationary level in terms of the electroosmotic mobility, u_s:

$$\Delta u_o = \frac{4u_s r}{R^2} \Delta r \qquad \text{(vi)}$$

If the fractional error in u_e due to fluid flow is defined as δ, then:

$$\delta = \frac{\Delta u_o}{u_e} = \frac{4u_s r}{R^2 u_e} \Delta r \qquad \text{(vii)}$$

From expression (vii) one may calculate the maximum value of u_s which will produce fractional errors of less than δ in u_e at distances up to Δr from the stationary level.

Materials and Methods

Corning #7740 borosilicate particles were used as a model system for screening the effectiveness of polysaccharide derivatives as low zeta potential surface coatings for glass. All chemicals were reagent grade and the water was distilled twice in pyrex ware.

The particles were prepared by grinding Corning #7740 glass tubing with water in an aluminum oxide ball mill for 16 hours. Particles of suitable size for electrophoretic measurements were obtained by repeated sedimentation followed by removal of particles having a sedimentation rate greater than 1 mm/3-4 min. Particles thus obtained were $\leq$ 5 μm in diameter. These were transferred to glass centrifuge bottles and cleaned with hot aqua regia followed by hot 6N HCl. Possible organic contaminants were removed by sequential treatment of the particles with toluene and isopropanol. The particles were then washed six times with 0.015 M NaCl at 60-70°C.

Further cleaning treatments failed to increase the electrophoretic mobility of the particles which was monitored throughout the cleaning procedure.

Coating Materials. Diethylaminoethyl-methylcellulose (DEAE-methylcellulose) was prepared from methylcellulose (Dow Methocel MC, 8000 cps Premium, Lot MM-110786) by a scaled down modification of the method of Peterson and Sober (6). Air oxidation of carbonyl groups on the methylcellulose was prevented by the addition of sodium borohydride (0.5 g per 100 g of methylcellulose). Acetone was used to precipitate the polymer after the addition of the 2 M NaCl. The product was then filtered and exhaustively dialysed against water to pH 7 in Union Carbide dialysis tubing. The resulting DEAE-methylcellulose was a clear gel which was lyophilized and dispersed in a Waring blender.

The extent of methylcellulose modification was controlled by adjusting the amount of diethylaminoethyl chloride used in the alkylation reaction (6). Two batches of 60 grams each of methylcellulose were modified with 7 grams and 35 grams, respectively, of diethylaminoethyl chloride. These preparations will be referred to as 'lower' and 'higher' degree amino substituted methylcellulose derivatives.

The nitrogen contents of the DEAE derivatives were determined for the lyophilized materials by the micro-Kjeldahl method (7). The effectiveness of methylcellulose and its derivatives as low zeta potential coating agents was examined by suspending glass particles in a 0.015 M NaCl solution containing a 0.1% (w/v) concentration of a given polymer and measuring the electro-

phoretic mobility of the glass particles at 25°C in an all glass
cylindrical chamber apparatus equipped with either grey platinum
electrodes (5) or with reversible Ag/AgCl electrodes as described
by Seaman and Heard (8).

As a test of the tenacity of the coating, the particles were
washed one to three times in dilute saline media. The effect of
the washes was assessed by measuring the electrophoretic mobility
of the particles in 0.015 M NaCl-NaHCO$_3$, pH 7.2 ± 0.2 and noting
any changes.

In attempts to improve the stability of the coating, epichlorohydrin was employed as a potential crosslinking agent for
DEAE-methylcellulose. A modification of Flodin's procedure (9)
was used. A 0.2% (w/v) solution of the higher modified DEAE-
methylcellulose was prepared to which glass particles were added
to make a milky white suspension. The particles were subsequently centrifuged, the supernatant decanted and the particles
suspended in 1.4 M sodium hydroxide to which sodium borohydride
(0.5 g per 100 g DEAE-methylcellulose) had been added to prevent
oxidation of the polymer. This suspension was mixed with redistilled epichlorohydrin, 0.2% w/v (Eastman Kodak), and incubated
at 70°C in a water bath for 4 hours. The particles were subsequently centrifuged, washed three times in 0.015 M NaCl-NaHCO$_3$,
pH 7.2 and subjected to electrophoresis. The effect of further
washes on the electrophoretic mobility of the particles was
monitored.

Finally the internal surface of an all glass cylindrical
electrophoresis chamber was coated with DEAE-methylcellulose and
then treated with epichlorohydrin. The chamber was filled with a
0.2% w/v solution of DEAE-methylcellulose in 1.4 M sodium hydroxide containing 0.5 g sodium borohydride per 100 g DEAE-methylcellulose and 10% (v/v) epichlorohydrin. The chamber was
incubated in a 70°C water bath overnight. The chamber was then
thoroughly rinsed out with 1 M KCl and set up in the usual manner
(10). The effect of the coating on the electroosmotic flow was
assessed by measuring the electrophoretic mobility of native
glass particles in media of varying ionic strengths at a series
of distances from the inner chamber wall.

Results

In analytical particle electrophoresis mobility measurements
are in principle made at the stationary level of the electrophoresis chamber where no electroosmotic fluid flow influences
the measurements. The pattern of electroosmotic fluid flow has
been described for closed systems of various geometries (11) so
that it is possible to calculate the location of the stationary
level for ideal conditions. In practice, mobility measurements
are only made in the vicinity of the real stationary level for
several reasons. Typically, the location of the real stationary
level is not experimentally defined, but rather is calculated for

ideal conditions from the electroosmotic flow equations and used
in the adjustment of the apparatus.   A lack of appropriate optical refraction or aberration corrections, small errors in
instrument settings, or shifts of the location of the real stationary level from the theoretical location due to nonideal
conditions such as contamination of the chamber surface all lead
to errors in locating the real stationary level.   Even if the
apparatus is optimally focused, the particles under examination
will not remain at the stationary level for the duration of the
measurements for the reasons outlined in the theoretical section,
so that electroosmosis when present is a significant source of
experimental error.

The magnitudes of these errors in the mobility measurements
when the particles are not at the stationary level may be estimated for small distances from the stationary level by means of
equation (vii).   The results of these estimations are shown in
Figure 2 which is interpreted as follows.   If the electrophoretic
mobility, u_e, is measured for particles at 20 μm from the stationary level in a 2 mm diameter capillary, there will be a
fractional error, δ, in the mobility value obtained, the magnitude
of which will depend on the relative magnitudes of the electroosmotic mobility, u_s, and u_e.   Thus for a particle with $u_e = 2.0$,
and $u_s = 2.1$, the measured electrophoretic mobility at 20 μm from
the stationary level would deviate from the true mobility by a δ
of 0.06 or 6%.   In a cleaned borosilicate glass capillary at
neutral pH, u_s ranges from 5-6 μm sec^{-1} $volt^{-1}$ cm at an ionic
strength of 0.015 and from 2 to 3 at 0.15 ionic strength.   It can
be seen that electroosmotic mobilities of such magnitude give
rise to appreciable errors in the u_e values obtained for particles at distances from the stationary level likely to be encountered experimentally.

The estimation in Figure 2 is particularly useful in esta-
blishing what reduction of u_s will reduce errors to an acceptable
degree for mobility measurements made near the stationary level.
For mobility measurements made at the calculated stationary level
it is not necessary to reduce u_s to zero in order for the errors
to be small unless the electrophoretic mobilities to be measured
are near zero.   In practice, any substantial reduction of u_s
facilitates data collection, reduces operator errors in judging
whether particles are in focus, and minimizes the influence of
changes in the location of the stationary level due to contamination of the wall surface by adsorbed substances or to heterogeneous distribution of charge on the uncontaminated wall.   In
preparative particle electrophoresis or in automated analytical
techniques where measurements are made at locations other than
near the stationary level, it is necessary to reduce u_s to zero.

Methylcellulose and two preparations of DEAE-methylcellulose
which each contained different levels of amino group substitution
were tested as coating agents for cleaned borosilicate glass.
The nitrogen content of the low and higher amino substituted

methylcellulose derivatives were found to be 0.37 and 1.28 mg
per gram, respectively, by micro-Kjeldahl assay.

The electrophoretic mobilities of bare borosilicate glass
particles and particles which have been subjected to various
treatments are presented in Table I. It should be noted that
the relatively high zeta potential ($\sim$ -70 mV) calculated from
equation (ii) for the bare glass particles will result in an
average pH at the surface of the glass particles of about 1.2
units lower than the bulk value (12). The washing experiments
show that the methylcellulose is readily washed off the particles
whereas the DEAE-methylcellulose is adherent and survives washing. The weak adsorption of the methylcellulose results in a
significant increase in zeta potential of the borosilicate
particles, a phenomenon which has been observed for other weak
and reversibly adsorbed neutral polymers such as dextrans (13).

In Figure 3 the electrophoretic mobility versus pH relationships are presented for native borosilicate glass particles and
epichlorohydrin crosslinked DEAE-methylcellulose coated particles. Evidence for an appreciable degree of crosslinking is
provided by the formation of a water-insoluble gel-like material
which develops after addition of epichlorohydrin to the reaction
mixture. In contrast the precipitate of the polymer which is
produced by the addition of sodium hydroxide alone is water
soluble. The pH versus mobility relationship for bare glass
suggests the presence of two or more surface ionogenic groups.
The crosslinked DEAE-methylcellulose coated particles exhibit a
positive branch to the pH versus mobility relationship at low
pH and an increase in mobility between pH 9 and 10 consistent
with the presence of a positive group of pK ca. 9.5.

Figure 4 shows the electrophoretic velocity data on bare
glass particles as a function of their radial position within
the electrophoresis tube coated with epichlorohydrin crosslinked
DEAE-methylcellulose. Consistent electrophoretic velocities
were obtained for up to 15 hours. Calculation of the electroosmotic flow at the wall of the chamber by means of equation
(iv) yielded values which ranged between -0.09 and +0.12 μm
sec^{-1} $volt^{-1}$ cm over the range of ionic strengths investigated.
The dashed line in Figure 4 represents the electrophoretic
velocity of the bare glass particles in an uncoated glass
chamber. Use of equation (iv) for the results obtained from a
bare glass chamber yield an average value for the electroosmotic
flow of +5.9 μm sec^{-1} $volt^{-1}$ cm corresponding to an electrophoretic mobility of -5.9 which agrees well with the observed value of
-5.6 μm sec^{-1} $volt^{-1}$ cm for bare borosilicate glass particles.
After twenty-four hours the chamber coated with crosslinked DEAE-
methylcellulose showed an increase in negative surface charge
which resulted in an electroosmotic flow of +1.21 μm sec^{-1} $volt^{-1}$
cm for the aqueous 0.015 M NaCl medium as a result of slow
desorption of the coating.

TABLE I

ELECTROPHORETIC MOBILITIES OF BOROSILICATE GLASS
PARTICLES FOLLOWING VARIOUS TREATMENTS

Treatment	pH	Observed[a] $\bar{u} \pm s$ (n)	Corrected[b] $\bar{u}$
None	7.2	-5.55 ± 0.25 (80)	
Methylcellulose 1 x wash[c]	7.2	-3.65 ± 0.22 (30) -5.67 ± 0.22 (20)	-8.40
DEAE Methylcellulose[d] 1 x wash[c] 3 x wash[c]	8.1 7.2 7.2	$+0.60 \pm 0.04$ (20) $+0.76 \pm 0.05$ (20) $+0.70 \pm 0.04$ (20)	$+1.23$
DEAE Methylcellulose[e] 1 x wash[c] 3 x wash[c]	8.0 7.2 7.2	$+1.21 \pm 0.04$ (20) $+0.93 \pm 0.05$ (20) $+0.96 \pm 0.05$ (20)	$+2.65$

a. Electrophoretic mobility, $\bar{u}$, in μm sec^{-1} $volt^{-1}$ cm ± sample standard deviation (n = no. of measurements) for particles at 25°C in 0.015 M NaCl only or with added polymer.
b. Corrected for the bulk medium viscosity effects of the polymer derivatives which were present at a concentration of 0.1% (w/v).
c. Washed samples had been suspended in > 100 volumes of 0.015 M NaCl-NaHCO$_3$ pH 7.2 ± 0.1 containing no coating material, centrifuged and resuspended in fresh saline the number of times indicated.
d. Low level amino group substituted methylcellulose.
e. Higher level amino group substituted methylcellulose.

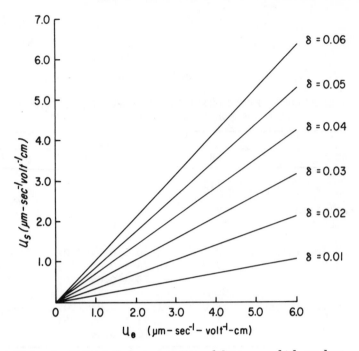

Figure 2.   Estimated electroosmotic mobilities, u_s, which produce fractional errors, δ, in the absolute values of electrophoretic mobility, u_e at 20 μm from the stationary level in a 2.0-mm bore capillary tube.

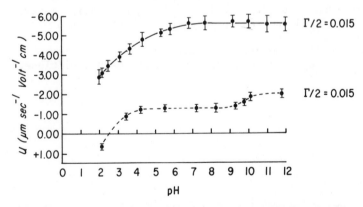

Figure 3.   The electrophoretic mobility of bare borosilicate glass particles (——) and crosslinked DEAE–methylcellulose coated particles (– – –) as a function of pH of the suspending medium. Error bars indicate the standard deviation. Migration toward the cathode is positive. The ionic strength of the suspending medium ($\Gamma/2$) is 0.015 g ions l.$^{-1}$.

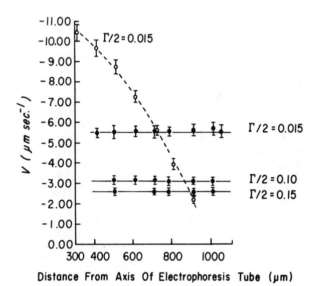

Figure 4.   Electrophoretic velocities of bare glass particles (——) in a crosslinked DEAE–methylcellulose coated cylindrical glass electrophoresis chamber at various radial positions inside the tube. The dotted line represents the same system for an uncoated glass chamber. The ionic strengths ($\Gamma/2$) at which each series of measurements were carried out are indicated.

Discussion

　　　The aim of the work reported in this paper is the development of a stable coating with low or zero zeta potential for the
surface of electrophoresis chambers in order to minimize or
eliminate the perturbing influence of electroosmosis. In surface
modification or coating work it is important that the material to
be coated be well defined and capable of being cleaned by some
standard procedure so that baseline or control electrokinetic
data are readily obtainable. The use of plastics such as poly-
methylmethacrylate (plexiglas) or polycarbonates in coating
studies is complicated by the various anti-oxidants and other
additives which result in wide variations in the surface properties for ostensibly the same product obtained from various
sources or lots. In comparison to glasses, cleaning methods for
plastics and their evaluation for success are much more difficult to devise. Taking these factors into consideration as well
as the extensive use of glass in the fabrication of electrophoresis chambers, efforts were focused on coating pyrex glass
tubes and particles. The nature of borosilicate glass surfaces
and the physical and chemical criteria required for an optimum
coating designed to eliminate electroosmosis determine the
selection of potential coating agents.
　　　The negative charge of borosilicate glass surfaces is
reported to originate from the ionization of surface groups
which are an integral part of the glass. Two types of sites
are present, one with a pK_a of approximately 7 and the second
with a pK_a of 5.1 (14). The weaker acidic site is due to the
silanol group $\geqslant$Si-OH, while the stronger acidic site is $\geqslant$B-OH.
It should be noted that the electrophoretic mobility of a particle and electroosmotic fluid flow in a closed electrophoresis
apparatus are both manifestations of surface charge, in the
former case that of the particle and in the latter that of the
chamber.
　　　The geometry of the chamber precludes the use of adhesive
films to eliminate the surface charge which leaves two alternative general approaches, namely:
　　　a) covalent bonding to surface groups to eliminate the
　　　　　negative groups by chemical modification (total charge
　　　　　zero)or to introduce positive groups such that the net
　　　　　charge is zero, or a combination of these types of
　　　　　surface modification.
　　　b) to use substances, usually macromolecules, which physi-
　　　　　cally adsorb to the glass surface, thereby shifting the
　　　　　electrophoretic plane of shear out from the original
　　　　　glass surface to the new macromolecular surface. By
　　　　　this means the charge on the original surface will have
　　　　　a negligible influence on ion distribution at the new
　　　　　plane of slip during electrophoresis (13).
The vast number of surface reactions encompassed by the

above two approaches can be severely restricted by an examination of the criteria required for a satisfactory surface coating or modification. The following properties of the surface coating should be considered in judging its general suitability and application limitations:

i) Physical properties:
   a) transparent to light, at least over the same range of wavelengths as the original glass.
   b) small coating thickness relative to the radius of the electrophoresis tube.
   c) capable of being applied uniformly.
   d) an electrical conductance not significantly greater than that of the particle suspending media.
   e) low fixed or acquired electrostatic charge in buffers.
   f) poor surface for gas bubble nucleation and formation.
   g) nonadhesive to particles or biological cells to be examined in the coated tube.

ii) Chemical and biological properties:
   a) hydrophilic to minimize adsorption of components from samples.
   b) compatible with system to be examined, e.g., non-toxic towards living cells.
   c) not subject to attack or degradation by any biological specimens under test.
   d) does not desorb or change the electrokinetic properties of the particles under examination.

iii) Electrokinetically stable (15) for duration of studies to:
   a) buffers up to physiological ionic strengths ($\sim$ 0.15 g ions liter^{-1}).
   b) pH range, 2 to 11.
   c) temperature range $\sim$ 0-50°C.
   d) shear rates encountered when filling or cleaning the electrophoresis tube.

Taking the above criteria into consideration possible coating materials or surface modifiers include:

A. Polysaccharides [agarose; dextran; ficoll; glycogen; methylcellulose; and starch].

B. Methacrylates and acrylamides [2-hydroxyethylmethacrylate (HEMA) (16); 2,3-dihydroxypropylmethacrylate (DHPMA) (16); and polyacrylamide].

C. Polymeric alcohols [polyethylene glycol (PEG); and polyvinyl alcohol (PVA)].

D. Silane derivatives as either sub-layer agents or primary coatings [γ-glycidoxypropyltrimethoxysilane (Dow Corning, Z6040) (2); γ-aminopropyltriethoxysilane (Union Carbide, A-1100) (2); γ-methacryloxypropyltrimethoxysilane (Dow Corning, Z6030) (2); and vinyltrimethoxysilane (Dow Corning (2)].

To date, polysaccharides have offered many promising features

as coatings. They are hydrophilic, have small numbers of charged
groups and are generally biocompatible. As a class they may be
crosslinked and their residual charge groups eliminated or
chemically reduced by procedures such as those used by Porath et
al. (17) on agarose. In addition polysaccharides may be deriva-
tized or oxidized and mixtures used to produce coatings with
either a net negative, net positive or net zero charge.

   The work reported has been directed at developing a surface
coating material which may be easily applied directly to a
cleaned borosilicate glass surface under mild conditions to form
a very thin stable layer which neutralizes or masks the glass
surface charge. Unmodified dextran (molecular weight 2 x 10^6
daltons) and methylcellulose (molecular weight 110,000) were
easily washed from model glass particles and conditions were not
found for applying agarose to the particles to form a uniform
thin layer, even though Hjertén has reported the successful use
of formaldehyde crosslinked methylcellulose to eliminate electroosmotic flow in his free zone electrophoretic equipment. These
considerations prompted the synthesis of diethylaminoethyl
derivatives of these substances with the rationale that the
introduction of cationic groups into the polymers should enhance
their binding to the glass surface where subsequently the thin
adsorbed layer could be chemically crosslinked to provide further
stabilization if necessary. Each DEAE-derivative was found to
adsorb to borosilicate glass particles and upon two to three
washes in 0.015 M NaCl, their mobilities were positive at neutral
pH. Work on DEAE-dextran and on DEAE-agarose was not carried
beyond this point, even though they appeared to be potentially
useful.

   The electrophoretic mobility versus pH for the clean borosilicate glass particles (Figure 3) is similar to that reported
in the literature. However extensive comparisons are difficult
because of variations in suspending media composition and ionic
strength from worker to worker (19, 20, 21). The decrease in
mobility which occurs below about pH 7 is consistent with the
presence of boranol and silanol groups (14) and the constant
mobility with increasing pH above a value of 7, typical behavior
for an anionogenic hydrophilic surface. The failure of the
crosslinked DEAE-methylcellulose coating to reverse the net
charge of the glass particles at neutral pH is indicative of
incomplete coverage, insufficient amino group substitution, and
perhaps also some carboxyl group contamination. The pK of about
9.5 is in the expected range for DEAE-derivatives (22). The small
increase in mobility as the ionization of the positive group is
suppressed indicates a low degree of substitution in the methylcellulose. Coverage of the glass surface appears to have been
sufficiently extensive to eliminate any obvious dependence of
mobility on pH over the range 4 to 7. The decrease in zeta
potential if it were to occur at a chamber wall is sufficient to
reduce errors in the electrophoretic mobility of particles with

mobilities of $\sim$ 3 µm sec^{-1} volt^{-1} cm from about 11% to 2% in
0.015 ionic strength media based on the conditions given in
Figure 2.

The transformation which occurs on coating a chamber with
epichlorohydrin treated DEAE-methylcellulose is depicted in
Figure 4 where the dashed line shows the extreme dependence of
particle velocity on position in the chamber for an uncoated
glass chamber. The full lines demonstrate for several different
ionic strengths the lack of dependence of particle velocity on
position in the chamber for a DEAE-methylcellulose coated glass
chamber. A complicating factor at present is the slow desorption of the coating which leads to the gradual reappearance of
significant electroosmotic flow and the possibility of contamination of any indicator particles during the course of electrophoretic measurements. The best conditions for satisfactory
coating with several of the polysaccharide derivatives have yet
to be worked out but these studies point to the probability of
developing single stage coatings which are easy to apply, do not
require a sub-layer nor elaborate pretreatment procedures and
which will remain electrokinetically stable for appreciable
periods of time.

Acknowledgement: This investigation was supported by
Contract No. NAS8-30887 from the National Aeronautics and Space
Administration.

Abstract

Physical, chemical and practical criteria are discussed for
glass surface treatments or coatings designed to minimize
electroosmosis as a source of error in analytical particle
electrophoresis. A rapid method for screening potential coating
agents was developed in which the electrophoretic mobility was
measured for cleaned borosilicate glass particles before and
after coating treatments. DEAE-derivatives of agarose, dextran
and methylcellulose adsorb directly to the particles and significantly reduce their negative surface charges to small net positive charges at neutral pH. A coating procedure involving
epichlorohydrin treatment of DEAE-methylcellulose is described
which produces a coating of reasonable stability for glass
electrophoresis chambers and significantly decreases errors in
electrophoretic mobility measurements due to electroosmotic flow.

Literature Cited

(1) Strickler, A., and Sacks, T., Ann. N.Y. Acad. Sci., (1973),
    209, 497-514.
(2) Lee, L.H., J. Coll. Inter. Sci., (1968), 27, 751-760.
(3) Van Oss, C.J., Fike, R.M., Good, R.J., and Reinig, J.M.,
    Anal. Biochem., (1974), 60, 242-251.
(4) Smoluchowski, M., Bull. Acad. Sci. Cracovie, (1903), 182.

(5) Bangham, A.D., Flemans, R., Heard, D.H., and Seaman, G.V.F., Nature, (1958), 182, 642-644.
(6) Peterson, E.A., and Sober, H.A., J. Amer. Chem. Soc., (1956), 78, 751-755.
(7) "Micro-Analyses in Medical Biochemistry 3rd Edition", p. 50, ed., E.J. King and I.D.P. Wootton, J. & A. Churchill Ltd., London, 1959.
(8) Seaman, G.V.F., and Heard, D.H., Blood (1961), 18, 599-604.
(9) Flodin, P., Dextran Gels and their Application in Gel Filtration, Ph.D. Dissertation, Uppsala, Sweden, 1962, pp. 14-24.
(10) Seaman, G.V.F., in "The Red Blood Cell, Vol. II", ed. D. MacN. Surgenor, pp. 1135-1229, Academic Press, New York, 1975.
(11) Abramson, H.A., Moyer, L.S., and Gorin, M.H., "Electrophoresis of Proteins and the Chemistry of Cell Surfaces", pp. 43-57, Van Nostrand-Reinhold, Princeton, New Jersey, 1942.
(12) Hartley, G.J., and Roe, J.W., Trans. Faraday Soc., (1940), 36, 101-109.
(13) Brooks, D.E., and Seaman, G.V.F., J. Coll. Inter. Sci., (1973), 43, 670-686.
(14) Hair, M.L., and Altug, I., J. Phys. Chem., (1967), 71, 4260-4263.
(15) Heard, D.H., and Seaman, G.V.F., J. Gen. Physiol., (1960), 43, 635-654.
(16) Andrade, J.D., Med. Instrm., (1973), 7, 110-120.
(17) Porath, J., Janson, J.C., and Låås, T., J. Chromat., (1971), 60, 167-177.
(18) Hjertén, S., "Free Zone Electrophoresis", p. 51, Almqvist and Wiksells Boktryckeri AB, Uppsala, 1967.
(19) Abramson, H.A., J. Gen. Physiol., (1929), 13, 169-177.
(20) Chattoraj, D.K., and Bull, H.B., J. Amer. Chem. Soc., (1959), 81, 5128-5133.
(21) Mehrishi, J.N., and Seaman, G.V.F., Biochim. Biophys. Acta, (1966), 112, 154-159.
(22) Peterson, E.A., "Cellulosic Ion Exchangers", p. 236, North Holland/American Elsevier, 1970.

Streaming Potential Studies on Gel-coated Glass Capillaries

SHAO M. MA, DONALD E. GREGONIS, RICHARD VAN WAGENEN, and JOSEPH D. ANDRADE

Department of Materials Science and Engineering, University of Utah, Salt Lake City, Utah 84112

The separation of living cells in zero gravity environments appears to offer significant advantages over analogous separations on earth. The zero gravity environment eliminates sedimentation and density fluctuation problems. As demonstrated in Apollo 16 experiments (1), a major problem with free zone electrophoresis in a closed system is the differential fluid movement due to the surface electrical properties of the container (electroosmosis). As a consequence, the resolution of the separation and the analysis involved are grossly compromised by the resulting "bullet" shape flow patterns. A solution to this problem is to produce surfaces which exhibit essentially no electrokinetic properties when in equilibrium with their environment. This can be achieved by neutralizing the surface charge or by shifting the hydrodynamic shear plane away from the interface by use of hydrophilic surface coatings.

In this study, glass tubes were subjected to various coatings and surface treatments. In particular, it was felt that uncharged hydrophilic methacrylate polymer coatings should decrease the electroosmosis. The overall effects of these treatments upon streaming potential is discussed.

Apparatus

The streaming potential apparatus has been described in detail (2,3). It is an all-glass system utilizing Ag/AgCl electrodes. Streaming fluid is forced back and forth from one reservoir to the other through a glass tube using pressurized pure nitrogen gas. A glass pH electrode and a thermometer are positioned in one of the reservoirs. A Keithley model 616 digital electrometer measures streaming potential or streaming current.

All new and large glassware are cleaned initially in chromic sulfuric acid, thoroughly rinsed in running, distilled water, soaked in doubly distilled water, rerinsed in double distilled water and then air dried in a dust-free environment. Small

glassware is cleaned by radiofrequency glow discharge (4).
To carry out the streaming potential measurement each reservoir is filled with about 500 ml of phosphate buffer at pH 7.2
containing 0.01 M KCl. The electrodes are inserted. The potential is then measured as a function of driving pressure. The
streaming potential, E_{str}, is linearly dependent on the pressure
drop, P, across the tube, as given by (5)

$$\zeta = \frac{4\pi\eta K}{D} \frac{E_{str}}{P}$$

where ζ is the zeta potential, η and D are the viscosity and
dielectric constant, respectively, in the diffuse portion of the
electrical double layer, and K is the bulk specific conductivity
of the electrolyte (assuming surface conductance is negligible
relative to bulk conductance).

Ball and Fuerstenau (6) have reviewed the streaming potential
literature in regard to streaming potential data. They have
concluded that, due to as yet unexplained flow and symmetry
potentials common to a wide variety of electrode types, the
slope of the loci of E_{str} data at a number of driving pressures,
P, in opposite flow directions, $\Delta E_{str}/\Delta P$, should be utilized in
the above equation. This has been the case in this study.
Streaming potential data was obtained by measuring streaming
potentials at a driving pressure of 2cm Hg, then reversing the
flow direction and repeating the measurement. The driving pressure was increased by 2cm Hg and streaming potentials were again
measured in both flow directions. This process was repeated
until the driving pressure reached 12cm Hg. The loci of streaming potential data as a function of driving pressure were then
fitted to a linear regression best fit straight line using a
Hewlett Packard (Model 9820A) programmable calculator.

Absolute values of zeta potential calculated from the above
equation may be erroneous due to assumptions concerning values
for double layer viscosity, dielectric constant, and entrance
flow effects resulting from the large ID of the tubes utilized
in this study (2,3).

Materials

Thick-walled "Pyrex" capillary tubes, approximately 0.2cm
ID and 15cm long, were used. Uncoated capillaries had $\Delta E_{str}/\Delta P$
of -0.357 mV/cm Hg.

Different silane adhesion promoters (7) were used to pretreat
the glass surfaces prior to gel coating. An investigation of
several silanizing reagents using different procedures showed a
slight decrease in $\Delta E_{str}/\Delta P$ for all the silane coatings. The
$\Delta E_{str}/\Delta P$ values were not sensitive to the organic functionality
of the silane.

A variety of hydrophilic polymers were used as coating
materials including methylcellulose (Dow Methocel MC, premium,

4000 cps); hydroxypropylmethylcellulose (Dow Methocel, 90Hg,
premium, 15000 cps); dextrin (Matheson, Coleman and Bell) and
agarose (Biorad Labs). Hydroxyethyl methacrylate was donated
by Hydro Med Sciences, Inc. Methoxyethyl methacrylate and
methoxyethoxyethyl methacrylate were prepared in our laboratories
by base catalyzed transesterification of methyl methacrylate
with the corresponding alcohol (8).

Methods and Results

The silane compounds were applied using standard procedures
(6). A silane solution was rinsed through the tubes, first in
one direction, then in the other, and the tubes were vacuum
dried overnight. The silane compounds used and the $\Delta E_{str}/\Delta P$
values are given in Table I.

TABLE I.

$\Delta E_{str}/\Delta P$ Values for Silane Coatings

Coating	$\Delta E_{str}/\Delta P$ (mV/cm Hg)
1,1,1,3,3,3-hexamethyldisilazine:triethylamine	-0.242
Chlorotrimethylsilane:triethylamine (1:1)	-0.221
Dichlorodimethylsilane:triethylamine (1:1)	-0.253
γ-methacryloxypropyltrimethoxysilane:triethylamine (1:1)	-0.234
γ-glycidoxypropyltrimethoxysilane:triethylamine (1:1)	-0.229
γ-glycidoxypropyltrimethoxysilane:n-propylamine (1:1)	-0.230
γ-glycidoxypropyltrimethoxysilane:pH 4 aq. acetic acid	-0.241

The γ-glycidoxypropyltrimethoxysilane:triethylamine (1:1)
silane coating was selected for further treatment with the
neutral polysaccharides. It was thought that coating the polysaccharides upon the epoxy-silanized glass was either an acid or
base catalyzed process resulting in covalent attachment of the
polysaccharide to the surface (Figure 1).
Standard solutions of various polysaccharides were prepared
by dissolving in either 0.5% HCl or 0.5% KOH. If the polysaccharide was soluble, the viscosity of the solution was "regulated"
by the addition of polysaccharide until it had the consistency of
a thick syrup. Otherwise the polysaccharide was added to complete
saturation. The polysaccharide solution was pulled through the
tubes in each direction by a water-aspirator partial vacuum.
The tubes were then placed in a vacuum oven at 120°C for 12 hours,
unless otherwise specified. The tubes were exhaustively rinsed
with distilled water and the $\Delta E_{str}/\Delta P$ values were obtained.
Consistently lower $\Delta E_{str}/\Delta P$ values were obtained when the
polysaccharides were applied in acidic solution as compared to

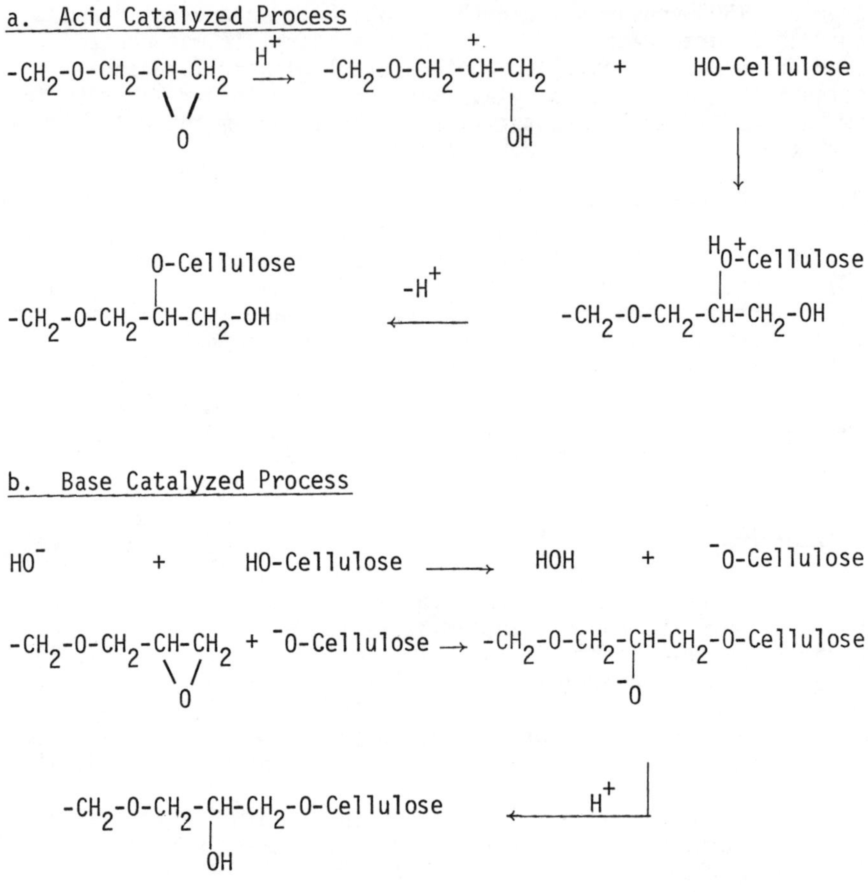

a. Acid Catalyzed Process

$-CH_2-O-CH_2-CH-CH_2 \overset{H^+}{\longrightarrow} -CH_2-O-CH_2-\overset{+}{C}H-CH_2$ $\quad + \quad$ HO-Cellulose
 \ /
  O

$\qquad\qquad\qquad\qquad\qquad\qquad\qquad\qquad\qquad$ OH

$\qquad\qquad\qquad\qquad\qquad\qquad\qquad\qquad\qquad\qquad\qquad\qquad$ ↓

$\qquad\qquad$ O-Cellulose $\qquad\qquad\qquad\qquad\qquad$ $\overset{H}{\underset{}{O}}\overset{+}{-}$Cellulose

$-CH_2-O-CH_2-CH-CH_2-OH \qquad \overset{-H^+}{\longleftarrow} \qquad -CH_2-O-CH_2-CH-CH_2-OH$

b. Base Catalyzed Process

$HO^- \qquad + \qquad$ HO-Cellulose $\longrightarrow$ HOH $\quad + \quad ^-$O-Cellulose

$-CH_2-O-CH_2-CH-CH_2 + {}^-$O-Cellulose $\rightarrow -CH_2-O-CH_2-CH-CH_2-O-$Cellulose
 \ / $^-$|
  O $^-$O

$-CH_2-O-CH_2-CH-CH_2-O-$Cellulose $\qquad \overset{H^+}{\longleftarrow}$
 |
 OH

Figure 1.   Reaction schemes for the attachment of polysacchardies to epoxy–silane surfaces. A, acid catalyzed; B, base catalyzed.

basic solution; however, when methylcellulose in distilled water
was applied to the epoxy-silanized glass, a $\Delta E_{str}/\Delta P$ value was
obtained comparable to the results obtained with methylcellulose
in acid solution. At the moment the mechanism that governs
attachment of methylcellulose to the epoxy-silanized glass is in
doubt. The treatment of glass surfaces with just methylcellulose
or γ-glycidoxypropyltrimethoxysilane does not give the low
$\Delta E_{str}/\Delta P$ values. This phenomenon is still being investigated.
    Further coating of methylcellulose onto a methylcellulose
coating showed no change in $\Delta E_{str}/\Delta P$. The use of 1,2,4,5,9,10-
triepoxydecane (TED) as a crosslinker in the methylcellulose
solutions (5% TED in 0.5% HCl-methylcellulose) was investigated.
Curing was accomplished by heating overnight at 120°C in vacuum.
No statistical change was noted for this treatment over normal
methylcellulose tubes (Table II).

TABLE II.

Effects of Silane and Polysaccharide Coatings Upon $\Delta E_{str}/\Delta P$ Values

Coating	$\Delta E_{str}/\Delta P$ (mV/cm Hg)	Standard Deviation
(1) Uncoated	-0.356	0.010
(2) γ-glycidoxypropyltrimeth-oxysilane:triethylamine (1:1)	-0.310	0.007
(3) Procedure 2, then 0.5% HCl	-0.226	---
(4) Procedure 2, then 0.5% KOH	-0.261	---
(5) Methylcellulose in distilled H_2O	-0.166	0.090
(6) Procedure 2, then methylcellulose in 0.5% KOH	-0.130	0.012
(7) Procedure 2, then methylcellulose in 0.5% HCl	-0.078	0.008
(8) Procedure 2, then methylcellulose in distilled H_2O	-0.058	0.002
(9) Methylcellulose coated tubes (Procedure 7) recoated with methylcellulose with 5% 1,2,4,5,9,10-triepoxydecane on 0.5% HCl	-0.103	0.003
(10) Procedure 2, then hydroxypropyl-methylcellulose in 0.5% HCl	-0.096	---
(11) Procedure 2, then dextrin in 0.5% HCl	-0.245	---
(12) Procedure 2, then agarose in 0.5% HCl	-0.250	---

A simple water-polyfluorocarbon surface contact angle test was used to detect surface-active extractables (9) from the coated capillaries. High contact angles were noted with all tubes. No significant change in the contact angle was evident with the methylcellulose-coated tubes, indicating that the methylcellulose is not readily extracted from the surface.

To obtain capillary tubes coated with methacrylate hydrogels, first soluble polymers were prepared. This was accomplished by radical initiation of the desired monomer at low dilution (1 to 10, v/v) in ethanol. A small amount of this polymer solution is allowed to flow through the glass capillary, and with a little care and patience, a very uniform coat of the polymer could be deposited on the inside of the capillary. The viscosity of the polymer solution was regulated by the addition or evaporation of solvent. The tubes were cured overnight at 60°C and then allowed to equilibrate in distilled water.

The monomers that were investigated were methoxyethyl methacrylate (MEMA), hydroxyethyl methacrylate (HEMA) and methoxyethoxyethyl methacrylate (MEEMA). It has been determined that these polymers swell in water to incorporate 3.5%, 40% and 63% water, respectively (9). The corresponding $\Delta E_{str}/\Delta P$ values are comparable to the best methylcellulose values (Table III).

TABLE III.

Effects of Hydrophilic Methacrylate Coatings Upon $\Delta E_{str}/\Delta P$ Values

Coating	$\Delta E_{str}/\Delta P$ (mV/cm Hg)	Standard Deviation
(1) Uncoated tubes	-0.290	0.045
(2) Hydroxyethyl methacrylate (HEMA)	-0.046	0.029
(3) Methoxyethyl methacrylate (MEMA)	-0.058	0.050
(4) Methoxyethoxyethyl methacrylate (MEEMA)	-0.038	0.012
(5) HEMA with 1% methacrylic acid (MAA)	-0.080	0.024
(6) HEMA with 3% MAA	-0.100	0.024
(7) HEMA with 10% MAA	-0.113	0.035
(8) HEMA with quaternized 1% dimethyl-aminoethyl methacrylate (DMAEMA)	+0.070	0.024
(9) HEMA with quaternized 3% DMAEMA	+0.074	0.004
(10) HEMA with quaternized 10% DMAEMA	+0.087	0.014

To observe the effect that charged groups in the hydrophilic polymers had upon $\Delta E_{str}/\Delta P$ values, HEMA was polymerized at low dilution with various amounts of methacrylic acid (MAA) and dimethylaminoethyl methacrylate (DMAEMA). In the streaming solution buffered to pH 7.2, methacrylic acid exists as the charged salt, but dimethylaminoethyl methacrylate exists as an uncharged species. After polymerization of the HEMA-DMAEMA copolymer, methyl iodide was added to the polymer solution to react with the free amino groups forming quaternary ammonium iodides (Figure 2). The capillary tubes were coated as described for the other methacrylate polymers, cured at 60°C overnight and then allowed to swell to equilibrium in either distilled water or streaming potential buffer.

The MAA-HEMA copolymer coatings exhibited average $\Delta E_{str}/\Delta P$ values which increased negatively with increasing percentage of methacrylic acid in the polymer: -0.080 for 1% MAA, -0.100 for 3% MAA, and -0.113 for 10% MAA. The quarternary DMAEMA-HEMA copolymer coatings produced a net positive electrokinetic surface as indicated by increasingly positive $\Delta E_{str}/\Delta P$ values with increasing percentage of DMAEMA; +0.070 for 1% DMAEMA, +0.074 for 3% DMAEMA, and +0.087 for 10% DMAEMA (Table III).

Discussion

According to Brook's model (10) the zeta potential in the presence of adsorbed polymer (hydrophilic, uncharged) could be higher or lower than that in the absence of polymer. The effect depends on the relative magnitude of the thickness of the adsorbed polymer layer, d, and the thickness, d_f, within which nonzero fluid flow occurs during an electrokinetic experiment. Three areas have been cited:

1. For a totally free draining adsorbed layer, the location of the shear plane is unaffected by the presence of the adsorbed layer, i.e. $d = d_f$. In this case, polymer adsorption causes an increase in zeta potential.
2. For a partially free draining adsorbed layer, the shear plane is shifted to a position within the adsorbed polymer layer ($0 < d_f < d$). In this case polymer adsorption may cause either an increase or a decrease in zeta potential.
3. When flow is totally excluded from the adsorbed layer, i.e., the shear plane is shifted outside the adsorbed layer ($d_f = 0$), polymer adsorption causes a decrease in zeta potential.

Our data was obtained from a large variety of neutral polymer coatings; all show a decrease in $\Delta E_{str}/\Delta P$. Our neutral polymer coatings were cast on glass capillary surfaces and are thicker than those obtained by adsorption. Although one cannot directly compare our coatings with the three cases presented by

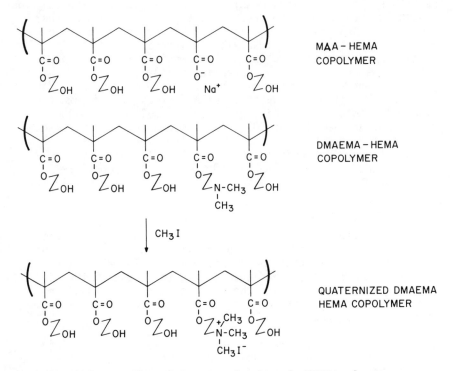

Figure 2.   Charged groups introduced into the HEMA polymer

Brook's (10) it does appear that in our case the shear plane has been shifted far away from the interface. Consider the $\Delta E_{str}/\Delta P$ values given in Table III for positive or negative groups co-polymerized with HEMA. At low streaming potential values, increased experimental error is introduced into the measurements. At a given P, E_{str} changes with time, partly due to a change in the relative fluid levels in the two reservoirs which causes a change in P and partly due to electrode drift with time of unknown cause (2-3). It was observed that the ΔE_{str} readings were more stable for surfaces with high streaming potentials. As a consequence, the error is larger when the absolute $\Delta E_{str}/\Delta P$ value is small. Although different samples of similar coating or the same sample measured on different days do show large varia-tions in their $\Delta E_{str}/\Delta P$ values, when the same sample was measured three times within a short period, its $\Delta E_{str}/\Delta P$ value was quite consistent. Also, in each series of measurements, the $\Delta E_{str}/\Delta P$ value for samples 5-10 followed the same trend as given in Table III. Therefore, we would like to attribute the large standard deviation values to the variations in experimental conditions, surface roughness, uneven coatings, etc., and consider the effect of increasing absolute streaming potential values with increasing charge as real.

Conclusions

A decrease in $\Delta E_{str}/\Delta P$ occurs when the glass surface is silanized or coated with a neutral polysaccharide such as methylcellulose. Silanization followed by methylcellulose treatment gives an almost ten-fold decrease in $\Delta E_{str}/\Delta P$ over uncoated "Pyrex" glass. Both acid and neutral methylcellulose solutions coated upon a γ-glycidoxypropyltrimethoxysilane base gives the lowest $\Delta E_{str}/\Delta P$ values for the polysaccharides examined. Base-containing methylcellulose deposited on a γ-glycidoxypropyltrimethoxysilane surface gives slightly higher $\Delta E_{str}/\Delta P$ values, perhaps due to the corrosive action of the base. No additional improvement in $\Delta E_{str}/\Delta P$ values was obtained when the methylcellulose tubes are recoated with methylcellulose containing a crosslinker.
Hydrophilic methacrylate polymer coatings also lower $\Delta E_{str}/\Delta P$ values about ten-fold over untreated glass capillaries. By added quaternary ammonium groups to these polymer coatings, the sign of $\Delta E_{str}/\Delta P$ is changed from negative to positive. It is felt that by adjusting the amount of charged co-monomers, a zero $\Delta E_{str}/\Delta P$ value could be obtained.

Acknowledgements

This work has been supported by the U. S. National Aeronautics and Space Administration, Marshall Space Flight Center, Huntsville, Alabama, under Contract No. 8-30253, Dr. J. Patterson, Contract Monitor. The assistance of R. Middaugh is gratefully acknowledged.

Abstract

Streaming potential techniques were used to measure the interfacial electrical potential of glass capillaries. Different silanizing reagents were used to coat the glass capillaries and decrease the streaming potential values by a comparable amount, i.e., about half that of the uncoated glass capillaries. A ten-fold decrease in streaming potential is observed when methylcellulose is coated onto a γ-glycidoxypropyl silanized glass capillary. A comparable decrease is noted when hydroxyethyl methacrylate (HEMA), methoxyethyl methacrylate (MEMA) and methoxyethoxyethyl methacrylate (MEEMA) polymers are coated on the glass capillary. By adding anionic or cationic charged groups to the poly-HEMA coating, streaming potential values of opposite sign are obtained.

Increasing the amount of methacrylic acid copolymerized with the HEMA monomer negatively increases the average $\Delta E_{str}/\Delta P$ values. Increasing the amount of quaternized dimethylaminoethyl methacrylate in the copolymer positively increases the average $\Delta E_{str}/\Delta P$ values.

Literature Cited

1. Snyder, R. S., Bier, M., Griffin, R. N., Johnson, A. J., Leidheiser, Jr., H., Micale, F. J., Vanderhoff, J. W., Ross, S., van Oss, C. J., Sep. Purific. Methods (1973) $\underline{2}$ (2), 259.
2. Van Wagenen, R., Ph.D. Dissertation, University of Utah, September 1975.
3. Van Wagenen, R., Andrade, J. D., and Hibbs, J. B., Jr., Submitted to J. Electrochem. Soc.
4. Hollahan, J. R., and Bell, A. T., Eds., "Techniques and Applications of Plasma Chemistry," Wiley, New York, 1974.
5. Davies, J. T., and Rideal, E. K., "Interfacial Phenomena," 2nd Ed., Academic Press, 1963.
6. Ball, B., and Fuerstenau, D. W., Miner. Sci. Eng., (1973) $\underline{5}$, 267.
7. Bascom, W. D., Macromolecules (1972) $\underline{5}$, 792.
8. Gregonis, D. E., Chen, C. M., and Andrade, J. D., "The Chemistry of Some Selected Methacrylate Hydrogels," this symposium.

9. Baier, R. E., Gott, V. L., and Feruse, A., Trans. Amer. Soc. Artificial Internal Organs (1970) 16, 50.
10. Brooks, D. E., J. Colloid and Interface Science, (1973) 43, 687.

19

Water Wettability of Hydrogels

FRANK J. HOLLY and MIGUEL F. REFOJO

Eye Research Institute of Retina Foundation, 20 Staniford Street, Boston, Mass. 02114

The matrix of hydrogels consists of hydrophilic macromolecules which are crosslinked to form a three-dimensional network.
Thus, a gel spontaneously imbibes water until the gel reaches
equilibrium hydration. This equilibrium water content of the gel
depends on crosslink density and on the amount and nature of hydrophilic sites in the macromolecular matrix.

With the exception of poly(hydroxyethyl methacrylate)
[PHEMA], which is not soluble in water, the matrix of hydrogels
is usually made up of water-soluble polymers. At equilibrium
hydration a large fraction of hydrogels consist of water. Hydrogels containing over 95% water are not uncommon. This is probably
why it has been implicitly assumed that hydrogels have a hydrophilic surface. In other words, water is expected to spread
spontaneously over the surface of hydrogels, at least when the
surface is fully hydrated.

Past Work on Hydrogel Wettability

Acrylic hydrogels are fairly new and their wettability has
only been examined recently. However, there are several reports
in the literature in which the wettability of gelatin gels are
considered. The first systematic investigation of the wettability
of hydrated and air-dried gelatin gels appears to have been made
by Pchelin and Korotkina (1). They found a water contact angle of
98° on hydrated gelatin and a contact angle of 115° on the air-
dried gelatin film. These authors also observed that the wettability of the gelatin gels depended on the polarity of the material adjacent to the gel surface while the gel was being formed.
Thus gelatin formed against paraffin or air was hydrophobic,
while gelatin formed against a clean glass surface was hydrophilic. Garrett (2) also found gelatin to be hydrophilic when formed
against a glass surface.

Braudo and coworkers (3) studied the wettability of gelatin
gels as a function of the gelatin concentration. The water content of the gel varied from 14% for the air-dried gels up to 86%

for the most dilute gels studied. They found that the water contact angle on these gels varied from the surprisingly high value of 123° for the most hydrated gel to 87° for the dehydrated gels. These authors offer an explanation for the anomalous wettability of gelatin gels based on the preferred orientation of the water molecules in a surface free-water layer of the gel. It is hard to see, however, why an oriented water layer having the characteristics of the surface layer of pure water would not be wetted by water. It is suspected that the high, obtuse water contact angles were obtained because of surface dehydration and these values were further increased due to the roughness of the surface.

More recently the agarose gel was the subject of an investigation as to water wettability (4). If the gel was formed under water or had been exposed to water for long periods of time, the surface was hydrophilic. On the other hand, if the surface was formed while exposed to clean air, the surface was hydrophobic. When this hydrophobic layer was shaved off, the gel again had a hydrophilic surface.

We have recently found (5) that gels made of the polymer PHEMA appear to have a hydrophobic surface even when the gel is fully hydrated. The wettability of PHEMA gels having equilibrium water contents between 31 and 42% has been found to be surprisingly low: advancing contact angle values for water between 60 and 80° have been obtained by both the sessile-drop and the captive-bubble techniques. The receding contact angle values measured were found to be much lower but still larger than zero.

Description of the Hydrogels Studied

The wettability of several types of hydrogels, consisting of macromolecules more hydrophilic than PHEMA, was measured (Figure 1). Gels were prepared by simultaneous polymerization and crosslinking of purified monomers in aqueous solutions in molds consisting of two glass plates separated by a silicone gasket (6). The copolymer of glyceryl methacrylate and methyl methacrylate [P(GMA/MMA)] was obtained from Corneal Sciences, Inc. Boston, MA. P(GMA/MMA) was made by bulk polymerization. The gel surface was polished and then equilibrated with water.

Poly(glyceryl methacrylate) [PGMA] forms gels that have twice the number of hydroxyl groups per monomer unit than does the PHEMA gel. Poly(hydroxyethyl acrylate) [PHEA] forms gels that are similar to PHEMA in hydrophilic sites but this polymer does not have the hydrophobic, bulky methyl side group on the acrylic backbone. The preparation of the PHEMA gels was described previously (5). The crosslinking agents for PHEMA, PGMA, and PHEA are the diesters which are usually found accompanying the monoesters. Poly(acrylamide) [PAA] gels were prepared using N,N'-methylene bisacrylamide as a crosslinking agent.

In addition to these synthetic materials, a polysaccharide gel, agarose, consisting of D- and L-galactopyranose, was includ-

ed in the study along with another material not strictly a hydrogel, poly(dimethyl siloxane) bulk-grafted with poly(vinyl pyrrolidone) [SIL-g-PVP] (7), which was obtained from the firm Essilor
(Paris, France).
     The gels, with the exception of P(GMA/MMA), were formed
against glass, which was made hydrophobic for some of the PHEMA
gels by Siliclad coating (5). The gels were thoroughly washed and
stored in distilled water (that was frequently changed) for at
least one month prior to the wettability measurements. Under
these conditions, the hydrophobic or hydrophilic nature of the
glass surface in contact with the gel during its formation did
not seem to affect gel wettability.

Method of Measurement of Wettability

     The contact angles were determined by a Ramé-Hart Goniomet-
er. The contact angle of the sessile droplets was determined in
a closed chamber (Ramé-Hart Environmental Chamber for Goniometer).
Clean air was circulated through two consecutive gas-washing
bottles filled with water and the chamber in a closed system by
a varistaltic pump in order to keep it saturated with respect to
water vapor. The captive-bubble technique was also employed occ-
asionally using a chamber designed for this technique in order
to guard against the slight possibility of surface dehydration.
As with PHEMA (5), this latter technique yielded contact angle
values that were consistently a few degrees higher (!) than the
data obtained with the sessile-drop technique.
     Both the advancing and the receding contact angles were det-
ermined. The sessile droplets (or captive bubbles) were slightly
increased or decreased in size after the contact with the gel
surface had been established, and until the angle remained constant upon additional change in volume. Thus, the contact angles
measured were static rather than dynamic values as the terms
seem to imply.

Gel Wettability as Characterized by Water Contact Angles

     Figure 2 displays both the advancing and receding contact
angle values obtained with water on the various gels as a function of the equilibrium water content of the gels. As a reference,
the average advancing and receding contact angle values for poly-
(methyl methacrylate) [PMMA] are also shown in the figure.
     It is clear from the graph that only the agarose gel of very
high water content (over 96%) is completely wetted by water. The
next most wettable gel is PAA as expected from its chemical composition. The wettability of PGMA and PHEA are similar indicating
that the additional hydroxyl group of PGMA just about cancels the
hydrophobic effect of the methyl group absent in PHEA.
     The wettability of the PHEMA gels, having considerably lower
equilibrium hydration, is widely variable, but all PHEMA gels are

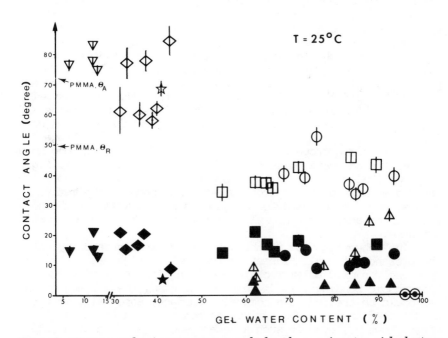

GEL OR SOLID	R₁-	R₂-
1. PMMA	$-CH_3$	$-OCH_3$
2. PHEMA	$-CH_3$	$-OCH_2CH_2OH$
3. PGMA	$-CH_3$	$-OCHOHCH_2OH$
4. PHEA	$-H$	$-OCH_2CH_2OH$
5. PAA	$-H$	$-NH_2$
6. P(GMA/MMA)	copolymer of 3. & 1.	
7. SIL-g-PVP	PVP grafted onto silicone	
8. Agarose	polysaccharide (galactopyranose)	

Figure 1. Chemical composition of hydrogels investigated

Figure 2. Contact angle of water on various hydrogels as a function of hydration. ▽, Sil-g-PVP; ◇, PHEMA; ○, PGMA; ☆, P(GMA/MMA); □, PHEA; △, PAA; ◉, agarose. Solid symbols indicate receding contact angles.

less wettable than PGMA or PHEA. P(GMA/MMA) is about as hydrophobic as PHEMA, while SIL-g-PVP having a rather low water content (6-12%) is only as hydrophobic as the less wettable PHEMA
gels. The average water wettability of the PHEMA gels as judged
by the magnitude of the advancing contact angle is comparable to
that of PMMA (1.5% water).

The receding contact angle values of water on the gels are
all much lower than the advancing contact angles. Hence, quite
large contact angle hysteresis was observed with the hydrogels.
Four factors or processes have been recognized as being capable
of causing contact angle hysteresis: surface roughness, physical
interaction between solid and liquid (e.g. dissolution), chemical
interaction between solid and liquid (e.g. hydrolysis), and
stereochemical changes at the solid-liquid interface.

All the hydrogels investigated had a smooth enough surface
to appear glossy when the excess water had been blotted off their
surface. Thus, the small degree of roughness present cannot
account for the pronounced hysteresis. All the gels were washed
repeatedly for prolonged periods of time prior to contact angle
measurements, so water-soluble, surface-active contaminants as a
possible cause of contact angle hysteresis can be eliminated. The
hydrogels were in thermodynamic equilibrium with water, so chemical interaction could not take place either.

Since hydrogels are not rigid solids, it is expected that
the vertical component of the water surface tension would deform
the solid at the drop periphery resulting in local stress and
distortion of the gel shape. We could observe no indication that
such an effect occurs, but it cannot be ruled out.

The only probable explanation left is the possible stereochemical changes that may take place at the gel boundary. We have
postulated (5) that the large contact angle hysteresis observed
with PHEMA, and this could hold true for the other gels as well,
is due to the relatively high mobility of the polymer chains and
segments at the gel surface. By changing orientation and/or conformation, the polymer network at the gel surface can change
character from hydrophobic to relatively hydrophilic depending on
whether the adjacent phase is water vapor or liquid water. Figure
3 schematically depicts such an occurrence. At the gel-vapor
interface, most of the hydrophobic groups, such as the methyl
groups, would be exposed to the gaseous phase, while at the gel-
water interface, the hydroxyl and other hydrophilic groups would
dominate.

The schematic view of the gel surface in Figure 3 indicates
the presence of small "lakes" of free water which are separated
by closely packed polymer chains held in close proximity by
hydrophobic bonding (8). If such is the case, the surface of the
gel could be treated as a composite surface, especially if the
liquid used to probe the surface is nonpolar. In the case of
water, however, even if such a heterogeneous surface structure
existed, the polymer chains at the surface are probably mobile

enough to shift so that no free-water lakes are expected to cross
the periphery of the sessile water droplet.

The data contained in Figure 2 seem to indicate that for a
given type of gel, the equilibrium water content of the gel has
no direct effect on surface wettability. However, there is some
vague indication, especially with the gels of higher water content
that hydrophobicity may increase with increasing water content. As
already mentioned, a similar phenomenon was observed for gelatin
by others (3).

The main factor determining water wettability appears to be
the chemical structure of the polymeric network at the interface.
The surface structure and orientation of the polymer matrix is
most likely affected by the polarity of the adjacent phase boundary as well as by some uncontrolled events relating to gel matrix
formation.

Water Wettability and Relative Contact Angle Hysteresis

One way to circumvent the difficulties introduced by the
unpredictable nature of the gel surface is to characterize it by
the advancing contact angle of water rather than by the bulk parameter; the degree of equilibrium hydration. Furthermore, one may
define the relative contact angle hysteresis (H_R) as the difference between the advancing (Θ_A) and the receding (Θ_R) contact
angles normalized with respect to the advancing contact angle.
That is,

$$H_R = (\Theta_A - \Theta_R) / \Theta_A$$

Then, the maximum value of hysteresis will be equal to 1
whenever the receding angle is zero, irrespective of the magnitude of the advancing contact angle.

The value of the relative hysteresis of the contact angle is
shown in Figure 4 as a function of the advancing contact angle of
water. It is apparent from the graph that the value of H_R increases approximately in a linear manner with the increasing
contact angle, i.e. with decreasing wettability.

This observation, common for all the gels investigated, can
be readily explained by our hypothesis. We have proposed that
both the hydrophobic nature of the gel surface and the convertibility to a more hydrophilic surface when in contact with water
are due to the segmental mobility and the amphipathic character
of the polymer chains at the gel boundary. The more mobile the
segments are for a given gel, the higher the advancing contact
angle of water and the greater the difference between the advancing and receding angles will be. Apparently this is so, because
at higher segmental mobility more hydrophobic groups can be exposed and lined up at the surface when it is exposed to water

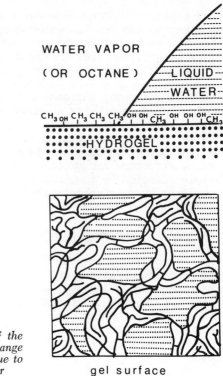

Figure 3. Schematic view of the hydrogel surface and of the change in surface layer orientation due to the presence of liquid water

gel surface

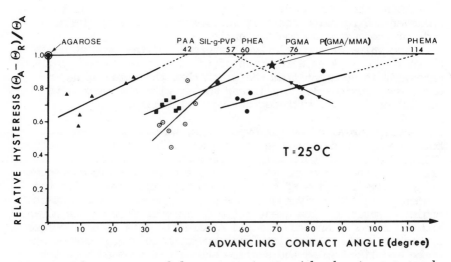

Figure 4. Relative contact angle hysteresis as a function of the advancing contact angle of water. ⊙, agarose; ▲, PAA; ▼, Sil-g-PVP; ⊙, PHEA; ■, PGMA; ●, PHEMA.

vapor (resulting in a higher advancing contact angle of wetting), and more hydrophilic sites can be exposed to the adjacent liquid water phase (resulting in lower receding contact angles). In other words, increased segmental mobility of the surface polymer chains for a given gel would be manifested by both decreased wettability and enhanced contact angle hysteresis.

It is possible to obtain a value for a "limiting advancing contact angle" for each gel by fitting a straight line to the data points by the least square method and extrapolating it to the maximum hysteresis value: $H_R = 1$, $\Theta_R = 0$. While the empirical nature of the relationship and the considerable scatter in the data make this limiting value of questionable importance and accuracy, it is possible to assign a physical meaning to it. This limiting value of Θ_A corresponds to the water wettability of a hypothetical gel of the same composition, which would have a segmental mobility sufficiently high to exhibit a zero receding contact angle. Since this limiting contact angle is 114° for PHEMA, which value is larger than can be obtained with water on the closely packed methyl groups in the freshly cleaved surface of a hexatriacontane single crystal (111°) (9), it is unlikely that a zero receding contact angle of water can ever be obtained on a PHEMA gel, no matter how much the segmental mobility may be increased by some suitable manipulation of the gel-forming technique.

The extrapolated value of the advancing contact angle appears to be of a more realistic magnitude for the other gels. It must be remembered, however, that these gels are considerably more hydrated than PHEMA, and this fact is likely to decrease the surface density of the hydrophobic (and hydrophilic) groups and chain segments at the gel surface. It is of interest to note, however, that the maximum hysteresis observed with gels of different types is approximately the same, 0.8 - 0.9, and the highest was observed with the gel of the copolymer, P(GMA/MMA).

It is apparent from Figure 4 that the material SIL-g-PVP exhibits a behavior just the opposite to that of the hydrogels. The seemingly anomalous behavior of this material actually strengthens the foregoing argument based on the segmental mobility of the surface polymer chains in the gels. In PVP grafted silicone the convertibility of the surface from hydrophobic to hydrophilic depends on the poly(vinyl pyrrolidone) content. The more PVP there is, the more hysteresis is expected. However, more PVP at the surface would result in higher wettability, i.e. lower advancing contact angle, since silicone is inherently hydrophobic. This is why the straight line, which fits the data exceptionally well, has a negative slope in the graph.

Critical Surface Tension of the Hydrogels

We were curious to find out the wettability of these hydrogels by pure organic liquids which are immiscible with water. The

eight liquids chosen and their surface tensions are shown in Table
I. They all have a negative spreading coefficient on water with
the exception of tricresyl phosphate.

Table I

Organic Liquids Used in Critical Surface Tension
Determination

DIAGNOSTIC LIQUIDS	SURFACE TENSION dyne/cm	SPREADING COEFFICIENT ON WATER
Diiodomethane	48.8	negative
1,1,2,2-Tetrabromoethane	48.3	negative
1-Bromonaphthalene	43.0	negative
Tricresylphosphate	40.4	positive
1-Methylnaphthalene	38.7	negative
Nitromethane	36.1	negative
Dicyclohexylamine	33.8	negative
n-Hexadecane	27.1	negative

The contact angle data obtained on four different hydrogels
were plotted as the cosine of the advancing angle versus the
liquid surface tension (Zisman Plot) and also as the adhesion
tension (the product of the liquid surface tension and the cosine
of the contact angle), W_T , versus the liquid surface tension
(Wolfram Plot). This latter manner of contact angle presentation
has been used by some investigators (10-13). It has yielded a
straight line of reasonably good fit for nonpolar solids and
either aqueous solutions of surfactants or for surfactants dissolved in a two-phase immiscible liquid system.

Wolfram (11,12) and later Lucassen-Reynders (13) have shown
that the slope of the straight line described by the data in the
adhesion tension versus surface tension plot is equal to the
ratio of the Gibbs' surface excess concentration at the solid-
liquid interface (Γ_{sl}) and at the liquid-vapor interface (Γ_{lv}):

$$\text{slope} = -\ \Gamma_{sl}/\Gamma_{lv}$$

Figure 5 shows the Zisman plot obtained for the PHEMA gel of 40% equilibrium water content and Figure 6 shows the Wolfram plot for the same gel. These figures also include data obtained with poly(methyl methacrylate), which served as the control surface.

As can be seen, in both cases the scatter of the data is considerable. The least square method has been used to obtain the straight lines shown in the graph. These were used to obtain the critical surface tension value by extrapolating to $\cos \Theta = 1$ or $W_T = \gamma_{lv}$, for the Zisman plot and the Wolfram plot, respectively. These critical surface tension values are contained in Table II for several acrylic hydrogels as well as for poly(methyl methacrylate). Whenever possible, the gels representative of each type were chosen so that the equilibrium water contents were similar.

Table II

Critical Surface Tension of Acrylic Hydrogels

GEL OR SOLID	WATER CONTENT IN WT %	CRITICAL SURFACE TENSION* ZISMAN PLOT	WOLFRAM PLOT
P M M A	1.5	38.5	38.9
P G M A	73.3	37.6	38.2
P H E M A	40.0	36.0	36.9
P H E A	71.9	34.1	35.6
P A A	77.7	32.9	34.8

* in dyne/cm

All the gels appear to have a critical surface tension value below that of PMMA. These critical surface tension values seem to decrease with increasing water wettability. The only exception to this rule is PGMA, which has a somewhat higher critical surface tension than does PHEMA. This may be due to the presence of the two adjacent hydroxyl group on the glyceryl side chain. However, the difference in the critical surface tension values are rather small and the scatter of data about the straight lines are rather large, so the reverse order of PHEMA and PGMA may not be real.

The tendency of the critical surface tension values of the gels to decrease with increasing water wettability, however, is

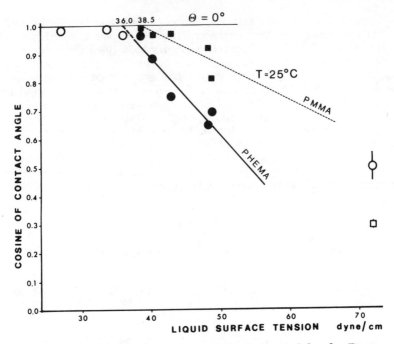

Figure 5. Critical surface tension of PHEMA obtained by the Zisman method using organic liquids. Solid symbols indicate values used to determine the straight lines.

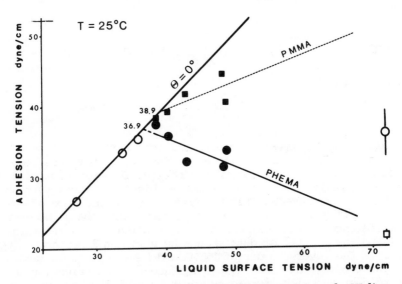

Figure 6. Critical surface tension of PHEMA obtained from the Wolfram plot using organic liquids. Solid symbols indicate values used to determine the straight lines.

significant in our opinion. It probably indicates the existence of lakes of free water at the surface of the gels in increasing proportion. Such water would have a low critical surface tension toward hydrophobic liquids, since the dispersion force component of water surface tension is only about 22 dyne/cm (14).

Criterion of Hydrophilicity Using Two-Liquid Systems

The common definition of a hydrophilic solid surface is that water spontaneously spreads over it, or that the contact angle of water on the solid is zero. This implies that the energy of adhesion of water to the solid is at least as high as the energy of cohesion of water. This definition is based on systems, where only one type of liquid is assumed to be present, so the solid cannot display a preference to another, nonpolar condensed phase.

It has been suggested (15) that the characterization of a solid or a gel surface should be made by water contact angle measurements in the presence of another liquid which is nonpolar. The selection of n-octane as the nonpolar phase has an advantage. Hamilton demonstrated (16) that the change in interfacial tension due to polar interaction across the solid-water interface can be calculated from the two-phase contact angle provided that the nonpolar liquid is n-octane. When there is no polar interaction between the solid and water, the water-octane contact angle is independent of the solid composition and is given by

$$\cos \theta = (\gamma_O - \gamma_W)/\gamma_{OW}$$

where γ_O = 22.4 dyne/cm, γ_W = 71.9 dyne/cm, and γ_{OW} = 51.4 dyne/cm, as obtained in our laboratory, are the surface tension of water-saturated octane, that of octane-saturated water, and the interfacial tension at the water-octane boundary, respectively. The upper limit for the water-octane contact angle obtained on an ideally nonpolar solid is thus equal to 164°.

Water-in-octane contact angle values were obtained for the gels in Table II as well as for PMMA and polyethylene. Table III contains the results together with the water-in-air contact angles obtained in water-vapor-saturated air. All the contact angles in Table III are the advancing type, i.e. the water droplet was increased in size while resting on the surface and the angle is taken through water, the denser phase as is customary. The standard deviations are also included.

It is reasonable to choose $\theta = 90°$ for the water-in-octane contact angles as the arbitrary dividing line between hydrophilic and hydrophobic surfaces, since at this angle the solid-liquid interfacial tensions are the same both with octane and water. As can be seen from the table, all the hydrogels investigated are hydrophilic to various degrees, since for all the gels this

contact angle is less than 90°. PHEMA is near the borderline
dividing the hydrophilic and hydrophobic solids. Conversely, PMMA
and polyethylene are hydrophobic.

Table III

Water-in-Octane Contact Angle Values for Hydrogels and Solids

GEL OR SOLID	WATER CONTENT IN WT %	CONTACT ANGLE OF WATER IN AIR (°)	CONTACT ANGLE OF WATER IN OCTANE (°)
P E	0.0	94.0±1.0	153.2±2.2
P M M A	1.5	72.6±1.5	120.0±1.1
P H E M A	38.9*	59.5±2.3	88.0±1.3
P H E A	71.9	44.0±2.5	79.3±2.1
P G M A	73.3	41.3±1.3	63.3±1.6
P A A	77.7	10.1±0.9	11.5±1.5

* This PHEMA gel had a slightly lower equilibrium water content
than the one included in Table II.

In a basic sense, this criterion of hydrophilicity is more
meaningful than the water advancing (or receding) contact angle,
since two liquid-solid interfacial tensions are compared simultaneously. The water-in-air contact angles, on the other hand,
provide information of practical importance. As a criterion of
hydrophilicity, the critical surface tension obtained by organic
liquids is rather a useless quantity as shown by the fact that
both polyethylene and poly(acrylamide) have similar critical surface tension values.

Conclusions

In conclusion, it was found that the surfaces of methacrylic
and acrylic hydrogels are all hydrophobic to a certain degree
(even at high equilibrium water content) as determined by the
magnitude of the advancing contact angle of water on the fully-
hydrated gel surface. The wettability is primarily determined by

the nature of the polymer and also by the segmental mobility of the polymer chains at the gel boundary.

A large relative contact angle hysteresis was found for all the gels, which increased with decreasing wettability for each series of gels. The gel water content at equilibrium hydration was not found to have a pronounced effect on gel wettability, especially for gels of the lower hydration range.

The critical surface tension obtained by hydrophobic organic liquids after the method of Zisman has not been found to be a useful parameter for the characterization of the water wettability of hydrogels, nevertheless it showed a consistent if reversed relationship with water wettability as determined by the advancing contact angle of water.

Although the water contact angle on gels is of practical importance, the distinction between a hydrophilic and a hydrophobic surface perhaps should be made by the preference indicated by the solid surface toward water in the presence of an inert, nonpolar liquid. By such a criterion, all the gels investigated can be classified as hydrophilic.

Acknowledgement

We wish to thank Professor J.J. Bikerman for calling our attention to the references on gelatin wettability. Ms. Fee-Lai Leong and Ms. Cheryl J. DeVine provided valuable technical assistance. This work was supported by PHS Research Grants No. EY-00208 and No. EY-00327, both from the National Eye Institute, National Institutes of Health.

ABSTRACT

The water wettability of various methacrylic and acrylic hydrogels were determined by the sessile-drop and the captive-bubble techniques. Surprisingly large advancing contact angles were obtained for all the gels, even though the surfaces were fully hydrated. The receding contact angles were much smaller than the advancing contact angles. The magnitude of the contact angle hysteresis appeared to increase with decreasing water wettability. It is proposed that large segmental mobility and amphipathic character of the polymer chains at the gel surface are responsible for the convertibility of the gel surface from a relatively hydrophilic to a hydrophobic one. The water contact angle measured in octane was less than 90° for all the gels studied. By this criterion all the hydrogels investigated have "hydrophilic" surfaces.

Literature Cited

1. Pchelin, V.A. and Korotkina, I.I., Zh.Fiz.Khim.,(1938)12:50

2. Garrett, H.E.,"Aspects of Adhesion", Vol. 2, p. 18 (ed. Alner D.J.) University of London Press Ltd., London, 1966
3. Braudo, E.E., Tolstoguzov, V.B., and Nikitina, E.A., Colloid J. (1974) 36:191
4. Padday, J.F., personal communication
5. Holly, F.J. and Refojo, M.F., J. Biomed. Mater. Res. (1975) 9:315
6. Refojo, M.F., J. Appl. Polymer Sci. (1965) 9:3161
7. Laizier, J. and Wajs, G., U.S. Patent 3,700,573 (1972)
8. Ratner, B.D. and Miller, I.F., J. Polymer Sci. A-1 (1972) 10:25
9. Shafrin, E.G. and Zisman, W.A.,"Contact Angle, Wetting, and Adhesion", p. 145 (ed. Fowkes, F.M.) Adv. Chem. Series #43, Am. Chem. Soc., Washington, D.C. 1964
10. Bargeman, D. and Van Voorst Vader, F., J. Coll. Interface Sci. (1973) 42:467
11. Wolfram, E., Plaste u. Kautschuk (1962) 12:604
12. Wolfram, E., Kolloid Z. u. Z. f. Polymere (1966) 211:84
13. Lucassen-Reynders, E.H., J. Phys.Chem., (1963) 67:969
14. Fowkes, F.M., Ind.Eng.Chem. (1964) 56:40
15. Glazman, Yu.M. and Fapon, I.D.,"Researches in Surface Forces" Proc. 2nd Surface Conf., Acad. Sci. USSR, Moscow, (1956) 2:232
16. Hamilton, W.C., J. Coll. Interface Sci. (1972) 40:219

Water Wettability of Proteins Adsorbed at the Hydrogel–Water Interface

FRANK J. HOLLY and MIGUEL F. REFOJO

Eye Research Institute of Retina Foundation, 20 Staniford Street, Boston, Mass. 02114

The hydrogels in general, and crosslinked poly-(hydroxyethyl methacrylate) [PHEMA] in particular, are presently being used as contact lens materials and are considered as biomaterials or coatings for biomaterials. The water-soluble surface-active substances in the blood plasma and tissue fluids including tears are proteinaceous in character. Therefore, the characteristics of the interface between hydrogels and water and its interaction with dissolved proteins are of importance.

Past Work

Little is known of the energetics of the hydrogel-water interface. Water has an unexpectedly high advancing contact angle (60-80°) on PHEMA gel surfaces (1) and even gels of more hydrophilic polymers such as poly(glyceryl methacrylate) [PGMA], and poly(hydroxyethyl acrylate) [PHEA] with twice the equilibrium water content of PHEMA exhibit advancing contact angles as high as 40° (2). A solid with such a low water wettability is expected to have an interfacial tension of considerable magnitude against water. The low receding contact angle value on hydrogels, however, indicates that this is not the case. Due to the considerable segmental mobility of the surface polymer chains of the gel matrix, orientational changes at the gel-water interface seems to lessen the interfacial tension significantly by increasing the density of the hydrophilic sites at the surface. Tension at the gel-water interface has been estimated to be fairly low by others (3). If such is the case, then the water-soluble proteins are expected to interact little with such an interface and adsorption would probably entail little if any irreversible conformational changes or denaturation of

the protein molecules.
  Protein adsorbed at hydrogel-water interfaces has
not been characterized in detail. To our knowledge, no
systematic attempts have been made to study the energetics including wettability of proteins adsorbed at
the hydrogel-water interface.

Materials and Methods

  We have studied the effect of protein adsorption
on the water contact angle on PHEMA gels having either
38.9 or 40.0% equilibrium water content. Some measurements have also been made on a crosslinked PHEA gel
containing 89.9% water at equilibrium hydration. These
gels are described elsewhere in this book (2). Occas-
ionally poly(methyl methacrylate) [PMMA] and polyethylene [PE] were also used as adsorbents for comparison.
  The protein used in this study was bovine serum
albumin [BSA] (Cohn Fraction V, Sigma Chemical Co., St.
Louis, MO) with a molecular weight of 68,000 g/mole. A
high molecular weight, commercially available glyco-
protein, bovine submaxillary mucin [BSM] (type I, MW=
4.1×10^6 g/mole, Sigma Chemical Company, St. Louis, MO)
and egg lysozyme [LYZ] (as chloride, MW=14,800 g/mole,
Miles Laboratories, Kankakee, IL) were also used in the
study. For comparison purposes, synthetic, water-soluble polymers were also used in the preliminary measurements.
  Contact angles were measured by using a Ramé-Hart
goniometer with an environmental chamber so that the
hydrogel sample could be kept in an atmosphere saturated with water vapor. Surface tension of the aqueous
solutions was measured by the Wilhelmy method using a
roughened platinum blade and a Cahn Electrobalance Model RM-2. Both the surface tension and the contact
angle were monitored for thirty minutes. Measurements
made at longer times indicated that such a time interval was sufficient to approximate equilibrium at least
for solutions with concentrations 10^{-2} wt% and above.

Effect of Synthetic and Biopolymers on Adhesion Tension

  The advancing (but static) contact angle for aqueous polymer solutions was determined on the PHEMA gel,
the PHEA gel, PMMA, and polyethylene (2). In addition
to the solutions of the three biopolymers already described, the following synthetic or modified natural
polymers were employed: Poly(vinyl alcohol) (Gelvatol
20/90), poly(ethylene oxide) (WSR-205), hydroxyethyl-
cellulose, and poly(vinyl pyrrolidone). The polymers

were dissolved in distilled water to a solution concentration of 0.1% by weight. Aqueous solutions of sodium
sulfosuccinate and methyl cellulose (USP grade) at 0.1%
concentration were also used on PMMA and PE. See Table
I for the surface tension of these solutions.

Table I

Surface Tension of Aqueous Solutions of Synthetic and

Biopolymers and Surfactant*

SOLUTE	SURFACE TENSION (dyne/cm)
Poly(vinyl alcohol)	52.6
Poly(vinyl pyrrolidone)	67.5
Methyl cellulose	55.4
Hydroxyethyl cellulose	63.5
Poly(ethylene oxide)	60.7
BSM	41.2
BSA	52.8
LYZ	58.0
Sodium sulfosuccinate	30.8

*
at 0.1% concentration and at 30 minutes

Figure 1 shows the results where the adhesion tension of the solutions to the gel or solid are plotted
as a function of the solution surface tension. For each
substrate, the polymer solutions yielded data which
appear to exhibit a linear dependence of the adhesion
tension on the solution surface tension. The 45° broken
line on the graph represents zero contact angle (adhesion tension = solution surface tension). Its intercept with the straight line, calculated by the least
square method for each substrate, represents a kind of

"critical surface tension" obtained with these aqueous
solutions for the particular solid or gel. The slopes
of the straight lines for PE, PMMA, PHEMA, and PHEA are
equal to -0.76, -0.18, +0.16, and +0.46, respectively.
It has been reported (4-6) that aqueous solutions
of surfactants, when placed on a nonpolar solid such as
paraffin, form contact angles that exhibit a general
relationship with respect to the surface tension of the
solution. The product of the surface tension of the
solution and the cosine of the contact angle (adhesion
tension, W_T) varies linearly with solution surface tension (γ_{lv}) such that

$$W_T = \gamma_{lv} \cdot \cos\Theta = A\gamma_{lv} + B \quad \text{where } A < 0$$

For low molecular weight surfactants, for which
the Gibbs adsorption equation is valid, the slope, A,
has been shown (4,6) to be equal to the negative ratio
of the surface excess concentration of the surfactant
at the solid-water interface (Γ_{sl}) and that at the
solution-air interface (Γ_{lv}). Assuming that the solute
surface excess concentration at the solid-vapor interface is zero, we may write:

$$A = -\Gamma_{sl}/\Gamma_{lv}$$

When A = -1, the extent of adsorption at the two
interfaces is the same and the adhesion energy of the
solution to the solid becomes independent of the solution surface tension. Such a situation has actually
been observed (7). If A = 0, the concentration of the
solute is the same at the solid-solution interface as
in the bulk, i.e. the interfacial excess concentration
is zero, $\Gamma_{sl} = 0$. Such a surface tension dependence
of adhesion tension was demonstrated by Lucassen-Reynders (6) using the data of Smolders (8) obtained in a
hydrogen gas - aqueous anionic surfactant solution -
mercury system at the electrocapillary maximum (zero
interfacial charge density). Positive slopes would mean
that the surface excess concentration of the solute(s)
is negative, i.e. the solute concentration is lower at
the interface than in the bulk.
Figure 1 demonstrates that the surface-active
polymer solutions behave similarly to simple surfactant
solutions on nonpolar solids. While the validity of the
Gibbs equation does not extend to macromolecules, it is

reasonable to suppose that the magnitude of the slope A reflects in some way the magnitude of the interfacial adsorption of the polymeric solute or at least its effect on the interfacial tension.

If this is so, then the results indicate that considerable adsorption of the polymeric solute takes place at the polyethylene-water interface. This is expected in view of the large interfacial tension at the polyethylene-water boundary. This effect is considerably less for the more polar PMMA surface as shown by the smaller absolute, but still negative, value of A for PMMA. This is no doubt due to the greater interaction of PMMA surface with water which competes with the polymeric solutes for the available adsorption sites.

It is quite interesting that the PHEMA gel is characterized by a straight line with a slightly posi-tive slope, which may indicate no difference between interface and bulk concentration or even lower concentration at the interface than in the bulk. The slope of the line for PHEA is also positive and even greater than that for PHEMA.

The scatter of the data is considerable for each solid and it is especially large for the gels. There are indications (9) that the type of the solutes employed also affect the position of the straight line in this Wolfram plot (adhesion tension vs. surface tension) especially for multicomponent polymeric solutions. Thus, considering the wide spectrum of the polymeric solutes used in this study, the results fit the straight lines reasonably well.

Film Pressure of Biopolymers Adsorbed at the Interface

Attempts have been made to obtain the film pressure of the biopolymers at the water-PHEMA gel and water-polyethylene interfaces as a function of the solution concentration. The decrease of interfacial tension, due to the adsorption of the biopolymer, is the film pressure of the adsorbate at the interface. It is given by the difference between the solution-solid and the solvent-solid adhesion tensions (10):

$$\Pi_i = \gamma_{sol'n} \cdot \cos\theta_{sol'n} - \gamma_w \cdot \cos\theta_w$$

The film pressures of albumin calculated for the PHEMA gel - water and the PE - water interfaces, together with the film pressures at the water-air inter-

interface, are shown in Figure 2.
The results show that the film pressure of albumin
is similar at the water-air and water-polyethylene interfaces, especially near the physiologic concentration
range (0.1 - 6%). On the other hand, the film pressure
of albumin at the PHEMA-water interface is not significantly different from zero within the rather large
deviation of the data, although it appears to decrease
with decreasing solution concentration.
The film pressure of mucin and lysozyme show similar variation with solution concentration at the PE-
water interface. At the PHEMA gel - water interface,
however, the film pressure is negative at high solution concentrations and approximates zero as the solution concentration is lowered.

Wettability of Biopolymers Adsorbed onto PHEMA and PE

The contact angle of the sessile droplet of the
biopolymer solutions on PHEMA and PE substrates that
had been exposed to a biopolymer solution of the same
composition and concentration was determined. By using
the same solution for the adsorbing medium and for the
wettability measurement, we avoided gross changes in
the interfacial tension at the droplet-substrate boundary due to dissolution or further deposition of the
biopolymers.
After the substrate had been in the biopolymer
solution of a given concentration for 1 hour, it was
gently blotted with a grease-free filter paper to remove the excess liquid. Then the contact angle of a
sessile droplet was determined in the environmental
chamber 30 minutes after the droplet of the same solution was deposited on the substrate. Afterwards, the
substrate surface was rinsed with distilled water,
blotted, and the contact angle was determined using the
same biopolymer solution. This measurement was repeated
at five different biopolymer concentrations for both
substrates.
Figures 3 and 4 show the advancing contact angles
obtained with bovine serum albumin solution on albumin
adsorbed on PHEMA gel and polyethylene as a function of
solution concentration. These graphs also contain the
data obtained with albumin solution on the clean, al-
bumin-free surfaces. The results obtained with mucin
and lysozyme are not shown here but are qualitatively
similar (11).
Albumin adsorbed from solutions containing the
protein at concentrations of the same order of magnitude as that of physiologic (tear and blood plasma)

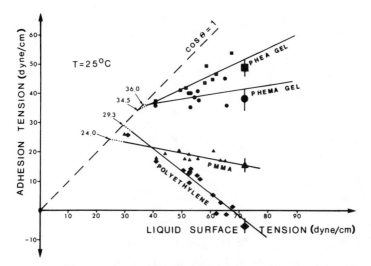

Figure 1. *Adhesion tension of aqueous polymer solutions to gels and solids as a function of the solution surface tension. (Large symbols indicate values obtained with pure water also showing standard deviation.)*

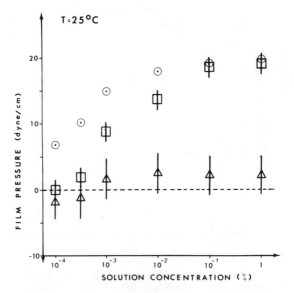

Figure 2. *Film pressure of bovine serum albumin at three interfaces: ⊙, water–air; □, water–polyethylene; △, water–PHEMA gel. (Vertical lines show standard deviation.)*

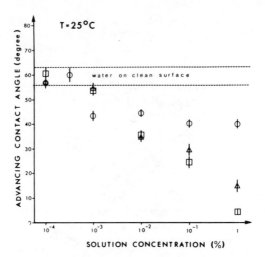

Figure 3. Wettability by albumin solutions of bovine serum albumin adsorbed onto PHEMA gels. ◯, clean surface. Surface exposed to albumin solution for 1 hour and blotted only (☐) then rinsed in distilled water and again blotted (△).

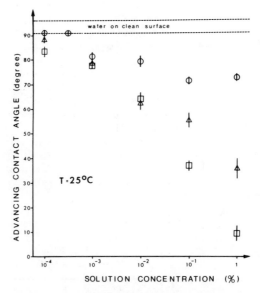

Figure 4. Wettability by albumin solutions of bovine serum albumin adsorbed onto polyethylene. ◯, clean surface. Surface exposed to albumin solution for 1 hour then blotted only (☐) then rinsed in distilled water and again blotted (△)

levels; 0.1 and 1%, makes both types of surfaces more wettable. At the solution concentration of 1%, no difference is discernible between the wettability of albumin-coated PHEMA and polyethylene. Rinsing decreases the wettability somewhat, but at these concentrations, it is still greater than that of the respective clean surface.

At the solution concentrations of 10^{-2} and 10^{-3} %, rinsing does not decrease wettability, i.e the adsorbed albumin layer is thinner but it is more tightly bound to both PHEMA and PE. At the concentration of 10^{-2} %, both surfaces are still more wettable than the respective clean surfaces. At the concentration level of 10^{-3} %, however, albumin adsorption does not seem to change wettability for the polyethylene and it actually decreases wettability for PHEMA.

At the lowest concentration level studied, 10^{-4} %, the wettability by albumin solution is the same for the blotted, rinsed, or clean surfaces of PHEMA and it is approximately the same as the water value on the clean gel surface. For polyethylene, only the exposed and blotted surface is somewhat more wettable than the clean surface, while the rinsed and the clean surfaces are wetted by this dilute albumin solution to about the same extent as clean polyethylene is wetted by water.

The receding contact angle of the respective solutions was found to be essentially zero for mucin, albumin, and lysozyme adsorbed on PHEMA and polyethylene from aqueous solutions having a biopolymer concentration 10^{-2} % or higher.

Critical Surface Tension of Adsorbed Biopolymers

In the past, we have determined the critical surface tension of mucin (12) and the two other biopolymers (13) using hydrophobic, diagnostic liquids according the well-known method of Zisman (14). However, we believe that such a parameter has little relevance to wettability in an aqueous system, especially if the segmental mobility of the surface polymer structure is large as it is with hydrogels (2) which also appears to be the case with adsorbed biopolymers. Therefore, we used the contact angle data obtained for clean, blotted, and rinsed surfaces that were discussed in the previous section, and plotted them in two ways: 1. cosine of the advancing angle (Zisman plot), and 2. the adhesion tension (Wolfram plot) both against the solution surface tension. Then - assuming a straight line fit - we extrapolated to zero contact angle using the least square method.

The results obtained with albumin on PHEMA are
shown in Figures 5 and 6, in a Zisman and Wolfram plot,
respectively. The albumin data on polyethylene are
shown in Figures 7 and 8. The pure water value with the
standard deviation shown are also indicated on these
graphs.
    For the albumin solution on a clean surface, the
contact angle values obtained at the five different
solution concentrations define a straight line reasonably well. For the albumin-coated surfaces, only the
contact angle values obtained at solution concentrations 1%, 0.1%, and 0.01% fall on the same straight
line.
    Similar graphs (not shown here) can be obtained
for mucin and lysozyme, except that for these biopolymers, even on a clean surface, only the values obtained
with the three most concentrated solutions gave a reasonably good fit to a straight line (11).
    Table II contains the critical surface tension
values obtained with the biopolymer solutions for the
clean substrates, PHEMA and PE. For comparison, the
table also contains the critical surface tension obtained with hydrophobic organic liquids (2).

Table II

Critical Surface Tension of PHEMA and PE as

Obtained by Aqueous Biopolymer Solutions

| DIAGNOSTIC LIQUIDS | CRITICAL SURFACE TENSION* | |
	PHEMA	POLYETHYLENE
Organic liquids	36.9 ; 36.0	34.7 ; 32.5
BSM sol'n (3 conc.)	21.8 ; 21.7	33.8 ; 32.4
BSA sol'n (5 conc.)	45.8 ; 42.5	36.4 ; 23.3
LYZ sol'n (3 conc.)	37.0 ; 27.8	29.8 ; 6.5

---

* in dyne/cm (Wolfram ; Zisman)

    Due to the small number of data points used to
calculate the straight lines, the exact numerical value

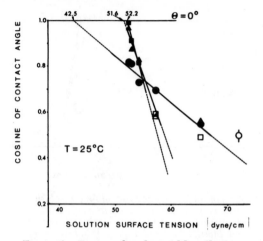

Figure 5. *Zisman plot obtained by albumin solution on clean and albumin-coated PHEMA gel surfaces. ○, clean surface. Surface exposed to albumin solution then blotted only (□) then rinsed in distilled water and again blotted (△). Solid symbols indicate values used to obtain the straight lines. Water value is also shown.*

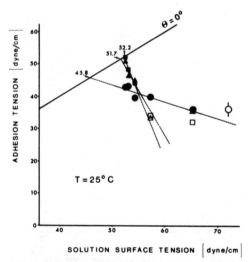

Figure 6. *Wolfram plot of adhesion tension obtained by albumin solutions on clean and albumin-coated PHEMA gel surfaces. ○, clean surface. Surface exposed to albumin solution for 1 hour then blotted only (□) then rinsed in distilled water and again blotted (△). Solid symbols indicate values used to obtain the straight lines. Water value is also shown.*

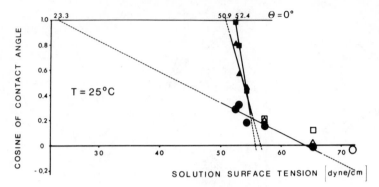

Figure 7. Zisman plot obtained by albumin solutions on clean and albumin–coated polyethylene. ○, clean surface. Surface exposed to albumin solution for 1 hour and blotted only (□) then rinsed in distilled water and again blotted (△). Solid symbols indicate values used to obtain the straight lines. Water value is also shown.

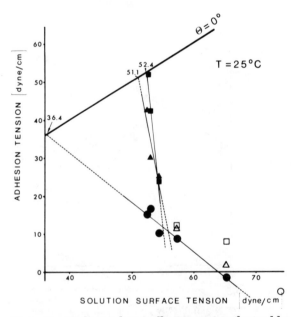

Figure 8. Wolfram plot of adhesion tension obtained by albumin solutions on clean and albumin-coated polyethylene. ○, clean surface. Surface exposed to albumin solution for 1 hour then blotted only (□) then rinsed in distilled water and again blotted (△). Solid symbols indicate values used to obtain the straight lines. Water value is also shown.

of the "critical surface tension" parameter is of limited significance. However, a few interesting observations can be made: 1. The critical surface tension of
polyethylene is approximately the same irrespective of
whether the hydrophobic liquids or any of the biopolymer solutions is used provided that it was obtained
from the Wolfram plot. When the Zisman method is used,
only the mucin solutions yield a critical surface tension value similar to that obtained with the hydrophobic liquids. 2. For the PHEMA surface, the highest
critical value (Wolfram) is obtained with the albumin
solutions. The lysozyme solutions yield a value (Wolfram) of critical surface tension practically identical
to the value obtained with hydrophobic liquids. The
critical surface tension value for PHEMA obtained by
the mucin solutions is as low as the dispersion force
field component of the water surface tension.

Table III contains the critical surface tension
value for the adsorbed biopolymers as determined by the
aqueous biopolymer solutions both before and after rinsing. The critical surface tension values obtained by
hydrophobic liquids are not shown. In a previous study
(12), the critical surface tension value for mucin adsorbed on polyethylene or glass was found to be about
38 dyne/cm, while the critical values for albumin and
lysozyme adsorbed on polyethylene were found to be
slightly higher, between 39-41 dyne/cm (13).

Table III

Critical Surface Tension of Biopolymers Adsorbed

on PHEMA and Polyethylene

BIOPOLYMER	CRITICAL SURFACE TENSION OF ADSORBED BIOPOLYMER ON*	
	P H E M A	POLYETHYLENE
BSM Solutions (3 conc.)	38.8 ; 38.7	37.8 ; 37.6
" " after rinsing	38.4 ; 38.3	33.7 ; 34.7
BSA Solutions (3 conc.)	52.2 ; 52.2	52.4 ; 52.4
" " after rinsing	51.7 ; 51.6	51.1 ; 50.9
LYZ Solutions (3 conc.)	45.2 ; 43.4	41.8 ; 31.1
" " after rinsing	37.4 ; 30.0	35.0 ; 21.6

* in dyne/cm (Wolfram ; Zisman)

With the exception of lysozyme, there is a good
agreement between the critical values obtained from the
Wolfram plot and those obtained from the Zisman plot.
Since the critical surface tension values obtained from
the Wolfram plot appear to be more consistent, it appears that the former values are more relevant for
aqueous solutions.

It may be mentioned here that these wettability
data were also treated according to the equation of
Good and Girifalco (15):

$$\cos\Theta = 2\Phi(\gamma_{sv}/\gamma_{lv})^{\frac{1}{2}} - 1$$

By assuming that the solvent interaction factor,
Φ, is constant, the fitting of a straight line to the
data in a $\cos\Theta$ vs. $(\gamma_{lv})^{-\frac{1}{2}}$ plot yielded critical surface tension values by extrapolating to $\cos\Theta = 1$. These
critical surface tensions compared well with those obtained from the Wolfram plot (see Tables I and II) except that the latter values appeared to be somewhat
more consistent (11).

It is of considerable interest that the critical
surface tension values for the unrinsed surfaces are
the same for both the PHEMA gel and the polyethylene
substrates. This observation is in line with the accepted view concerning the short range of surface forces.
The adsorbed albumin layer has the highest critical
surface tension, 52 dyne/cm, while the adsorbed mucin
layer has a critical value of about 38-39 dyne/cm. It
may be significant that these values are approximately
the same as the surface tension of the more concentrated aqueous solutions of the respective biopolymers. The
critical surface tension of the adsorbed lysozyme seems
to fall between these two critical values.

Conclusions

In conclusion, it was found that the two hydrogels,
PHEMA and PHEA with equilibrium water content 40% and
90%, respectively do not exhibit a high adsorptivity
at the water interface for a number of synthetic and
biopolymers. The water-gel interfacial tension at least
does not seem to change much in the presence of these
solutes and may even increase in certain cases.

The film pressure of albumin at the PHEMA gel -
water interface is small and approaches zero with decreasing solution concentration. On the other hand, the

film pressure of mucin and lysozyme at the PHEMA-water interface appears to be somewhat negative at high solution concentration, approximating zero with decreasing solution concentration. In other words, the decrease of water contact angle on the gel resulting from the presence of the biopolymer in the droplet may be accounted for solely by the change in the surface tension of water due to the adsorption of the biopolymer at the solution-air interface.

Nevertheless, biopolymers adsorbed on PHEMA gels from aqueous solution increase wettability. While the solution contact angle increases somewhat after rinsing at high solution concentration, the critical surface tension value obtained by these solutions show little change with rinsing for BSA and BSM adsorbed on PHEMA.

The critical surface tension values obtained by the aqueous biopolymer solutions from the adhesion tension vs. solution surface tension plot appear to be the most consistent among the different systems compared. Numerically, for albumin, mucin, and lysozyme respectively, it is equal to 52, 39, and 45 dyne/cm, when adsorbed on PHEMA. The corresponding critical values for these biopolymers adsorbed on polyethylene are 52, 38, and 42 dyne/cm.

Acknowledgement

Ms. Fee-Lai Leong and Ms. Cheryl J. DeVine provided vauable technical assistance. This work was supported by PHS Research Grants No. EY-00208 and No. EY-00327, both from the National Eye Institute, National Institutes of Health.

ABSTRACT

The surface tension of synthetic and biopolymer solutions and the contact angle on two hydrogels, poly-(hydroxyethyl methacrylate) and poly(hydroxyethyl acrylate), and two hydrophobic solids, polyethylene and poly(methyl methacrylate) were measured. The adhesion tension vs. the solution surface tension plot yielded an approximately straight line for each of the gels and solids. From the slope of the lines it was concluded that the effect of polymer adsorption on the gel-water interfacial tension is small as compared to that at the hydrophobic solid - water interface.

Adsorbed biopolymers increased the wettability of PHEMA gel and polyethylene. Critical surface tension values for the adsorbed biopolymers were obtained from

the adhesion tension vs. surface tension plot using the
aqueous contact angle data. They were found to be independent of the substrate, unaffected by rinsing (for
PHEMA) and was similar in magnitude to the surface tension of the concentrated biopolymer solutions for albumin and mucin on PHEMA and polyethylene.

Literature Cited:

1. Holly, F.J. and Refojo, M.F., J. Biomed. Mater.
    Res. (1975)9:315
2. Holly, F.J. and Refojo, M.F., These Proceedings.
3. Andrade, J., Lee, J.B., Jhon, M.S., Kim, S.W., and
    Hibbs, J.R.,Jr., Trans. Amer. Soc. Int. Organs,
    (1973)19:1
4. Wolfram, E., Plaste u. Kautschuk,(1962)12:604
5. Bargeman, J., J. Coll. Interface Sci.(1973)42:467
6. Lucassen-Reynders, E.H., J. Phys. Chem.(1963)67:
    969
7. Wolfram, E., Kolloid Z. u. Z. f. Polymere,(1966)
    211:84
8. Smolders, C.A., Rec. trav. chim. (1961)80:699
9. Holly, F.J., ARVO Abstracts, p. 109, Spring, 1975
10. Harkins, W.D.,"Rec. Adv. in Surf. Chem. and Chem.
    Phys.", p. 19 (ed. Moulton, F.R.), The Science
    Press, Lancaster, PA, 1939
11. Holly, F.J. and Refojo, M.F., manuscript in prep-
    aration.
12. Holly, F.J. and Lemp, M.A., Exp. Eye Res. (1973)
    11:239
13. Holly, F.J., unpublished results.
14. Fox, H.W. and Zisman, W.A., J. Coll. Sci.(1952)7:
    428
15. Good, R.J. and Girifalco, L.A., J. Phys. Chem.
    (1960)64:561

Radiation-Induced Co-Graft Polymerization of 2-Hydroxyethyl Methacrylate and Ethyl Methacrylate onto Silicone Rubber Films

TAKASHI SASAKI,* BUDDY D. RATNER, and ALLAN S. HOFFMAN

Department of Chemical Engineering and Center for Bioengineering, University of Washington, Seattle, Wash. 98195

A number of hydrogel systems have demonstrated good tissue compatibility and low thrombogenicity in in vivo and in vitro tests (1). The poor mechanical properties of hydrogels have encouraged the development of techniques for reinforcing hydrogels to make them suitable for use in biomedical devices. By radiation grafting monomers such as 2-hydroxyethyl methacrylate (HEMA), N-vinyl-2-pyrrolidone, and acrylamide onto strong, inert polymeric supports, materials have been produced which combine the desirable biocompatibility properties of the hydrogel systems with the good mechanical properties of the substrate polymer (2-8).

A clear correlation between the surface hydrophilicity of radiation grafted hydrogels and their biocompatibility has not yet been firmly established. Many useful biomaterials demonstrate a balance between hydrophilic and hydrophobic sites which might be important for biocompatibility. Some of these materials are summarized in Table I.

In order to systematically investigate the interrelationship between the hydrophilic-hydrophobic composition of a polymeric material and biological interactions with that material, a series of well characterized radiation graft copolymers of poly(HEMA) and poly(ethylmethacrylate) (EMA) have been prepared. A preliminary report on the preparation and characterization of these graft copolymers is presented here.

Experimental

Silicone rubber films, 1.9 cm. x 3.8 cm. x 10 mils thick (Silastic, type 500-3, Dow Corning Corp.) were cleaned with five minute sonications in 0.1% Ivory soap solution and then in distilled water (3 times). After washing they were equilibrated at

*On a leave of absence from the Takasaki Radiation Chemistry Research Establishment, J.A.E.R.I., Takasaki, Japan.

Table I

Thromboresistant Surfaces Which May Demonstrate a Balance
Between Hydrophilic and Hydrophobic Sites.

Material	Hydrophobic Site	Hydrophilic Site	Reference
blood vessel wall	lipid	protein, poly-saccharide	(9)
copolyether-polyurethanes	polyurethane blocks	polyether blocks	(10)
perfluorobutyryl ethylcellulose	perfluorobutyryl groups	cellulose rings	(11)
Stellite 21	tallow	steel	(9)
Poly(HEMA)	backbone methyl groups	hydroxyl groups	(12)
poly(vinyl acetate-crotonic acid)	vinyl acetate groups	carboxylic acid groups	(8)

52% R.H. The initial weight of the silicone rubber (W_s) was
measured at 52% R.H.

Highly purified HEMA was generously supplied by Hydron
Laboratories, Inc. and was used as received. EMA monomer was used
after distilling at a reduced pressure. Reagent grade solvents
and chemicals were used as received.

The films were immersed in monomer solutions and the system
was deoxygenated by bubbling nitrogen gas through the solution
for 30 minutes. Irradiation was performed in a ca. 20,000 Ci
Co-60 source. The irradiation dose used was 0.25 Mrad unless
otherwise specified.

After the irradiation, the films were washed twice (30 min.
and 2 hrs.) with acetone-methanol 1:1 mixture, and then with
distilled water for 24 hours changing the water twice during this
period. The wet weight (W_w) of the grafted films was measured by
blotting between two sheets of filter paper (Whatman #1) for ten
seconds under a constant presssure and then weighing. Films were
then dried in an evacuated desiccator containing magnesium
perchlorate for more than 12 hrs. at which point the dry weight
(W_d) was measured. The degree of graft and the water content in
the graft were defined as:

$$\text{wt. \% graft} = [(W_d - W_s)/W_d] \times 100$$

$$\text{water content (\%)} = [(W_w - W_d)/(W_w - W_s)] \times 100$$

Results

A preliminary experiment on the radiation induced bulk copolymerization of HEMA and EMA showed that the polymerization rate decreases abruptly as the fraction of EMA in a HEMA–EMA monomer mixture increases. Similar behavior was noted in a co-graft polymerization of HEMA and EMA onto silicone films in acetone or acetone–methanol (1:1 mixture), as is shown in Fig. 1. When the HEMA portion in the monomer mixture is more than 90 vol. %, precipitation of homopolymer in acetone was observed.

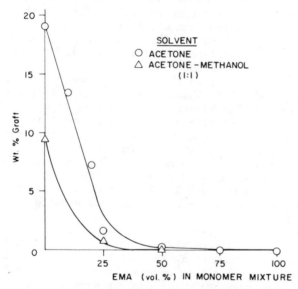

Figure 1. Co-graft polymerization of HEMA–EMA at various monomer compositions. Total monomer, 20 vol %.

In surveying appropriate solvent systems for the co-graft polymerization an accelerative effect by water on the graft polymerization, especially in alcoholic systems, was noted. Fig. 2 shows the effect of water on the grafting rate of 1:1 (by volume) mixture of HEMA and EMA in methanol (MeOH), ethanol (EtOH) and acetone. It can be seen from the figure that the addition of water to any of these solvents increases the grafting rate. Pure EtOH appears to be somewhat superior to pure MeOH for obtaining higher levels of graft, but the accelerative effect of water is more marked in the methanolic system.

Co-graft polymerization with various monomer compositions was carried out in a MeOH-H_2O or an EtOH-H_2O system. Fig. 3 and 4 show the relationship between the degree of graft and monomer ratio in these systems.

In Fig. 3, it should be noted that the wt.% graft is almost constant in the range of 25 to 75 vol. % of EMA in the monomer mixture at two monomer concentrations. This tendency is quite different from that shown in Fig. 1. In the solutions consisting of only EMA, some precipitation of homopolymer occurred though the viscosity of the solutions themselves remained rather low.

Fig. 4 shows another aspect of the copolymerization behavior of this system. The cografting in pure EtOH has a somewhat similar tendency to that seen in acetone-MeOH (Fig. 1). That is, the wt. % graft decreases with an increasing proportion of EMA in the monomer. The addition of water to the monomer solution, however, changes the situation entirely. Thus, with 7.5 vol. % of water in the solution, the wt. % graft is nearly constant in the range of 25 to 100 vol. % EMA in the monomer mixture. Addition of more water leads to practically the reverse trend from that seen in pure EtOH.

The effect of water on the grafting rate of EMA or HEMA in MeOH-H_2O and EtOH-H_2O is shown in Figure 5 which is a crossplot of some of the data in Figure 4. The degree of graft of pure EMA markedly increases with an increase of water in the solution. Precipitation of pure EMA homopolymer occurred when the fraction of water was more than 5% in the MeOH-H_2O system or 15% in the EtOH-H_2O system. In the ethanolic system, the addition of water to the HEMA solution shows a retarding effect. However, very little effect on HEMA graft level is noted upon the addition of water to the methanolic system in the concentration range studied.

Fig. 6 shows the graft water content for those samples which were prepared with various HEMA-EMA mixtures in an EtOH-H_2O solvent system (7.5% of water in the solution). The wt. % graft for these samples are almost constant ($\sim$22%). It can be seen from this figure that the water content decreases with increasing EMA portions in the monomer mixture. Upon dehydration and then rehydration of grafts contains larger portions of EMA, water contents are found to drop even further. After this treatment these films gave constant water contents, ca. 2-3% for 1/3 HEMA-EMA grafted films and > 1% for pure EMA grafted films, regardless of the graft level.

The water content of the graft is usually independent of graft level at a given monomer ratio if more than 50% HEMA is present in the monomer mixture. However, this is no longer observed for larger EMA portions. In Fig. 7 the water contents of a number of HEMA-EMA grafted polymers are plotted against the wt. % graft. In cases where more than 50% of the monomer

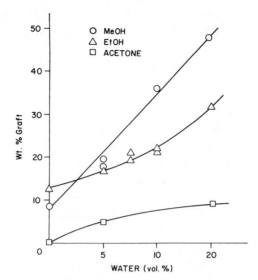

Figure 2.   *Effect of water on the grafting of a 1:1 mixture of HEMA and EMA in various solvents. Dose, 0.25 mrad; total monomer, 25 vol %.*

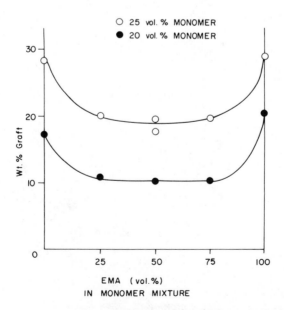

Figure 3.   *Co-graft polymerization of HEMA–EMA in MeOH–H₂O (5 vol % H₂O)*

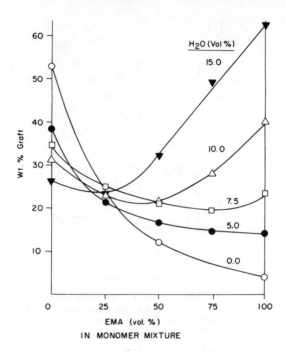

Figure 4.   Co-graft polymerization of HEMA–EMA
(25 vol %) in EtOH–H₂O

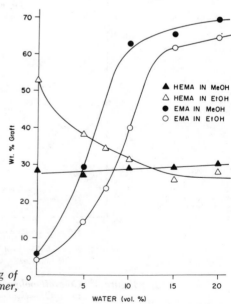

Figure 5.   Effect of water on the grafting of
HEMA or EMA. Dose, 0.25 Mrad; monomer,
25 vol %.

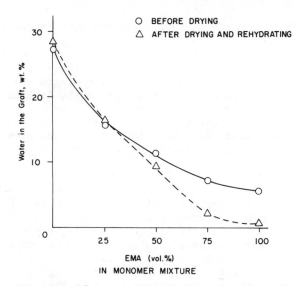

Figure 6. *Effect of monomer composition on the graft water content. Samples were prepared in EtOH–H₂O. Monomer, 25 vol %. H₂O = 7.5%; dose, 0.25 Mrad or 0.16 Mrad (HEMA).*

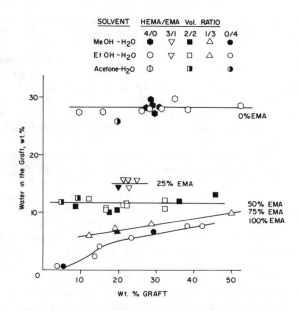

Figure 7. *Plot of water content vs. wt % graft. Total monomer, 25 vol %.*

mixture is EMA the water content increases with increasing graft level.

Figure 8 shows the results of a 2 hour vena cava ring test for a series of HEMA-EMA co-grafted rings. The graft levels of these samples are different and the water contents are larger than those for the films investigated above, probably due to differences in the formulation of silicone rubbers. The results, however, clearly show that HEMA-EMA co-grafted rings as well as pure HEMA grafted rings have much lower thrombogenicity than pure EMA grafted or ungrafted silicone rubber rings.

Discussion

The pronounced effect of small amounts of water on the grafting of HEMA-EMA mixtures to silicone rubber can be mainly attributed to the Trommsdorff effect. That is, the addition of the non-solvent water to an alcoholic solution of EMA will cause the growing chains to adopt a less expanded conformation which hinders the chain termination step to produce higher grafting rates. The situation is the reverse for pure HEMA systems as EtOH or MeOH mixtures with water are good solvents for HEMA polymer. This explanation could account for the copolymerization behavior shown in Fig. 3 and 4. Changes in the partitioning of the monomer between the silicone rubber and the grafting solution may also influence the grafting behavior demonstrated by this system upon the addition of water.

The variation in water content with degree of graft for a given monomer ratio only at fractions of EMA greater than 50% in the monomer mixture may be due to alterations in the porosity of these grafted films. The water in the solvent system may cause precipitation of EMA polymer as it is formed. This proces occurs simultaneously with the surface grafting. Such precipitated polymers often have an open cellular, porous structure. Thus, the water content in the gel should be related to the porosity of the graft as well as to the HEMA-EMA ratio and should therefore be highly dependent on the solvent conditions at the time of grafting. Upon drying, such a porous structure may collapse and, due to hydrophobic interactions between the poly(EMA) chains, be resistant to rehydration. This situation would account for the behavior seen at high EMA ratios in Figure 6 in which the water contents of grafted films which have been dried and then re-hydrated are found to be lower than they were before drying.

The in vivo evaluation of the thrombogenicity of silicone rubber rings by the vena cava ring test (figure 8) is consistent with the idea that a balance of hydrophobic and hydrophilic sites on surface might be important for the blood compatibility of materials. Two other conclusions can also be drawn from this experiment. First, low apparent thrombogenicity (as evaluated by the vena cava ring test) can be achieved using hydrogel systems which have low water contents (i.e. dow to ~ 12% H_2O).

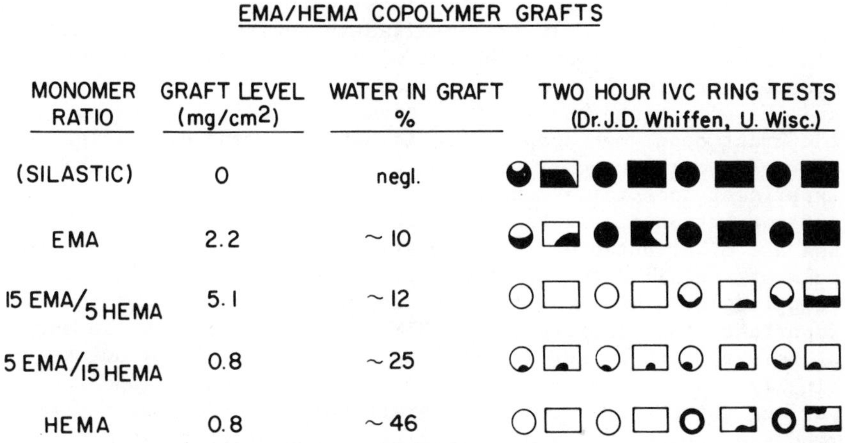

EMA/HEMA COPOLYMER GRAFTS

MONOMER RATIO	GRAFT LEVEL (mg/cm^2)	WATER IN GRAFT %	TWO HOUR IVC RING TESTS (Dr. J.D. Whiffen, U. Wisc.)
(SILASTIC)	0	negl.	
EMA	2.2	~10	
15 EMA/5 HEMA	5.1	~12	
5 EMA/15 HEMA	0.8	~25	
HEMA	0.8	~46	

Figure 8. Results of 2-hr vena cava ring tests for silastic and various HEMA/EMA grafted copolymers on silastic

Second, the vena cava ring test does not have the sensitivity to distinguish among most of the grafted polymers used in this experiment. As the series of grafted polymers differ dramatically in water contents and surface properties, some difference in performanace between grafted HEMA and HEMA-EMA copolymers would be expected.

The addition of 25% HEMA to the monomer mixture was found sufficient to produce a surface which, although possessing a low water content, is significantly less thrombogenic than the hydrophobic Silastic or Poly(EMA)-g-Silastic surfaces. The relatively linear decrease in water content with increasing EMA in the monomer mixture (Figure 6) is taken to indicate that both HEMA and EMA enter into the graft copolymer at similar rates and that the graft copolymer composition is not different by a large degree from the monomer solution composition. However, this experiment does raise a number of questions which effect the interpretation of the results. The degree to which the HEMA and EMA are distributed between the silicone rubber surface and the bulk of the silicone rubber is not presently known. The precise fractions of HEMA and EMA in the copolymer, and perhaps more important, the distribution of HEMA and EMA groups at the surface of the graft copolymer interfacing with the blood, is not known. Finally, the importance of the level of graft on the apparent surface thrombogenicity of the graft polymer cannot be determined from the preliminary vena cava ring implantation experiment. These problems are currently under study in an effort to clarify the interpretation of these results and to determine the importance of a hydrophilic-hydrophobic balance on the thrombogenicity of materials.

It should be emphasized that the vena cava ring test does not distinguish between non-thrombogenic materials and those which are only non-thromboadherent. Thus, the true thrombogenicity of the materials remains to be evaluated. Tests are underway which should allow a more realistic and quantifiable assessment of the thrombogenicity of HEMA-EMA grafted materials. An evaluation of the biological response to these grafted copolymers implanted in soft tissue is also underway.

Conclusions

A survey has been made of various solvent systems for use in the preparation of HEMA-EMA co-grafted silicone rubber films by the radiation grafting technique. The addition of water to the solvents investigated has a marked accelerative effect on the grafting rate of EMA. By the addition of appropriate amounts of water, it is possible to prepare co-grafted films with the same levels of graft throughout a wide range of HEMA/EMA ratios. The effect of water on the grafting process for this system has been discussed and it is suggested that the changes in graft levels noted might be due to the Trommsdorff efect or differences in the in the partitioning of the monomers between the silicone rubber and the monomer solution.

The water content of grafted films at a given monomer composition is independent of graft level at higher HEMA ratios, but increases with increases in the graft level at higher EMA ratios. This was concluded to be due to porosity affects in the graft.

HEMA-EMA grafted copolymers on silicone rubber and pure HEMA grafts on Silicone rubber show low apparent thrombogenicity by the vena cava ring test. Pure grafted EMA materials were found to be thrombogenic by this test.

An analysis of the exact copolymer composition of these films is planned for the near future. Also a wide range of additional studies on the interaction of these materials with biological systems are underway.

Acknowledgement

One of the authors (T.S.) would like to gratefully acknowledge the support of the Japanese Atomic Energy Institute, Takasaki, Japan. We would also like to thank the U.S. Atomic Energy Commission, Division of Biomedical and Environmental Research (Contract AT (45-1)-2225) for their generous support for these studies, and Dr. James D. Whiffen of the University of Wisconsin for the vena cava ring tests.

Literature Cited

1. Bruck, S. D., J. Biomed. Mater. Res., (1973), 7, 387.
2. Hoffman, A. S. and Kraft, W. G., ACS Polymer Preprints, (1972), 13(2), 723.
3. Hoffman, A. S. and Harris, C., ACS Polymer Preprints, (1972), 13(2), 740.
4. Lee, H. B., Shim, H. S. and Andrade, J. D., ACS Polymer Preprints, (1972), 13(2), 729.
5. Ratner, B. D. and Hoffman, A. S., Preprints - ACS Division of Organic Coatings and Plastics Chemistry, (1973), 33(2), 386.
6. Kearney, J. J., Amara, I. and McDevitt, M. B., Preprints - ACS Division of Organic Coatings and Plastics Chemistry, (1973), 33(2), 346.
7. Ratner, B. D. and Hoffman, A. S., J. Appl. Polymer Sci., (1974), 18, 3183.
8. Kwiatkowski, G. T., Byck, J. S., Camp, R. L., Creasy, W. S. and Stewart, D. D., "Blood Compatible Polyelectrolytes for Use in Medical Devices", Contract No. N01-HL3-2950T, National Heart and Lung Institute, National Institutes of Health, Bethesda, Maryland, Annual Report, July 1, 1972 - June 30, 1973, PB 225-636.

9. Baier, R. E., Dutton, R. C. and Gott, V. L., "Surface Chemical Features of Blood Vessel Walls and Synthetic Materials Exhibiting Thromboresistance" in "Surface Chemistry of Biological Systems", M. Black, ed., pp. 235-260, Plenum Press, N.Y., 1970.

10. Brash, J. L., Fritzinger, B. K. and Bruck, S. D., J. Biomed. Mater. Res., (1973), 7, 313.

11. Peterson, R. J. and Rozelle, L. T., "Ultrathin Membranes for Blood Oxygenators", Contract No. N01-HB-1-2364, National Heart and Lung Institute, National Institutes of Health, Bethesda, Maryland, Annual Report, December 1974, PB 231 324/5WM.

12. Ratner, B. D. and Miller, I. F., J. Polymer Sci., (1972), A-1, 10, 2425.

The Thermodynamics of Water Sorption in Radiation-Grafted Hydrogels

BERNARD KHAW,* BUDDY D. RATNER, and ALLAN S. HOFFMAN**

Department of Chemical Engineering and Center for Bioengineering,
University of Washington, Seattle, Wash. 98195

Hydrogels are lightly crosslinked, water swollen polymer networks and they have been suggested for or applied to a number of
biomedical applications (e.g., 1-3). Since bulk hydrogels display relatively poor mechanical properties when water swollen,
it is of interest to coat them onto stronger support materials.
Radiation graft copolymerization is a useful technique for preparing a covalently bonded coating of one polymer on or within
the surface region of another polymer and it has been used to
graft hydrogels to stronger polymer supports (3,4,5).

It has been hypothesized that both the content and "structural character" of the water which is sorbed in hydrogels can
be important determinants of the biological response at the
interface with a hydrogel (6,7,8). It has become important,
therefore, to study the manner in which water molecules are
taken up by these hydrogel-grafted films, in order to find a
correlation between the structure and character of the water in
these gels and their biocompatibility. From this knowledge it
may be possible to design improved biomaterials by adjusting the
water sorption behavior to some predetermined ideal.

One way of studying the mode in which water molecules are
taken up by hydrogels is by means of the thermodynamic parameters for the sorption of water vapor into the hydrogels. This
paper describes the thermodynamics of water sorption in three
different hydrogels, which were radiation grafted to silicone
rubber films.

Materials and Methods

Three monomers were used in these studies: Hydroxyethyl
methacrylate (HEMA), generously supplied by Hydro-Med Sciences
of New Brunswick, N.J. and used as received; N-vinyl pyrrolidone
(N-VP) purchased from Matheson, Coleman and Bell and purified by
double distillation; and ethylene glycol dimethacrylate (EGDMA)
purchased from Monomer-Polymer Laboratories, Philadelphia, PA.
and used as received. Films of Medical Grade Silastic were

* Present Address: ITT Rayonier, Shelton, Wash.
** To whom reprint requests should be directed.

purchased from Dow-Corning Corp. Their (nominal) thickness was
10mil.
   The method used for radiation grafting has been described in
previous publications (3,4). The radiation dose used here was
0.25 Mrad. A quartz spring vapor sorption apparatus was used to
obtain sorption isotherms at different temperatures for the three
different films. From these data, the standard differential
heats, ($\Delta H°$) entropies, ($\Delta S°$) and free energies ($\Delta G°$) of sorption
were calculated based on the method of Othmer, et.al. (9).

Results and Discussion

   The three films chosen for study are described in Table I.
They were chosen because of their wide range of water contents
and also because they were sufficiently grafted to yield readily
measured water sorption isotherms.

Table I. HYDROGEL/SILASTIC FILMS STUDIED

Hydrogel Grafted	Monomer Solution Composition (vol.%)[a]			Extent of Grafting (mg/cm^2)	$\%H_2O$ in Hydrogel[b]
	HEMA	NVP	EGDMA		
HEMA	20	0	1	4.40	26
HEMA/NVP	10	10	1	3.18	50
NVP	0	20	1	2.64	61

a)   Remaining 79% (vol.) is a 1/1 (vol.) mixture of Methanol/H_2O.
b)   As measured by blotting and weighing the grafted film after
     equilibration in H_2O (see Ref. 4).

   Typical isotherms for Silastic and two of the grafted films
are shown in Figure 1 and 2. It can be seen that water sorption
in the Silastic is negligible in comparison to that in the grafted
hydrogels. It is also clear that water sorption is always exothermic, as expected.
   The derived data for standard differential heats, entropies
and free energies of water sorption as a function of extent of
sorption are shown in Figures 3-5 for the three grafted films.
It should be noted that all the values for $\Delta H°$, $\Delta S°$, and $\Delta G°$
obtained for the three hydrogel films under study are more neg-
ative than those for the condensation of water vapor to liquid
water; the values for the condensation of water vapor to liquid
water at 26°C are: $-\Delta G°$ = 1.97 kcal/mole, $-\Delta S°$ = 28.29 e.u. (cal
$deg^{-1} mole^{-1}$) and $-\Delta H°$ = 10.48 kcal/mole. This indicates that
enthalpy, entropy and free energy changes more negative than those
for condensation are at work and that, as one would expect, the
sorption process is not merely water vapor condensing on the
films; that is, hydration or polar interactions between the hydrophilic groups in the hydrogels and the water molecules are no
doubt occurring. The same can be said for the curves obtained at

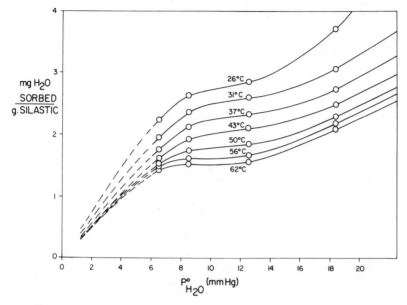

Figure 1. Sorption isotherms for water vapor in ungrafted Silastic films

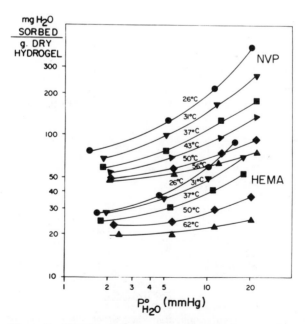

Figure 2. Sorption isotherms for water vapor in two radiation-grafted hydrogel/Silastic films

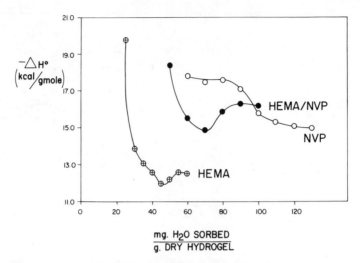

Figure 3. *Standard differential heat of sorption of water vapor as a function of the extent of water sorption in three radiation-grafted hydrogel/Silastic films at 26°C*

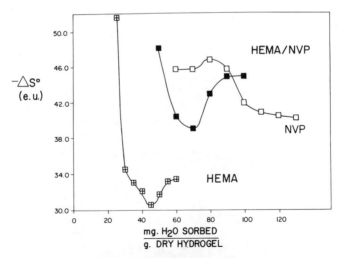

Figure 4. *Standard differential entropy of sorption of water vapor as a function of the extent of water sorption in three radiation-grafted hydrogel/Silastic films at 26°C*

higher temperatures of 31°, 37°, 43°, 50° and 56°C; the curves derived at these temperatures showed essentially the same behavior as at 26°C. Refojo and Yasuda (10) and later Warren and Prins (11) showed that when bulk HEMA hydrogels were immersed in water, they deswelled as temperature was raised up to about 55 to 60°C. These observations also support an exothermic process of water sorption below 55-60°C.

When the initial water molecules are sorbed in the grafted films they partake in highly exothermic interactions in all three hydrogel films. Probably these initial interactions in HEMA or HEMA/NVP films are localized at the most accessible -OH groups; the strong interaction and localization of a water molecule at such a site not only evolves heat (high $-\Delta H°$) but it also reduces the entropy of the system significantly (high $-\Delta S°$) as the water molecule loses randomness. It is also possible that the high $-\Delta H°$ and $-\Delta S°$ contain some contribution from increased hydrophobic bond formation upon sorption of water molecules, which could "plasticize" local chain motions. That is, the interaction of H_2O molecules with available -OH sites could permit a certain amount of increased hydrophobic group interactions, which would occur with $-\Delta H°$ and $-\Delta S°$ changes. Then, if any water molecules were involved, these could become more structured around such hydrophobic domains, contributing to $-\Delta H°$ and $-\Delta S°$ changes. It is most likely that the polar interactions dominate the initial sorption process in the HEMA-containing films and calculated clustering functions reflect this (see below). In the NVP film there is probably a somewhat weaker interaction and localization of the H_2O molecule around the pyrrolidone carbonyl or nitrogen groups, and the changes here with increasing sorption are less marked, over the range of water contents studied.

As more water is sorbed, the evolution of heat and decrease in entropy both decrease in magnitude until most of the readily accessible sorption sites are saturated, and the curves begin to level off. Eventually the water will begin to penetrate and swell the bulk of the HEMA hydrogels, opening up some new -OH sorption sites to the H_2O molecules; this leads to an increase in the (differential) exothermicity and decrease in (differential) entropy of sorption. This can explain the minima noted in both the $-\Delta H°$ and $-\Delta S°$ curves, for the HEMA containing hydrogels; these minima occur at approximately 1 $H_2O/3$ HEMA-OH for the HEMA system and 1 $H_2O/1$ HEMA-OH for the HEMA/NVP copolymer system. A model has been proposed for the structure of poly (HEMA) gels in which many of the OH-groups are tied up in strong intramolecular H-bonded crosslinks which form organized regions into which water cannot easily penetrate. (12) The ratio of 1 H_2O molecule per 3 HEMA-OH groups may indicate that two-thirds or more of the hydroxyl groups in poly(HEMA) are involved in these regions in the dehydrated gel. The upturn in the $-\Delta H°$ and

$-\Delta S°$ curves (Figures 3 and 4) for HEMA could indicate that although the gels are swelling and rearranging, a large number of free OH-groups are not being opened up for exothermic H-bonding with the water. In the HEMA/NVP copolymer films this HEMA organized structure may be partially broken up by the NVP rings and therefore not as many hydroxyl groups are occupied in the H-bond structure.

The NVP system does not show any minimum (and may in fact exhibit a small maximum) in the $-\Delta H°$ and $-\Delta S°$ curves. This may be due to the lack of a porous structure in this grafted hydrogel system (discussed below), coupled with sorption of relatively large amounts of H_2O.

Finally all curves should level off at some equilibrium value of $-\Delta H°$ and $-\Delta S°$ representing the combined process of condensation, and mixing (or dilution) of water in the swollen hydrogel/Silastic elastic network, where there is a balance between decreased free energy due to sorption vs. increased free energy due to network swelling. The gradual decrease in $-\Delta G°$ with increasing sorption (Figure 5) is a reflection of the approach of the overall sorption process to saturation equilibrium.

The results in this study for the three hydrogel films are comparable to those obtained by Masuzawa and Sterling (13) for the sorption of water vapor by agar, carboxymethyl cellulose, gelatin and maize starch and also to those reported by Bettelheim and Ehrlich (14) for the sorption of water vapor by mucopolysaccharides. In addition, the presence of maxima and minima and the close parallels between the shapes of the entropy and enthalpy curves seen here were also observed by these authors. Both groups also report a gradual decrease in $-\Delta G°$ with increase in mg. H_2O sorbed/g sorbant, as seen here in our data, and expected as saturation water sorption is approached.

The tendency for water molecules to cluster (or not to cluster) together within the grafted hydrogel matrices may be estimated by calculating the "clustering function" for water, after the technique of Zimm and Lundberg (15). These calculations have been made, assuming no change in volume on mixing, and results are shown in Fig. 6. It can be seen that all values of the clustering function are significantly negative, suggesting that the water molecules are strongly site bound at widely separated sites within the hydrogel matrix. This is especially true for the HEMA-containing polymers. Data of Zimm and Lundberg (15) for collagen are reproduced in Fig. 6 and show a similar behavior to the hydrogels studied here. To quote their discussion of these data: "The initial water is tightly bound" and as more water is absorbed "the sorption process changes . . . from one of sorption on a few highly specific sites to a diffuse swelling phenomenon". (15)

The diffusion coefficient for water in the hydrogel grafted films may be calculated from the initial slope of the sorption or desorption curve when it is plotted as (e.g., see Ref. 16):

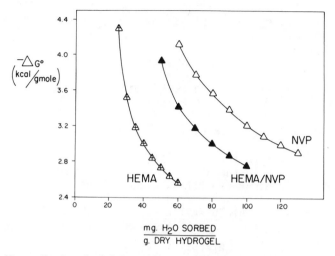

Figure 5. *Standard differential free energy of sorption of water vapor as a function of the extent of water sorption in three radiation-grafted hydrogel/Silastic films at 26°C*

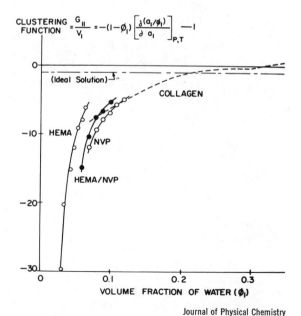

Figure 6. *Clustering functions for water as a function of water content in the three grafted hydrogel films and in collagen (15). Calculated by the technique of Zimm and Lundberg (15).*

$$\frac{M_t}{M_e} \text{ vs. } \frac{\sqrt{t}}{\ell}$$

Where M_t = amount of H_2O sorbed at time t.
        M_e = amount of H_2O sorbed at equilibrium (t = ∞)
        t = time
        ℓ = film thickness

Table II presents these results, and it can be seen that the diffusivity of H_2O in the HEMA film is significantly higher than in either of the other two films, even for the desorption process.

Table II. DIFFUSIVITIES OF WATER IN GRAFTED HYDROGEL/SILASTIC FILMS

Hydrogel Composition	DIFFUSIVITIES, cm^2/sec x 10^{10}		
	Sorption	Desorption	Average
HEMA	24.1	18.9	21.5
HEMA/NVP	5.4	7.4	6.4
NVP	9.2	6.9	8.1

The increased rate of diffusion of water vapor into and out of the HEMA grafted films may indicate a more porous and open structure for the graft regions of this material. The homopolymer surrounding the HEMA film after irradiation was a solid white gel. Such white HEMA gels formed in precipitating solvent systems are often macroporous in structure (10,17). Thus, the graft on the film may also be macroporous. The homopolymers surrounding the HEMA—N-VP copolymer films and the pure N-VP films were clear viscous liquids. In a non-precipitating polymerization medium one would not expect the formation of an open macroporous structure. Thus, the high diffusivity of water into the HEMA grafted film may be a reflection of the conditions under which it was prepared.

Conclusions

The thermodynamics of the sorption of water vapor by three hydrogel-grafted films have been studied. The films were prepared by the radiation-initiated grafting of Silastic backbones in monomer solutions containing 2-hydroxyethyl methacrylate (HEMA), N-vinyl pyrrolidone (NVP) and an equivolumetric mixture of these two monomers. A fused-quartz spring vapor sorption apparatus was used in this study. The standard differential heat of water vapor sorption by each film was calculated at different temperatures based upon the measured water vapor sorption data. The standard differential entropies and standard free energies of water vapor sorption were then calculated. Clustering functions for water were also calculated from the sorption isotherms. Diffusion coefficients were calculated from the sorption and desorption kinetic data. All of these data indicate that the grafted HEMA systems are more porous and more penetrable to water molecules and also site-bind water molecules strongly when compared to N-VP systems, and support a model for the HEMA polymer in which strong intramolecular H-bonds are important to its gel structure.

The water sorption thermodynamic and kinetics in hydrogel systems can yield valuable insight into the state or "character" of the imbibed water and its interactions with the hydrogel, as well as help to understand better the microscopic structure of the hydrogel. When these techniques are coupled with observations of biological interactions of these systems, one may be better able to design new and better materials for biomedical applications.

Acknowledgment

The support of the U.S.E.R.D.A. Division of Biomedical and Environmental Research (Artificial Heart Program) Contract AT(45-1)2225 and the NIGMS Bioengineering Center Grant No. GM-16436-07 are gratefully acknowledged.

Abstract

The thermodynamics of water sorption in three different radiation-grafted hydrogels have been studied. Diffusion coefficients for water in these films were also calculated from the sorption and desorption kinetic data. The thermodynamic and kinetic data indicate that the grafted hydroxyethyl methacrylate (HEMA) films are more porous and penetrable to water molecules and their OH groups interact more strongly with water molecules when compared to grafted N-vinyl pyrrolidone (NVP) films. Strong site binding of water molecules is also reflected by the relatively large negative values calculated for the water clustering function in all films, although the values are less negative for the poly NVP films than for the poly HEMA films. These data all support a model for the HEMA polymer in which strong intramolecular H-bonds are important to its gel structure. Such data may be relevant to the biological interactions of hydrogels in the living system environment.

Literature Cited

(1) (a) Wichterle, O. and Lim, D., Nature, (1960), 165, 117.
     (b) Levowitz, B.S., et al., Trans.Amer.Soc.Artif.Int.Organs, (1968), 14, 82.
(2) Andrade, J.D., et al., Ibid., (1971), 17, 222.

(3) Hoffman, A.S., et al., Ibid., (1972), 18, 10.
(4) Ratner, B.D. and Hoffman, A.S., ACS Org.Coatings and Plastics
     Chemistry Preprints, (1973), 33(2), 286; J.Appl.Polymer Sci.,
     18, 3183.
(5) Hoffman, A.S., et al., in "Permeability of Plastic Films and
     Coatings," ed. H.B. Hopfenberg, Plenum Press, N.Y., (1974),
     441.
(6) Drost-Hansen, W., Ind.Eng.Chem., (1969), 61, 10.
(7) Jhon, M.S. and Andrade, J.D., J.Biomed.Mtls.Res., (1973), 7,
     509.
(8) Hoffman, A.S., J.Biomed.Mtls.Res., Biomed.Mtls. Symposium No.
     5, Part 1, (1974), 8, 77.
(9) (a) Othmer, D.F., Ind.Eng.Chem., (1940), 32, 841.
     (b) Othmer, D.F. and Sawyer, F.G., Ibid., (1943), 35, 1269.
     (c) Othmer, D.F. and Sawyer, F.G., Ibid., (1944), 36, 894.
     (d) Jocefowitz, S. and Othmer, D.F., Ibid., (1948), 40, 739.
(10) Refojo, M.F. and Yasuda, H., J.Appl.Polymer Sci., (1965), 9,
     2425.
(11) Warren, T.C. and Prins, W., Macromolecules, (1972), 5, 506.
(12) Ratner, B.D. and Miller, I.F., J.Polym.Sci., Part A-1, (1972),
     10, 2425.
(13) Masuzawa, M. and Sterling C., J.Appl.Polymer Sci., (1968),
     12, 2023.
(14) Bettelheim, F. and Ehrlich, S.H., J.Phys.Chem., (1963), 67,
     1948.
(15) Zimm, B.H. and Lundberg, J.L., J.Phys.Chem., (1956), 60, 425.
(16) Crank, J. and Park, A.S., "Diffusion in Polymers," 16, Aca-
     demic Press, N.Y., 1968.
(17) Yasuda, H., Gochin, M. and Stone, W., Jr., J.Polymer Sci.,
     (1966), 4, 2913.

23

Biological and Physical Characteristics of Some Polyelectrolytes

E. O. LUNDELL, G. T. KWIATKOWSKI, J. S. BYCK, F. D. OSTERHOLTZ, W. S. CREASY, and D. D. STEWART

Union Carbide Corp., Chemicals and Plastics, Research and Development, Bound Brook, N. J. 08805

It has been recognized for some time that the surface charge of blood components plays a role in their interactions with the blood vessel wall.[1] Based largely on the results of Sawyer and coworkers[2], it has also become apparent that surface charge, and electrochemical phenomena in general, can significantly influence blood/material interactions and that their control may be a necessary condition for development of a truly blood compatible material. Research stemming from ths concept, involving blood compatibility studies with metals and polymers, has led to the conclusion that a limiting level of negative charge density or negative potential may be required to achieve nonthrombogenic character, provided that other necessary conditions are also satisfied. This has prompted extensive research directed at development of anionic polyelectrolytes which might be expected to be blood compatible.

As represented in Figure 1, the blood compatibility and, in particular, thromboresistance of carboxyl-containing copolymers would be expected to depend upon the interaction of several physical and chemical parameters. These include not only the total concentration of ionizable groups, the degree to which these are neutralized by various monovalent or divalent metal salts, but also the hydrophilicity of the total polymer system. This last property depends largely on the selection of the neutral comonomer. As was expected at the outset of this program, and has been confirmed by experimental results, none of these parameters can be considered to be a sufficient condition for blood compatibility. All must be controlled, however, as necessary conditions for compatibility.

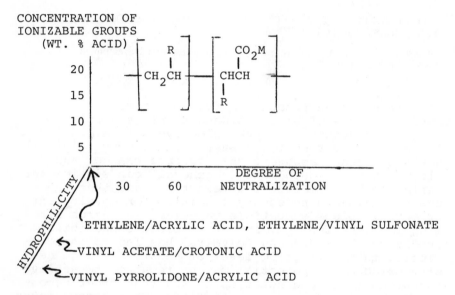

Figure 1. *Schematic of the surface charge density and hydrophilicity of some polyelectrolytes*

METHODS
BIOLOGICAL COMPATIBILITY TESTING

Biological test data described represent the results of evaluation programs conducted by other investigators under contract to the National Heart and Lung Institute of the National Institutes of Health. Methods employed by these testers are summarized in this section.

In Vitro Test Methods

Partial Thromboplastin Time.[a]

Polymer samples were submitted in the form of tubes and stirring paddles. Glass tubes (26 mm i.d. x 100 mm) were solution coated on the entire interior surface except for a 0.5"-1.0" rim at the orifice. This was accomplished by introducing 3 ml of polymer solution into the tube, fitting the tube to a rotary evaporator, tilting it so that the entire inner surface would be coated (except for the rim at the orifice), and applying heat to the exterior of the evacuated tube while being rotated. Stirring paddles were prepared by solution coating glass rectangles (3 mm x 18 mm x 120 mm) or by cutting equivalent size strips from compression molded polymer sheet.
To carry out the partial thromboplastin time test, 25 ml of native whole blood was drawn into a silicone-coated (G. E. Dri-Film SC-87) glass syringe and immediately transferred to the polymer coated tube. The paddle was employed to stir the blood sample for four minutes at 60 rpm. At the end of this time, the sample was anticoagulated by addition of sodium citrate and the cellular elements were removed by centrifugation (425 g) at 4°C. Partial thromboplastin time of the resultant platelet-poor plasma was determined according to the methods of Langdell[3] and Rodman. [4]

Stypven Time.[a]

Samples of platelet-poor plasma prepared for determination of partial thromboplastin time were also used to measure Stypven time according to the methods of Miale[5] and Hardisty. [6]

a. Dr. R. G. Mason, University of North Carolina, School of Medicine, Chapel Hill, North Carolina

In Vivo Test Methods

Canine Implantation Studies.[b]

Test samples were in the form of vena cava
("Gott") rings. These are 9 mm long and have an o.d.
of 8 mm. Both inflow and outflow orifices were streamlined. Ethylene/acrylic acid copolymer and ionomer
rings had a wall thickness of 1 mm (6 mm i.d.), whereas all remaining samples had a wall thickness of 0.5
mm (7 mm i.d.). The larger i.d. ring was the standard
size prescribed by Gott. Vena cava rings may be prepared by compression molding (i.d.) and machining
(o.d.), by injection molding, or by solution coating
of the test polymer onto injection molded high density
polyethylene rings. There was a single circumferen-
tial groove midway between the ends of each ring to
permit the ring to be tied in place after implantation.
Tests were conducted by inserting the ring, using
a special insertion device, into the canine inferior
vena cava through a small incision in the right atrium.
A wrapping was placed over the ring external to the
blood vessel to prevent cul de sac formation and the
ring was tied in place. Initially three samples of
polymer were each implanted for two hours, after which
time they were removed and examined for the presence
of adherent thrombi. Polymers which were rated highly
thromboresistant on the basis of this acute test were
then implanted in the same manner for a period of two
weeks.

Canine Implantation Studies.[c]

Implant samples were in the form of 30 mm long
tubes with an o.d. of 9 mm and a wall thickness of 1 mm
(7 mm i.d.). Both inflow and outflow orifices were
streamlined. Tubes may be fabricated by the several
methods noted above for vena cava rings. The tubes
were circumferentially grooved near each end.

[b.] Dr. V. L. Gott, Johns Hopkins University, Dept. of
Surgery, Baltimore, Maryland.

[c.] Dr. P. N. Sawyer, State University of New York,
Downstate Medical Center, Brooklyn, New York.

Implantation was in the canine inferior vena cava through an incision in that vessel. The tube is tied in place at each groove, with one tie proximal and one tie distal to the incision. For each polymer, one tube was employed as an acute (2 hour) test and two tubes as chronic (14 days) implants.

POLYMER SYNTHESIS

Ethylene/Acrylic Acid Copolymers and Ionomers

The family of ethylene/acrylic acid copolymers seemed an excellent experimental probe for accomplishing this study for several reasons. Copolymers containing 2-20% by weight of acrylic acid are well known materials and could be readily obtained commercially; conversion of these polymers to their partially neutralized alkali metal, alkaline earth, and organic amine salts had been thoroughly investigated and could be controlled over a range of 0-70% neutralization; and these polymers are tough, flexible thermoplastics capable of being fabricated from solution or in the melt to yield highly satisfactory test samples.

Ethylene/Vinyl Sulfonic Acid Copolymers and Ionomers

Ethylene/vinyl sulfonic acid copolymers were prepared by a two-step procedure. In the first step, ethylene was copolymerized with sodium vinyl sulfonate in a high pressure aqueous emulsion polymerization. The resulting fully neutralized sodium vinyl sulfonate ionomers were converted by treatment with anhydrous HCL to the desired sulfonic acid copolymers which could be titrated with base to yield partially neutralized ionomers.

N-Vinyl Pyrrolidone/Acrylic Acid Copolymers and Ionomers (NVP/AA)

In initial studies leading to preparation of N-vinyl pyrrolidone (NVP)/acrylic acid (AA) copolymers, synthesis was carried out as a two-step procedure. In the first step, sodium acrylate was copolymerized with NVP in buffered aqueous solution at pH 7-8 after which the resulting fully neutralized ionomer was acidified to yield the desired N-VP/AA copolymer. Potassium persulfate was employed as initiator in the polymerization step. Although the copolymerization could be carried out successfully, attempts to carry out acidification with concentrated or half-concentrated hydrochloric acid, followed by drying at elevated temperatures, resulted in decomposition and discoloration of the product. Although there had initially been concern that direct acidification might, therefore, not be possible, it was subsequently found that satisfactory colorless polymers could be obtained if two precautions were taken. First, acidification was carried out at temperatures below 10°C. Second, exhaustive purification of the polymer, by dissolving in water and reprecipitating with acetone, was required to remove all traces of acid before drying. It was also found that this purification procedure produced a drop in acrylic acid content, as determined by titration, presumably by removing impurities of acrylic acid homopolymer. Even after removal of these impurities, however, the products were consistently found to have higher AA/NVP ratios than in the original monomer feed. By running a polymerization to partial conversion, with continuous feed of monomers and continuous removal of reaction mixture, it was possible to conveniently control unreacted monomer ratios to prepare a uniform copolymer throughout the course of the reaction.

Vinyl Acetate/Crotonic Acid (VA/CA) Copolymers and Ionomers

Vinyl acetate and crotonic acid were polymerized under aqueous emulsion conditions at 65-69° C (slight vinyl acetate reflux), with potassium persulfate employed as initiator. Small amounts of sodium acetate were used as buffer for the crotonic acid. Emulsions were prepared with final solids content of approximately 50%. At low crotonic acid (CA) content (8%), the reaction was slightly exothermic, whereas with CA levels higher than 10% the reaction was more sluggish. Reaction times ranged from 20-30 minutes for 5% CA copolymers to 15-20 hours for copolymers containing ≥15% crotonic acid. The polymer emulsion obtained proved to be very stable and difficult to coagulate. Isolation was best accomplished by gradual addition of methanol to the thick latex at 0-5°C, with continuous removal of the slimy product. This was dried under vacuum to yield relatively brittle polymer of only modest molecular weight. High molecular weights are unobtainable because of the chain transfer activity of crotonic acid.

Coating of Polyelectrolyte Hydrogels onto Polypropylene

The hydrophobic PE surface can be made wettable
with water by treatment with chromic acid "cleaning
solution". The PE tubes were suspended in a mixture
of concentrated sulfuric acid containing solid chromi-
um trioxide and the mixture stirred at room temperature for one day. The tubes were rinsed well with distilled water followed by a rinse with methanol. After
this treatment the tubes were not allowed to dry and
were touched only with methanol rinsed clean rubber
gloves. They were stored until used in a jar containing methanol. The tubes were placed in a Ferris wheel
apparatus and coated with a 10% (wt/wt) solution of
Union Carbide A-174 silane adhesion promoter in methanol either by spraying or dipping. The tubes were
dried and heat treated (80°/0.5 hr) in an oven while
turning with their cylindrical axis parallel to the
axis of rotation.
    The tubes were then coated with a solution of
polyelectrolyte. After the desired amount of polyelectrolyte had been coated onto the PE ring the coating was heat cured (80°/1 hr). VA/CA solution containing 10% by weight polymer in methanol was used. Two
complete immersions of the tube in the solution, followed by drying, produced a defect-free coating of pro-
per thickness. After each dip, a droplet of solution
rolled around inside the tube as it is rotated in the
80°C oven until the solvent is evaporated. In the case
VP/AA, a solution containing 5% by weight polymer in
methanol was used. One immersion followed by shaking
to remove any excess drops is used to insure that the
entire tube has some finite coating of the polymer.
After evaporating off the remaining solvent a 0.15 ml
droplet of the solution is placed inside the tube followed by evaporation of solvent with rotation.
    For irradiation crosslinking the presence of a
controlled amount of water in the coating and the absence of oxygen both enhanced the efficiency of crosslinking of the coating. Water could be added to the
coating by exposing the tubes to a humid atmosphere.
The absorption of 2% by weight water by a 20 ml wit-
ness strip was used as a guide. The dampened tubes
were sealed under nitrogen or argon in a PE bag and
irradiated with the Van de Graaff. The dosages required were 35 megarads for VA/CA and 100 megarads for
NVP/CA. However, variation in hydrogel density were
attainable in the NVP/AA case over a range of 80-200
megarads.

Direct Graft Polymerization of Vinyl Acetate/Crotonic Acid from Polypropylene

The polypropylene objects to be coated with hydrogel were placed in a flask containing either an aqueous emulsion or an ethanol solution of vinyl acetate, crotonic acid, sodium crotonate and a surfactant (Tergitol 12-P-12). The concentration of monomers was either 25% or 50% by weight, and the surfactant was 1% by weight. The solution (emulsion) was deoxygenated with nitrogen, placed in a Co^{60} source and irradiated at a rate of 50,000 rad/hr while agitation was maintained by nitrogen. After irradiation the samples were removed from the flask and thoroughly washed with ethanol to remove any adsorbed polymer. The samples were then equilibrated with water.

RESULTS AND DISCUSSION

Initially, an attempt was made to elucidate the relationship of the surface charge density of a material to its hemocompatibility. Two systems, ethylene/acrylic acid (EAA) and ethylene/vinyl sulfonate (EVS), were chosen for study. No member of the EVS series of copolymers and ionomers showed significant thromboresistance[7], and only two members of the EAA series (the 19% acrylic acid copolymer and its 60% sodium ionomer) showed significant thromboresistance.[7-10]

To elucidate the thrombogenic dependence on hydrophilicity, two systems, N-vinyl pyrrolidone/acrylic acid (NVP/AA) and vinyl acetate/crotonic acid (VA/CA) copolymers and ionomers, were chosen for study. While no member of the NVP/AA series displayed any thromboresistance[11,12], members of the VA/CA series showed various degrees of thromboresistance. The 2% crotonic acid 60% sodium ionomer, in particular, exhibited a high degree of thromboresistance in vena cava implant studies.

Ethylene/Acrylic Acids Polyelectrolytes

A series of copolymers and ionomers was prepared which represented four levels of total carboxyl concentration (4.0, 9.8, 14.7, and 19.2% acrylic acid). At each level, unneutralized copolymers were compared with copolymers which had been converted to approximately 30% and 60% neutralized ionomers. Counteranions investigated included lithium, sodium, potassium, magnesium, calcium, and dimethylethanolamine. The

polymers which comprise the series of twenty-one ethyl-
ene/acrylic acid copolymers and ionomers were expected
to differ from one another in a number of ways. These
differences included variation in measurable physical
chemical parameters by which such surface properties
as wettability and electrochemical behavior can be
characterized, as well as variation of chemical interactions with blood on the molecular level. The contact
angles of distilled water on the polymer surfaces ranged
from a high of 87 (surface free energy = 19.9 ergs/cm)
for the unneutralized 9.8% acrylic acid copolymer to a
low of 71 (31.6 ergs/cm) for highly neutralized lithium
and sodium ionomers of the 19.2% acrylic acid copolymer.
Zeta potentials determined by Dr. Sawyer ana coworkers[13] (Table I) at SUNY-Downstate Medical Center were
large negative values, suggesting that the electrochemical conditions proposed for blood compatibility were
satisfied.

Initial evaluations of the thrombogenicity of these
materials was performed by Mason, et al. (8), at the
University of North Carolina School of Medicine and
Pennington (9) at Battelle Memorial Institute, using
Stypven time and partial thromboplastin time as a measure of blood compatibility (Table II). No member of
the series of twenty-one ethylene/acrylic acid copolymers and ionomers appeared to be highly thromboresistant on the basis of the data. The variations in polymer structure appear to have no significant effect on
the Stypven time, since all polymers tested yielded
values within experimental error of values obtained for
the siliconized glass controls. In the case of the
partial thromboplastic times, however, there was far
greater variability. In virtually all cases, the polymers appeared to be more thrombogenic than the stan-
dards. In only two cases — the 9.8% acrylic acid copolymer and its 29.5% neutralized sodium ionomer — were
values seen which were significantly higher than the
values for siliconized glass. In neither case, however, were high values registered by both test groups.
In fact, there appeared to be little obvious correlation between the results of the two parallel sets of
tests.

As an additional portion of the basic in vitro
blood compatibility studies performed at Battelle
Memorial Institute (9), adhesion of erythrocytes and
platelets to the experimental polymers was also investigated. The results of these tests indicated fairly
uniform behavior throughout the series of polymers.
In all cases, adherence of erythrocytes was very light,
with some showing no adherence at all.

Platelet adherence ranged from light to moderate, with slight aggregation in some cases. The only significant exception was the dimethylethanolamine salt, which showed heavy platelet adherence with heavy aggregation.

TABLE I

ZETA POTENTIALS ACROSS SOME POLYELECTROLYTES - SALINE/1000 SOLUTION INTERFACES AS OBTAINED FROM STREAMING POTENTIALS

% AA	% Ionomer	dE/dP* mv/cm Hg	Zeta Potential
9.8	0	325	-17.2
19.2	0	350	-18.5
19.2	27.0% Mg	355	-18.8
19.2	25.5% Ca	315	-16.8
19.2	61.0% Na	390	-20.7
19.2	61.5% Mg	340	-18.0
19.2	61.5% Ca	365	-19.5
19.2	34.5% DMEA	325	-17.5

* dE/dP is the average slope of the streaming potential versus pressure relation from which the zeta potential is calculated.

In vivo blood compatibility studies were also performed on many of the ethylene/acrylic acid copolymers and ionomers by Sawyer and coworkers[14] at State University of New York Downstate Medical Center and Gott[18] and coworkers at Johns Hopkins University (Table III). Sawyer's results indicate that two of the polymers exhibited a moderate level of thromboresistance. These were the 19% acrylic acid copolymer and its highly neutralized sodium ionomer. In both cases, two hour and two week implants were either thrombus-free at the end of the implant period or contained only minimal junctional thrombi. Based on the observations in these experiments[16], these two polymers appeared to be moderately thromboresistant, whereas copolymers at lower neutralization or lower acrylic acid content, as well as calcium ionomers, were thrombogenic. The highest level of thrombogenic activity was found in the magnesium and dimethylethanolamine ionomers. In Gott's tests, all of the polymers displayed significant levels of thrombogenic activity, with only limited variation

TABLE II

IN VITRO BLOOD COMPATIBILITY OF
ETHYLENE/ACRYLIC ACID POLYELECTROLYTES

Acrylic Acid Cor ent	% Ionomer	Stypven Time*	Partial Thromboplastin Time*	
			University of North Carolina	Battelle
4.0%	0	--	--	109
"	24.0% Na	--	--	99
"	55.0% Na	--	--	101
9.8%	0	98	76	137
"	29.5% Na	98	115	83
"	71.0% Na	97	95	89
14.7%	0	97	106	101
"	32.0% Na	98	85	99
"	62.0% Na	98	74	86
19.2%	0	100	74	92
"	27.0% Li	98	78	70
"	28.0% Na	103	62	94
"	24.0% K	101	59	83
"	27.0% Mg	99	79	91
"	26.5% Ca	103	68	78
"	34.5% DMEA	104	84	103
"	65.0% Li	98	78	--
"	61.0% Na	96	89	--
"	40.5% K	100	94	109
"	61.5% Mg	107	65	99
"	61.5% Ca	97	74	104

* % of siliconized glass control

TABLE III

IMPLANTATION OF ETHYLENE/ACRYLIC ACID COPOLYMERS AND IONOMERS

% AA	% Neut.	Cation	Johns Hopkins 2 hr. Implant	Duration	Downstate Medical Center	% Occlusion
9.8	0	—		14 days		clean
				14 days		50%
				2 hours		80%
9.8	29.0	Na		14 days		20%
				14 days		10%
				2 hours		90%
9.8	71.0	Na		14 days		5%
				14 days		10%
				2 hours		60%
14.7	0	—		14 days		junctional
				14 days		30%
				2 hours		90%
14.7	62.0	Na		14 days		50%
				18 hours		30%
				2 hours		90%
19.2	0	—		14 days		junctional
				14 days		clean
				2 hours		junctional

TABLE III (Continued)

19.2	27.0	Mg	14 days	80%	
			14 days	60%	
			2 hours	60%	
19.2	26.5	Ca	14 days	5%	
			6 days	10%	
			2 hours	40%	
19.2	34.5	DMEA	14 days	60%	
			14 days	80%	
			2 hours	20%	
19.2	61.0	Na	14 days	5% (inl.)	
			14 days	clean	
			2 hours	clean	
19.2	60.5	K			
19.2	61.5	Mg	4 hours	90%	
			14 days	30%	
			2 hours	junctional	
19.2	61.5	Ca	31 days	50%	
			14 days	clean	
			2 hours	50%	

within the series. Several of the polymers displayed
some thromboresistance, including the highly neutralized sodium ionomer of the 19% acrylic acid copolymer.

Ethylene/Vinyl Sulfonate Polyelectrolytes

     The chemistry of the anionic moiety of a poly-
electrolyte must also be considered as a factor regulating blood compatibility. Although carboxyl-containing polymers have been the primary subjects in our
investigation, comparison of their physiological activity with the blood compatibility of analogous sulfonic acid-containing copolymers and ionomers was also
an essential element of the program. Like carboxyl
groups, sulfonic acid moieties ($-SO_3H$) are present in
many naturally occurring polyelectrolytes. Most nota-
bly, they are present in heparin, in high concentration, as sulfate half-esters and are presumed to play
a role in the biological activity of that compound.
However, sulfonic acids are stronger acids than their
carboxylic acids analogues and would be expected to
differ in their chemical reactions with plasma. In
order to assess the significance of this factor to the
development of blood compatible polyelectrolytes,
ethylene/vinyl sulfonic acid copolymers and ionomers
were prepared and biologically evaluated.
     Initial tests of two copolymers in this series,
and an ionomer of each, suggested that no significant
differences could be detected between the in vitro
blood compatibility of these polymers (partial thromboplastin time; Stypven time, Table IV), and analogous
data for similar acrylic acid copolymers and ionomers.
The poor blood compatibility of these polyelectrolytes
was confirmed by in vivo studies of two members of the
series by Gott and coworkers (Table V).(13)

N-Vinyl Pyrrolidone/Acrylic Acid Polyelectrolytes

     Ethylene/acrylic acid and ethylene/vinyl sulfo-
nate copolymers and ionomers represent a series of
polyelectrolytes which are hydrophobic or, at most,
only slightly hydrophilic, depending on the content of
the hydrophobic comonomer ethylene and the degree of
neutralization. It is known, however, that naturally
occurring polyelectrolytes, which impart negative surface charge to blood vessel walls or to blood cells,
are considerably more hydrophilic than even the most
hydrophilic ethylene/acrylic acid ionomers, due to the
presence of large numbers of polar groups along the

polymer chain.[16,17] It was also found that certain hydrogels[18] and ionic interpolymers[19] exhibit good physiological properties on exposure to blood. An appropriate continuation of studies of the relationship of anionic groups and negative surface charge to the blood compatibility of polymers, therefore, seemed to be investigation of polyelectrolytes more hydrophilic than ethylene/acrylic acid copolymers and ionomers. In order to accomplish this, vinyl acetate and N-vinyl pyrrolidone were employed instead of ethylene as the neutral comonomer to prepare carboxyl-containing copolymers. N-vinyl pyrrolidone/acrylic acid copolymers were selected as a system to be investigated recognizing that the presence of high concentrations of N-vinyl pyrrolidone would produce water soluble polyelectrolytes.

Vinyl pyrrolidone/acrylic acid copolymers and ionomers were crosslinked by Van de Graaff irradiation to form water insoluble hydrogels which swell significantly in aqueous media by imbibing large quantities of water. The degree of crosslinking was increased by increased dosage. Among the features controlled by the density of crosslinking, the tightness of the gel determines the maximum size of solute molecules capable of penetrating the polyelectrolyte coating as water is imbibed.

Samples of N-vinyl pyrrolidone/acrylic acid copolymers and sodium ionomers were coated on high density polyethylene. The 8% ionomers were studied by Gott[17] and were shown to be highly thrombogenic. The 2% ionomers were studied at the Utah Biomedical Test Laboratory and were also highly thrombogenic (Table VI)[12]. In both series of ionomers, the implant results showed total or near total occlusion, often accompanied by ascites or pulmonary embolism.

TABLE IV

IN VITRO BLOOD COMPATIBILITY OF ETHYLENE/VINYL SULFONIC ACID COPOLYMERS AND IONOMERS*

% Vinyl Sulfonic Acid	% Na Ionomer	Stypven Time**	Partial Thromboplastin Time**
4	0	91	67
4	40	99	87
8	0	94	100
8	40	100	90

* University of North Carolina
** % of siliconized glass control

TABLE V

IN VIVO BLOOD COMPATIBILITY OF
ETHYLENE/VINYL SULFONIC ACID SODIUM IONOMERS*

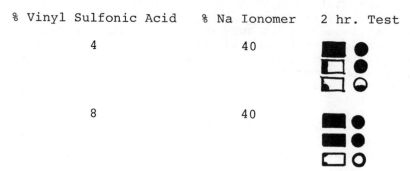

% Vinyl Sulfonic Acid	% Na Ionomer	2 hr. Test
4	40	
8	40	

* John Hopkins University

TABLE VI

THROMBORESISTANCE OF N-VINYL PYRROLIDONE/ACRYLIC
ACID COPOLYMERS AND IONOMERS. VENA CAVA
IMPLANTATION OF 8% ACRYLIC ACID-CONTAINING POLYMERS

% Neut.	Duration	Johns-Hopkins
0%	2 hours	
30%	2 hours	
60%	2 hours	

Vinyl Acetate/Crotonic Acid Polyelectrolytes

Since the N-vinyl pyrrolidone/acrylic acid copolymers imbibed such large amounts of water, the investigation of polyelectrolyte systems which had more modest water uptake was undertaken. Vinyl acetate, being less hydrophilic than N-vinyl pyrrolidone, was chosen as the comonomer in an attempt to prepare a hydrogel with a modest water uptake. Vinyl acetate and acrylic acid could not be satisfactorily copolymerized because of widely differing reactivity ratios, so vinyl acetate/crotonic acid copolymers were, therefore, adopted as an acceptable substitute. As it was found that copolymers above 15% crotonic acid level were too brittle for convenient handling, polymers were prepared for blood compatibility studies at 2.3%, 7.6%, and 16.2% crotonic acid monomer ratios. The crotonic acid content in these materials was determined by titration of aliquot samples with N/10 NaOH. In analogy to the E/AA polyelectrolytes, sodium ionomers were prepared at approximately 30% and 60% conversions .

Radiation treatment, using the Van de Graaff electron accelerator, was investigated as the method of choice for crosslinking these copolymers and ionomers to convert them to water-swellable hydrogels. Experiments were carried out in which compression molded plaques, 0.02" thick, were exposed to various radiation doses ranging from 5 to 75 megarads. Immersion of the treated samples in water showed that doses of 30 to 40 megarads were sufficient to convert 30% neutralized

TABLE VII

IN VITRO BLOOD COMPATIBILITY OF
VINYL ACETATE/CROTONATE COPOLYMERS AND IONOMERS

% Crotonic Acid	% Na Salt	Stypven Time	Partial Thrombo- plastin Time
5%	0 (dry)	106	131
5%	0 (wet)	105	140
5%	30	82	109
5%	60 (dry)	105	112
5%	60 (wet)	110	142
15%	0	105	103
15%	30	94	94
15%	60	90	74

TABLE VIII

THROMBORESISTANCE OF VINYL ACETATE/CROTONIC ACID COPOLYMERS AND IONOMERS

VENA CAVA IMPLANTATION OF 2% CROTONIC ACID-CONTAINING POLYMERS

% Neut.	Duration	Johns Hopkins		SUNY-Downstate
0	2 hours	▢ ◯ ▢ ◯ ▢ ◯		10% occl. Junct. thrombus - Sacrif.
	4 days	◼ ● ▢ ◯	died died of pneumonia	
	5 days	▢ ◯	died of pulmonary embolism	
	7 days			90% occl. Junct. and laminar thrombus
	14 days			Junct. thrombus - Sacrif.
30% Na	2 hours	▢ ◯ ▢ ◯ ▢ ◯		Junct. thrombus Sacrif.
	4 days	◼ ●	died	
	6 days			80% occl. junct. and laminar thrombus
	14 days	▢ ◯ ▢ ◯		Junct. thrombus Sacrif.
60% Na	2 hours	▢ ◯ ▢ ◯ ◼ ● ▢ ◯ ▢ ◯		Clean - sacrif.
	14 days	▢ ◯ ▢ ◯ ◼ ●	ascites	Clean - sacrif. Clean - sacrif.

TABLE VIII (Continued)

IMPLANTATION OF VINYL ACETATE/CROTONIC ACID COPOLYMERS AND SODIUM IONOMERS

% Crotonic Acid	% Neutralization	Johns Hopkins 2 hr. Implant	Downstate Medical Center Duration		% Occlusion
7.6	0		2 hours		30%
			14 days		minimal
			14 days		10%
7.6	30		2 hours		90%
			18 hours		70%
			14 days		10%
7.6	60		2 hours		10%
			14 days		clean
			14 days		90%
16.2	0		2 hours		40%
			14 days		clean
			14 days		junctional
16.2	30		2 hours		clean
			7 days		30%
			5 hours		30%
16.2	60		2 hours		clean
			2 days		junctional
			14 days		clean

ionomers to water-insoluble materials, but that for the 60% neutralized ionomers, insolubilization was achieved only after exposure to doses of 75 megarads. In order to equalize radiation-induced changes in the polymers, 75 megarads was chosen as a standard dose for all vinyl acetate/crotonic acid copolymers and ionomers. Irradiation was carried out stepwise in 3-5 megarad doses to minimize side effects from overheating of samples.

Mechanical properties have been measured on five of the six materials after radiation and exposure to water for approximately 100 hours. All of the properties indicated very weak materials; thus, a solid support material was necessary to provide sufficient mechanical strength for test samples and end use applications (see Methods section).

The initial in vitro evaluations were the Stypven time and partial thromboplastin time performed by Dr. R. G. Mason[8] and Dr. C. J. Pennington. The results indicated that several compositions were more blood compatible under these conditions than siliconized glass controls (Table VII). Furthermore, it was also found that blood compatibility could be further enhanced by preswelling the polyelectrolyte hydrogels by 30 minutes immersion in 0.154 M NaCl.

Nine members of the vinyl acetate/crotonic acid series were evaluated by Dr. P. N. Sawyer[14] as coatings of polyolefin rings. These were copolymers containing 2, 8 and 16 percent crotonic acid, as well as the 30% and 60% sodium ionomers of these copolymers. Dr. V. L. Gott[15] implanted six of the nine members of the vinyl acetate/crotonic acid polyelectrolyte hydrogel series. Copolymers containing approximately two percent crotonic acid gave excellent two hour results at all levels of neutralization and provided fairly good two week results (Table VIII). In the unneutralized case, two of the animals had clean rings at four and five days, one dying of pneumonia and the other appearing to die of pulmonary embolism. The third ring in this series thrombosed at four days. The two ionomers looked surprisingly good at two weeks, with two of the three rings in each case being free of thrombus. Comparison of results from both groups suggested that the material which consistently demonstrated the most encouraging thromboresistance is the 60% sodium ionomer of vinyl acetate/2% crotonic acid copolymer.

Using Co^{60} radiation, it has been possible to graft polymerize vinyl acetate/crotonic acid/sodium crotonate directly onto the polypropylene. This has provided an opportunity to carefully control the coating thickness and water uptake of the hydrogel (see Table IX and X) in order to optimize biological results. In vivo studies have been performed on samples prepared with various coating thicknesses and water uptake values. The Gott vena cava implant and Kusserow renal embolus tests [20] show the poly(vinyl acetate-co-2%-crotonic acid) 60% sodium ionomer prepared at 0.1 Mrad (12 μ coating and 55% water uptake) to be the most consistently thromboresistant polyelectrolyte hydrogel in this series.

TABLE IX

COATING PARAMETERS AS A FUNCTION OF RADIATION DOSE-SURFACTANT PROMOTED IN ETHANOL SOLUTION

Dose (Megarads)	Monomer Conc. (%)	Coating Thickness (μ)	Eq. Water Abs. (%)
0.10	50	12[a]	50.0
0.20	50	30[a]	33.4
0.20	25	7[a]	53.8
0.20	25	13[a]	28.0
0.40	25	30	29.6
1.20	25	32	18.7
2.40	25	65	9.7

a. Approximate; calculated from coating weight

TABLE X

COATING PARAMETERS AS A FUNCTION OF RADIATION DOSE-SURFACTANT PROMOTED IN WATER[b]

Dose (Megarads)	Monomer Conc. (%)	Coating Thickness (μ)	Eq. Water Abs. (%)
0.05	25	6[a]	111.3
0.10	25	16[a]	55.6
0.20	25	30	16.4
0.40	25	70-100	3.3
0.80	25	130-180	3.2
1.00	25	140-200	3.3

a. Approximate; calculated from coating weight
b. Nitrogen gas agitation during irradiation to maintain emulsion

Results of vena cava implant tests on poly(vinyl-co-2%-crotonic acid) 60% sodium ionomer samples pre-pared at varying grafting doses are shown in Table XI. For those samples prepared by emulsion grafting pro-cedures, there appears to be a minimum dose (0.1 Mrad) below which the samples become significantly less non-thrombogenic. This may be due either to a physically different hydrogel produced at the lowest dose or to a non-uniform coating. At present, we cannot distin-guish between these two possibilities. It is note-worthy that those samples prepared at 0.07 Mrad show a 43% incidence of occlusion at two hours while those prepared at 0.1 and 0.4 Mrads show a 100% patency rate in the same period. The data also tend to indicate (with the exception of the samples prepared at 0.4 Mrad) that coatings with high water uptake values are more thromboresistant than those with low water up-take values. For samples prepared at 1.0 and 1.2 Mrad in ethanol (20% water uptake), there was a 35% inci-dence of complete occlusion at two hours.

TABLE XI
VENA CAVA IMPLANT TESTS

Grafting Dose (Megarads)	Water Uptake (%)	Results[a]		
		O	I	X
0.07[b]	75	2	2	10
0.1[b]	56	5	2	5
0.4[b]	3	3	2	1
1.0[c]	20	−	3	2
1.2[c]	19	2	−	4
60.0[d]	120	3	2	1

a. X = Occluded, I = Partial Occlusion,
    O = Patent.
b. Emulsion Polymerization.
c. Solution Polymerization in Ethanol.
d. Van de Graaff Irradiation of Preformed
    Hydrogel.

CONCLUSION

Systematic variation of the surface charge den-sity and hydrophilicity of polyelectrolyte systems has yielded important information regarding nonthrombo-genic surfaces. The results from experiments with ethylene/acrylic acid and ethylene/vinyl sulfonate polyelectrolytes have shown a thrombogenic dependence on the chemical nature of the ionic moiety bound to the polymer backbone and its counterion. Evaluation of vinyl acetate/crotonic acid and N-vinyl pyrrolidone/ acrylic acid polyelectrolyte hydrogels has revealed a thrombogenic dependence on both the surface charge density and the hydrophilicity of the polyelectrolyte.

In vivo and in vitro blood compatibility studies established that the vinyl acetate-co-2%-crotonic acid 60% sodium ionomer was the most thromboresistant material in its series. Furthermore, the vinyl acetate/ crotonic acid hydrogel system was significantly more thromboresistant than the N-vinyl pyrrolidone/acrylic acid hydrogel system. Also, within each series it has been shown that there is a thrombogenic dependence on the surface charge density. These results imply that the antithrombogenic character of a surface does not increase indefinitely as the hydrophilicity and surface charge density of the surface increases. Rather, an optimum hydrophilicity and charge density is reached, beyond which the surface again becomes thrombogenic.

Based on the results of our studies, we are concentrating our efforts on optimizing the promising vinyl acetate-co-2%-crotonic acid ionomer hydrogel system. Further results of studies on vinyl acetate-co-2%-crotonic acid ionomer hydrogel coatings prepared by the direct polymerization of monomers from surfaces will be reported subsequently.

LITERATURE CITED

1. Sawyer, P. N. and J. W. Pate, Surgery, (1953), 34, 191; Amer. J. Physiol., (1953), 103, 113, 175.
2. Sawyer, P. N., ed., "Biophysical Mechanisms in Vascular Hemostasis and Intravascular Thrombus", Appelton-Century-Crofts, New York, 1965.
3. Langdale, R. D., J. Lab. Clin. Med., (1953), 41, 637.
4. Rodman, N. F., An. J. Clin. Path., (1958), 29, 525.
5. Maile, J. P., "Laboratory Medicine-Hematology", 2nd ed., C. V. Mosby Co., St. Louis, 1962.
6. Hardisty, J., Brit. J. Haemat., (1966), 12, 764.
7. Gott, V. L. and R. E. Baier, "Materials Compatible with Blood", Contract PH43-68-84-2, April 1969.
8. Mason, R. G. and L. D. Ikenberry, "Antithrombogenic Surfaces: Platelet-Interface Reactions", Contract PH43-67-1416, September 1969.
9. Pennington, D. J., L. L. Peterson, J. H. Rotaru and J. B. Boatman, "Biological Evaluation of Prosthetic Materials III: In Vitro Evaluation of Cardiovascular Prosthetic Devices", Contract PH43-67-1404-F-3, April 1967.

10. Sawyer, P. N. and S. Srinivasan, "New Approaches
    in the Selection of Materials Compatible with
    Blood", Contract PH43-68-75, December 1969.
11. Gott, V. L. and R. E. Baier, "Materials Compatible
    with Blood, Vol. I.", Contract PH43-68-84, Septem-
    ber 1973.
12. Daniels, A. U. and J. D. Mortenson, "Vena Cava
    Ring Tests of Biomaterials in Dogs", Contract NOl-
    HB42979, November 1974.
13. Lucas, T., S. Srinivasan and P. N. Sawyer, Trans.
    Amer. Soc., Artif. Int. Organs, (1970) 16, 1.
14. Sawyer, P. N., and S. Srinivasan, "New Approaches
    in the Selection of Materials Compatible with
    Blood". Contract PH43-68-75, December 1970.
15. Gott, V. L. and R. E. Baier, "Materials Compatible
    with Blood", Contract PH43-68-84-2, April 1970.
16. Kirk, J. E., "Mucopolysaccharides and Thromboplas-
    tin in the Vessel Wall, In: Biophysical Mechanisms
    in Vascular Hemostasis and Intravascular Thrombo-
    sis", Ed. P. N. Sawyer, Appelton-Century-Crofts,
    New York, 1965.
17. Bruck, S. D., "Blood Compatible Synthetic Polymers
    - An Introduction", Charles C. Thomas, Springfield,
    Ill., 1974.
18. Halpern, B. D., R. Shibakawa, H. Cheng, and C.
    Cain, "Non-Clotting Plastic Surfaces", Report PB-
    178469, June 1967.
19. Nelson, L. M., "Development of Polyelectrolyte
    Complexes as Anti-Thrombogenic Materials", Report
    PB-177694, January 1968.
20. Kusserow, B. K., R. W. Larrow and J. E. Nichols,
    "Analysis and Measurement of the Effects of Mate-
    rials on Blood Leukocytes, Erythrocytes and Plate-
    lets", Contract NOl-HB-8-1427, March 1975.

ACKNOWLEDGEMENTS

    This work was supported by the Biomaterials Program of the
Division of Blood Diseases and Resources, National Heart and
Lung Institute, National Institutes of Health, Bethesda, Maryland
under Contract No. PH43-68-1388 and NOl-HB-3-2950. Valuable and
stimulating discussions were held with Dr. S. D. Bruck, Project
Officer of the Biomaterials Program. Contributions of Drs.
R. L. Camp, J. F. Gaasch and L. S. Gonsior are gratefully ac-
knowledged.

Fibrous Capsule Formation and Fibroblast Interactions at Charged Hydrogel Interfaces

J. J. ROSEN and D. F. GIBBONS

Department of Biomedical Engineering, Case Western Reserve University, Cleveland, Ohio 44106

L. A. CULP

Department of Microbiology, Case Western Reserve University, Cleveland, Ohio 44106

Precise measurement of the biological response to specific properties of an implanted material requires a well defined materials system and a method for accurately evaluating variations in the response. Ionogenic hydrogels have been used as a materials system in this study to evaluate the effect of charge on the early phases of fibrous capsule formation. Image analysis techniques were used to quantitatively measure the response. Specific aspects of fibroblast interactions with these materials were also investigated using cell culture techniques.

The fibrous capsule is a layer of connective tissue formed as part of the biological response to an implanted foreign material. It consists primarily of collagen fibers, but may also contain a variety of inflammatory cell types and new capillaries, depending on the degree of the response(1). It has been previously demonstrated in vivo, that the addition of charged functional groups to a hydrogel network can significantly alter the early phases of the response (2). An understanding of the mechanisms responsible for these changes is an essential step in the development of materials which can fulfill the varied requirements encountered in soft-tissue implant applications.

Although many attempts have been made to correlate cell attachment with properties of the substrate material in culture (3,4,5), they have generally been limited by an ability to uniquely vary the specific material parameter of interest. The hydrophilic gel polymers based on hydroxyethyl methacrylate (HEMA) represent a class of well-characterized materials which are particularly well suited to isolate the effect of surface charge on cell attachment kinetics.

The same hydrogel compositions were used in the implant study and for cell culture substrates. This allowed direct comparisons to be made between the effect of surface charge on the in vivo reaction, and on the cellular response observed in vitro.

Methods and Materials

Monomers. 2-Hydroxyethyl methacrylate as received from Polysciences, Inc., contained between 0.25 and 0.40 percent methacrylic acid as an impurity, 200-400 ppm MEHQ as an inhibitor and had a pH of 6.00 to 6.20. Adsorption with potassium carbonate (anhydrous) lowered the methacrylic acid content to below 0.15 percent and reduced the inhibitor slightly to 150-200 ppm. The monomer purity levels were determined using gas chromatography. The final pH was 6.85 to 6.90. Methacrylic acid (Polysciences, Inc.) (b.p. = 59°C, 10mm Hg.) and diethyl amino ethyl methacrylate (Rohm and Haas Co.) (b.p. = 90°C, 3 mm Hg.) were vacuum distilled under dry nitrogen and their final purity was checked using gas chromatography. Tetraethylene glycol dimethacrylate (Polysciences, Inc.) was the crosslinking agent and was used as received.

Polymerization. Sheets of hydrophilic gels were prepared by solution polymerization between glass plates with 1.0 millimeter glass spacers for the implant materials and 0.15 millimeter glass spacers for the cell culture substrates. Polymerization proceeded for 24 hours at room temperature. The reaction mix compositions are given in Table I. The gels were removed from the glass by soaking in distilled water. All hydrogels were washed in distilled water which was changed daily, for a minimum of three weeks. Equilibrium water content of the gels was determined on fully hydrated samples by measuring the percent weight loss after vacuum drying at 120°C until no further weight loss was observed.

In Vivo. Implant specimens, measuring 7 x 15 mm., were cut from the washed gels. The samples were reequilibrated in phosphate buffered saline solution for three days and boiled for 30 minutes prior to implantation. Sprague-Dawley male rats, weighing between 300 and 350 grams were used for all experiments. Subcutaneous implants were placed away from the incision in a pocket opened precisely at the facia layer of the dorsal surface. The implants were removed with the surrounding tissues after various times from one to twenty-one days, fixed in formalin and prepared for histology by standard techniques. Sham experiments in which the entire surgical protocol was performed, but with no material implanted, served as controls. The tissue response was evaluated qualitatively on the basis of cell types, degree of granulation, morphology of collagen organization and integrity of implant material. A modified tri-chrome stain differentiated the collagen fibers from surrounding features and provided the basis for quantitative analysis of the fibrous capsule. Computor-assisted image analysis using a Millipore particle counting microscope was used to measure the mean

distance in microns from the material interface to the edge of
the fibrous capsule, the percentage of the capsule cross-
sectional area actually filled with collagen fibers, and the
total collagen area in square microns per unit length along
the material interface. A modified nuclear stain was used to
count the number of cells present per unit area adjacent to the
material interface. All measurements were made at seven
locations adjacent to the interior surface of the implant, and
were repeated on two sections for each material and each implant
time. The results are reported as the mean of these measurements.

In Vitro. All cell culture experiments were performed
using the Swiss 3T3 cell line, contact inhibited mouse fibroblasts. Only cells between their 10th and 20th passage were
used for this study. Cells were routinely grown in Eagles'
minimal essential medium which was supplemented with four
times the normal concentration of amino acids and vitamins
(MEMX4), penicillin (250 units/ml), streptomycin (0.25 mg/ml),
and 10 percent donor calf serum. The DNA of the source cells
was labelled with ^{3}H-thymidine by growing cells for 24 hours
in media containing 0.1 μCi/ml of ^{3}H-thymidine followed by
48 hours of growth in non-labelled media (6). This technique
provided a method for counting the cells and determining the
relative ability of Swiss 3T3 fibroblasts to attach to substrates of various hydrogel compositions. The substrates were
prepared from washed hydrogel membranes equilibrated in phosphate buffered saline. Discs (15 mm diameter) of the gels
were placed in the attachment media for two hours prior to the
attachment assay. The attachment media contained phosphate
buffered saline supplemented with divalent cations (100 mg/l
$MgSO_4 \cdot 7H_2O$ and 100 mg/l $CaCl_2$) and 10% donor calf serum.
The attachment assay consisted of inocculating dishes
containing hydrogel substrates with a known aliquot (2.0 x 10^6
cells in one milliliter of attachment media) of the labelled
cells. At each of the time points, one disc was removed from
each dish (there were two dishes for each gel composition),
dipped three times in three separate washes of phosphate
buffered saline containing divalent cations (100 mg/l $MgSO_4 \cdot 7H_2O$
and 100 Mg/l $CaCl_2$), and then placed in scintillation vials
containing Bray's scintillation solution.

Results and Discussion
The biological reaction to the presence of an implanted
material, often referred to as the "foreign body response",
is influenced by a   complex set of variables. A recent review
(7) differentiated some major classes of these variables and
attempted to distinguish between those associated with material
properties and those involved in the "normal" wound healing
process.

The large number of potentially relevant factors has sometimes made previous investigations difficult to control and their results difficult to interpret. The issue of what constitutes a "biocompatible" material in the soft tissues has been further confused by a persistant notion that a thin fibrous membrane around an implant is always a "good" reaction and a thick capsule is indicative of a severe or "bad" response. This concept was furthered by Laing's early study (8) of implanted metals which demonstrated that an increase in metal ion concentration or ion toxicity correlated with an increase in the surrounding fibrous membrane thickness.

Although it is true that chronic inflammation, tissue necrosis, or continued cellular proliferation would be considered undesirable in almost any clinical application of an implanted material, it does not follow that a stable, thick fibrous capsule per se, would always be an undesirable consequence of the biological reaction. For example, when mechanical stability is desired to limit lateral movement, such as in a hydrocephalus shunt, a dense, well organized acellular, and nonadherent capsule would allow for longitudinal growth adjustments while providing the necessary lateral stability. In the same application, the formation of fibrous tissue in or around the drain section often leads to blockage and eventual device failure. A much better understanding of basic interactions between material parameters and biological mediators is needed before this high level of control over the reaction can be achieved.

During this investigation our primary objective was to determine if the charge of the material significantly affected the process of fibrous capsule formation. The hydrogel system was selected as the material model because the charge could be varied by changing the chemical composition of the functional groups without changing the main chain methacrylate structure. The initial monomer reaction solutions, as shown in Table I, all contained the same percent of crosslinking agent and the same amount of water. After swelling to equilibrium the highly acidic gels contained slightly more water (47%) than the most basic gels (41%). These differences may have contributed to the observed variations in the biological response, but they were considered to be small when compared with the wide range of charge density and sign that was possible with this system. Scanning microscopy of the polymers revealed fairly uniform, smooth surfaces with similar morphology for all compositions. It is now well established (7) that many other experimental parameters must also be considered as possible sources of artifact during the evaluation of implanted materials. To minimize the influence of implant edge and shape effects (9), only the tissue adjacent to the central region of implants was included in the analysis. The implant, itself, was placed in a "pocket" opened by blunt dissection away from the site of the initial incision. Variations due to anatomical location were reduced by always placing the implant directly over the muscle facia in approximately identical positions for each animal. By choosing this particular placement for the implant, new collagen formed as part of the fibrous capsule was easily differentiated from the "normal" tissues.

Figure 4 shows examples of the interface between the implant (on the left) and the surrounding tissues. Within the fibrous capsule, the collagen fibers filled only part of the tissue space. The remainder of the cross-sectional capsule area was filled with cells, fluid, or small vessels. The percentage of the capsule area actually filled with collagen, referred to in Tables II, III, and IV as "Collagen Density" varied with implant time and hydrogel composition and was related to the morphology of the newly formed collagen fibers. The thickness of the capsule multiplied by the collagen density was proportional to the total collagen produced per unit area along the capsule cross-section. This value was obtained from area measurements made on 176 micrometer segments of capsule for all samples and is reported as "Relative Amount Collagen" in Tables II, III, IV. This arbitrary but constant unit of capsule length represented approximately one microscopic field at 400x magnification.

A parallel measurement of total cellular activity in the vicinity of the implant was determined by counting the nuclei of all cells within 250 micrometers of the implant surface and normalizing to a constant unit area. This count was not limited to the capsule region because significant vessel development and other cell activity often occurred outside the actual fibrous capsule. The number of cells per unit area in normal connective tissue is relatively low (as indicated by the surgical control) and therefore, the value for total cellular activity is representative of cells that are directly involved in the response to the implant material.

The entire range of the biological response, for all materials tested, was within the limits of what has been considered "mild" (8). Contrary to previous work (2) none of the hydrogel materials consistantly elicited a "gross purulent exudate". Obvious infections of this type occurred rarely and did not correlate with the material compositions. The in vivo experiments of this study have demonstrated significant and reproducible changes in both the amount of collagen produced and its organizational morphology as a direct result of variations in the charge of the hydrogels.

The bar graphs in Figures 1, 2, and 3 show, for example, that the capsule formed around positively charged hydrogels during the first three weeks after implantation is thicker than the capsules around other hydrogels. This result agrees with

TABLE I
COMPOSITION OF HYDROGEL MONOMER SOLUTIONS

HYDROGEL MONOMER (Mole Percent)	COMPOSITION (Volume Percent)		
	HEMA	MAA	DEAEMA
100 HEMA	100.0		
95 HEMA, 5 MAA	96.0	4.0	
80 HEMA, 20 MAA	84.0	16.0	
60 HEMA, 40 MAA	67.5	32.5	
95 HEMA, 5 DEAEMA	94.0		6.0
80 HEMA, 20 DEAEMA	75.5		24.5
60 HEMA, 40 DEAEMA	51.0		49.0
95 HEMA, 2.5 MAA, 2.5 DEAEMA	94.5	1.8	3.6
80 HEMA, 10 MAA, 10 DEAEMA	78.5	7.0	14.5
60 HEMA, 20 MAA, 20 DEAEMA	57.0	14.5	28.5

HEMA= 2-Hydroxyethyl methacrylate, MAA= Methacrylic acid,
DEAEMA= Diethylamino ethyl methacrylate,
Crosslinker= Tetraethylene glycol dimethacrylate: 2.2 mole
percent, Distilled water: 33.0 volume percent,
Initiator: $(NH_4)_2SO_8$: 0.5 mole percent
Reducing Agent: $Na_2S_2O_5$: 0.5 mole percent.

TABLE II
IN VIVO RESPONSE TO HYDROGELS IMPLANTED FOR ONE DAY

HYDROGEL MONOMER (Mole Percent)	CAPSULE THICKNESS (Micrometers) MEAN (S.D.)	COLLAGEN[1] DENSITY (Percent) MEAN (S.D.)	RELATIVE[2] AMOUNT COLLAGEN MEAN (S.D.)	TOTAL[3] CELLULAR ACTIVITY MEAN (S.D.)
100 HEMA	27.1 (11.9)	54.4 (8.4)	14.2 (6.2)	192.1 (35.6)
95 HEMA, 5 MAA	27.8 (4.2)	48.9 (10.6)	14.9 (3.9)	190.0 (26.6)
60 HEMA, 40 MAA	41.8 (8.6)	60.9 (9.7)	24.0 (9.0)	133.9 (20.6)
95 HEMA, 5 DEAEMA	63.8 (19.8)	68.4 (9.4)	42.4 (11.5)	207.3 (57.9)
60 HEMA, 40 DEAEMA	77.1 (7.2)	55.9 (10.9)	36.7 (13.5)	361.7 (49.6)
95 HEMA, 2.5 MAA, 2.5 DEAEMA	37.8 (10.4)	48.9 (10.6)	17.1 (9.8)	223.1 (55.5)
60 HEMA, 20 MAA, 20 DEAEMA	19.2 (6.9)	60.9 (9.7)	11.5 (3.8)	227.4 (48.9)
SURGICAL SHAM [4]	--- ---	31.7 (10.6)	34.1 (11.7)	67.5 (17.7)

1. Percent of fibrous capsule composed of collagen fibers.
2. Area of collagen fibers proportional to new collagen formation.
3. Number of cells per unit area adjacent to implant.
4. Values obtained from connective tissue adjacent to surgically prepared implant site.

TABLE III
IN VIVO RESPONSE TO HYDROGELS IMPLANTED FOR ONE WEEK

HYDROGEL MONOMER (Mole Percent)	CAPSULE THICKNESS (Micrometers) MEAN (S.D.)	COLLAGEN DENSITY (Percent) MEAN (S.D.)	RELATIVE AMOUNT COLLAGEN MEAN (S.D.)	TOTAL CELLULAR ACTIVITY MEAN (S.D.)
100 HEMA	84.8 (11.5)	81.1 (5.8)	59.6 (15.1)	130.6 (26.7)
95 HEMA, 5 MAA	97.9 (16.3)	65.7 (7.8)	63.7 (8.4)	87.6 (14.3)
60 HEMA,40 MAA	79.4 (20.5)	61.4 (8.4)	46.9 (13.2)	159.8 (22.7)
95 HEMA, 5 DEAEMA	112.6 (23.5)	49.6 (8.7)	57.2 (10.8)	287.4 (25.9)
60 HEMA,40 DEAEMA	175.0 (15.1)	52.0 (7.9)	91.2 (11.3)	248.1 (39.8)
95 HEMA,2.5 MAA,2.5 DEAEMA	87.7 (16.6)	54.6 (14.8)	50.7 (8.9)	133.9 (25.5)
60 HEMA, 20 MAA, 20 DEAEMA	85.3 (18.1)	62.1 (5.1)	53.1 (7.3)	149.5 (26.2)
SURGICAL SHAM	--- ---	51.8 (7.1)	87.0 (11.9)	88.5 (17.8)

TABLE IV
IN VIVO RESPONSE TO HYDROGELS IMPLANTED FOR THREE WEEKS

HYDROGEL MONOMER (Mole Percent)	CAPSULE THICKNESS (Micrometers) MEAN (S.D.)	COLLAGEN DENSITY (Percent) MEAN (S.D.)	RELATIVE AMOUNT COLLAGEN MEAN (S.D.)	TOTAL CELLULAR ACTIVITY MEAN (S.D.)
100 HEMA	94.1 (11.6)	77.4 (6.4)	68.0 (11.6)	143.3 (25.0)
95 HEMA, 5 MAA	75.6 (7.6)	67.6 (6.2)	50.9 (5.6)	71.2 (6.7)
60 HEMA,40 MAA	50.2 (7.9)	79.2 (6.5)	39.8 (7.7)	81.2 (16.3)
95 HEMA, 5 DEAEMA	130.9 (28.0)	59.2 (4.9)	65.7 (7.1)	148.6 (19.5)
60 HEMA,40 DEAEMA	143.2 (17.9)	60.4 (5.5)	77.0 (15.4)	178.3 (14.2)
95 HEMA,2.5 MAA,2.5 DEAEMA	64.2 (14.6)	62.3 (10.8)	54.7 (11.2)	99.9 (16.9)
60 HEMA, 20 MAA, 20 DEAEMA	70.6 (5.8)	85.1 (7.3)	60.0 (6.4)	116.7 (21.9)
SURGICAL SHAM	--- ---	36.4 (10.1)	61.9 (17.2)	54.9 (16.0)

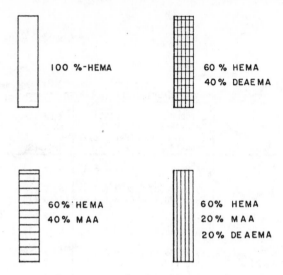

Bar configurations used in Figures 1, 2, and 3 to represent the indicated hydrogel compositions

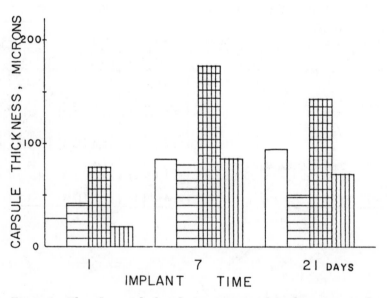

Figure 1.   The effect of hydrogel composition and implant time on the mean thickness of the fibrous capsule

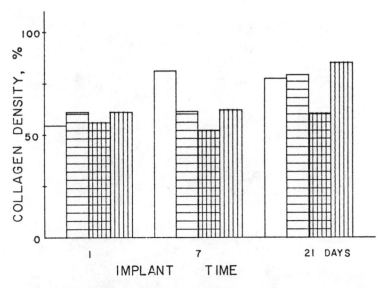

Figure 2.   *The effect of hydrogel composition and implant time on the percent of the fibrous capsule composed of collagen fibers*

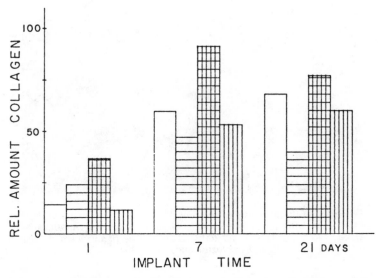

Figure 3.   *The effect of hydrogel composition and implant time on the relative amount of collagen produced during the formation of the fibrous capsule*

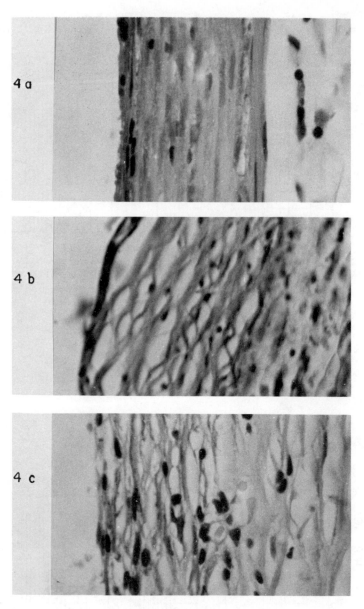

Figure 4. Fibrous capsule adjacent to hydrogel implants (at left) three weeks after implantation. 4a. 60% HEMA–40% MAA; 4b. 60% HEMA–40% DEAEMA; 4c. 100% HEMA. The tissues were stained with hematoxilin and eosin (340×).

the qualitative observations made by Barvic (2). Quantitative
evaluation provided by the image analysis techniques which have
been summarized here and described in detail (10) reveals that
although the total thickness is greater for positively charged
gels, the density of collagen fibers decreases and the actual
amount of collagen produced remains essentially constant during
the same three week period.

To begin to differentiate between possible models that could
account for these changes, fibroblast-hydrogel interactions were
studied under more controlled conditions. Cell cultures were
used to investigate the kinetics of fibroblast attachment to
hydrogel substrates. Substrate polymer compositions were chosen
from, and identical to those used for the in vivo experiments.
Only some of these compositions have been included in this paper
to illustrate the range of the observed reactions.

The results of the cell culture attachment experiments are
presented in Figures 5, 6, and 7. The values for attachment of
cells are reported as a percentage of the total cells available
for attachment. The substrates were equilibrated in phosphate
buffered saline containing divalent cations and calf serum
(see Methods). Fresh solutions of this composition also served
as the attachment media. For the short duration of these
experiments, the absence of vitamins and amino acids normally
found in growth media did not affect the rate or extent of
cell attachment (11).

The specific role of the serum proteins has not been elabor-
ated during this phase of the study, although it is not un-
reasonable to assume that these proteins interact differently
with hydrogel surfaces of different charges. The presence or
absence of serum has been shown to influence cell attachment in
other model systems (12) and is probably involved in the
mechanisms responsible for the observed variations in the hydrogel system. The initial objective of this study has been to
determine if fibroblasts respond differently to substrates of
different charge density and sign when there are no interactions
with other cell types, complement or clotting proteins.

The graphs in Figures 5, 6, and 7 show that fibroblasts do
alter their attachment behavior when the charge of the substrate
is varied. These changes occur during three distinct phases
of the attachment process. The first phase begins when the cells
first come in contact with the substrate material and ends when
cells begin to attach rapidly. This so called "lag" phase is
often interpreted as a time of cell "accomodation" or "adjustment" to the substrate. The second phase is a period of rapid
attachment characterized by a steeper slope on the attachment
curve. The third phase is the "plateau" or "leveling-off"
phase when there are no further significant increases in the
total cells attached. For the neutral gels that contained no
charge groups, the lag phase lasted between five and ten minutes.
The period of rapid attachment was over within forty-five

Figures 5, 6, and 7 are the results of an assay to determine the effect of hydrogel substrate composition on the attachment of Swiss 3T3 fibroblasts. The cells are labelled with ³H-thymidine and inocculated into dishes containing the hydrogel substrates. After times varying from 5 min to 2 hr, substrate discs are removed, washed, and counted using scintillation techniques to determine the percent of available cells that attached to the substrates.

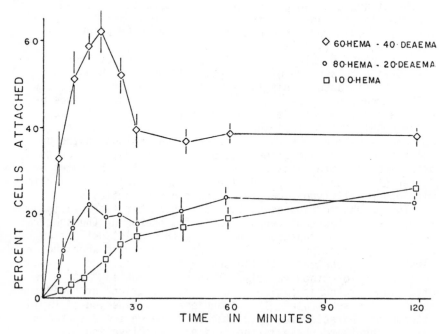

Figure 5.   Attachment of Swiss 3T3 fibroblasts to hydrogel substrates

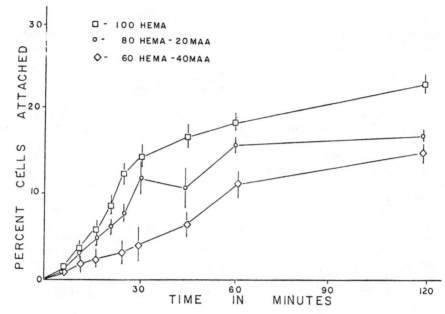

Figure 6. Attachment of Swiss 3T3 fibroblasts to hydrogel substrates

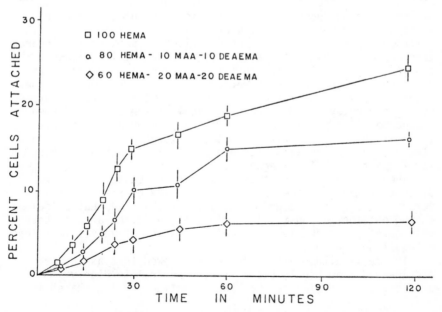

Figure 7. Attachment of Swiss 3T3 fibroblasts to hydrogel substrates

minutes and the plateau attachment value was approximately twenty-five percent of the available cells. As negative charges are added to the gel network, the lag phase remains essentially unchanged, but the slope of rapid attachment and the plateau value decrease. If positive charges are added, the lag phase goes toward zero and the slope of rapid attachment increases dramatically. The plateau value increases, but is not at the maximum percent attached due to some cells releasing from the substrate at around thirty minutes. The ionogenic neutral gels which contain equimolar amounts of positive and negative charge groups, have proportionately lower values for each phase than any of the other materials.

Conclusions

The ionogenic hydrogels have provided a useful model for evaluating the effect of charge on the early phases of the soft-tissue response. Quantitative evaluation and comparison of the capsule thickness, collagen density and the amount of collagen produced, and the relative degree of cellular activity were made possible by the image analysis of specially prepared histology sections of the in vivo response to the implanted hydrogels.

The neutral, homopolymer of hydroxyethyl methacrylate is progressively encapsulated during the first three weeks. The capsule contains increasing amounts of collagen which pack more efficiently with time to produce a moderate, densely packed structure. As positive charges are added by co-polymerizing with diethyl aminoethyl methacrylate, significant quantities of collagen are produced immediately after implantation and are accompanied by very high cellular activity. Although new collagen continues to be produced, efficient packing is hindered (perhaps by the high cell activity), and very thick, loose, capsules result. The fact that the capsule does not organize efficiently with time is characteristic of the response to these materials and is indicated by the consistantly low value for the collagen density. Figure 4b is typical of the capsule formed around positively charged gels. The acidic groups also have initially high cell activity, but this level decreases during the second week to allow for dense capsule formation without large quantities of collagen being produced.

When equimolar quantities of acidic and basic groups are present in the gel, the early cell activity decreases within a few days and remains significantly lower than the neutral gel throughout the response. The collagen production around these gels develops gradually and remains constant at values less than or equal to the neutral gel.

The cell culture attachment assay showed that as the gels are changed from highly basic through neutral to highly acidic: 1) the time for cell accomodation or "lag phase" is increased, 2) the slope of the rapid attachment phase decreases sharply, and 3) the maximum percent of attached cells decreases. Cell attachment to ionogenic, electroneutral gels is proportionately lower than for any other gels.

The implications of the cell culture experiments cannot be fully appreciated until further studies which are designed to elaborate the mechanisms of adhesion to hydrogel substrates are completed. Although direct correlation between cell culture and in vivo experiments is premature at this time, certain interesting trends are apparent. For example, the positively charged gels have a higher level of cell activity around implants and a much more rapid and extensive attachment of fibroblasts in culture. If the subsequent release of cells from positively charged substrates represents cell damage or death and this also occurs at the implant interface, the low density of collagen in the capsule may be related to increased numbers of cells and cell debris associated with the implant interface.

The combined information obtained from further cell culture and quantitative analysis of in vivo reactions will be useful in evaluating the role of specific mechanisms and biological pathways that control the formation of fibrous capsules around implanted materials.

Literature Cited

1. Kellermeyer, R.W., and K.S. Warren, J. of Exp. Med., (1970), 131, (1) 21-39.
2. Barvic, M., J. Vacik, D. Lim, and M. Zavadil, J. of Biomed. Mat. Res. (1971) 5, 225-238.
3. Grinnell, F., M. Milam, and P.A. Srere, Arch. of Biochem. and Biophys., (1972), 153, 193-198.
4. Rappaport, C., in "The Chemistry of Biosurfaces", M.L. Hair ed., Marcel Dekker, Inc., New York (1972).
5. Weiss, L., Exper. Cell Res., (1974), 83, 311.
6. Culp, L.A., J. of Cell Biol. (1974), 63, 71-83.
7. Coleman, D.L., R.N. King, and J.D. Andrade, J. of Biomed. Mat. Res. (1974), 8, (5) 199.
8. Laing, P.G., A.E. Ferguson, and E.S. Hodge, J. of Biomed. Mat. Res. (1967), 1, 135,
9. Wood, N.K., E.J. Kaminski, and R.J. Oglesby, J. of Biomed. Mat. Res., (1970), 4, 1.
10. Rosen, J.J., and D.F. Gibbons, (submitted for publication).
11. Mapstone, R. and L.A. Culp, (submitted for publication).
12. Witkowski, J.A. and W.D. Brighton, Exper. Cell Res., (1972) 70, 41.

INDEX